Exploring Space, Exploring Earth

Paul Lowman, a NASA scientist for over 40 years, describes the impact of space flight on geology and geophysics. A foreword by Neil Armstrong emphasizes that the exploration of space has led us to a far deeper understanding of our own planet. Direct results from Earth-orbital missions include studies of Earth's gravity and magnetic fields. In contrast, the recognition of the economic and biological significance of impact craters on Earth is an indirect consequence of the study of the geology of other planets. The final chapter presents a new theory for the tectonic evolution of the Earth based on comparative planetology and the Gaia concept. Extensive illustrations, a glossary of technical terms, and a comprehensive bibliography provide geologists and geophysicists with a valuable summary of research. The book will also serve as a supplementary text for students of tectonics, remote sensing and planetary science.

PAUL LOWMAN has been involved in a wide range of space research programs at the Goddard Space Flight Center. In 1963–4 he took part in planning for the *Apollo* missions. He was Principal Investigator for Synoptic Terrain Photography on the *Mercury*, *Gemini*, and *Apollo* Earth-orbital missions, an experiment that laid the foundation for *Landsat*. Between 1965 and 1970 he taught lunar geology at the University of California, Catholic University of America, and the Air Force Institute of Technology. Dr Lowman was also involved with the *Mariner 9* Mars mission, the *Apollo* X-ray fluorescence experiment and *Apollo 11* and *12* sample analysis among others. His main research interest was and still is the origin of the continental crust, as approached through comparative planetology.

In 1974, Dr Lowman received the Lindsay Award from the Goddard Space Flight Center. He was elected a Fellow of the Geological Society of America in 1975, and of the Geological Society of Canada in 1988. Drawing on his dual career in terrestrial and lunar geology, he authored *Space Panorama* (1968), *Lunar Panorama* (1970), and *The Third Planet* (1972). He also contributed to *Mission to Earth* (1976), the first NASA compilation of *Landsat* pictures, edited by N. M. Short.

Exploring Space, Exploring Earth

New Understanding of the Earth from Space Research

Paul D. Lowman Jr.
Goddard Space Flight Center

Foreword by Neil A. Armstrong

CAMBRIDGE
UNIVERSITY PRESS

University Printing House, Cambridge CB2 8BS, United Kingdom

Cambridge University Press is part of the University of Cambridge.

It furthers the University's mission by disseminating knowledge in the pursuit of education, learning and research at the highest international levels of excellence.

www.cambridge.org
Information on this title: www.cambridge.org/9780521890625

First published 2002

A catalogue record for this publication is available from the British Library

Library of Congress Cataloguing in Publication data

Lowman, Paul D.
Exploring space, exploring earth: new understanding of the Earth from space research / Paul D. Lowman; foreword by Neil A. Armstrong.
p. cm.
Includes bibliographical references and index.
ISBN 0 521 66125 0 (hb.) – ISBN 0 521 89062 4 (pb.)
1. Earth sciences – Remote sensing. 2. Astronautics in earth sciences. I. Title.

QE33.2.R4 L69 2002
550'.28'7 – dc21 2001052713

ISBN 978-0-521-89062-5 Paperback

To John A. O’Keefe
Founder of Space Geodesy

CONTENTS

FOREWORD

In the works of Homer, the Earth was portrayed as a circular disc floating on a vast sea and covered with a sky built from a hemispherical bowl. Critics soon noticed a flaw in this concept: the visible star field varied from place to place. From Greece, the Big Dipper was visible throughout its circle around the North Star, but southward along the Nile, it dipped below the horizon. Clearly, the surface of the Earth was somehow curved. Some thought the Earth was like the surface of a cylinder, curving to the north and south, but stretching in a straight line to the east and west. A student of Socrates, Parmenides, reasoned that the Earth must be a sphere, because any other shape would fall inward on itself. Plato also concluded that the Earth must be a sphere because a sphere was the most perfect shape for a solid body. Whether persuaded by the logic of Parmenides, or by loyalty to Plato, the Greeks came to accept a spherical Earth. A final argument, the most persuasive, was recorded by Aristotle. He noted that during an eclipse of the Moon, when the Earth's shadow fell on the surface of the Moon, the shadow was curved. The shape of the Earth would not be truly known, however, until the philosophers were replaced by the measurers.

In that category, one name stands above all others: Eratosthenes of Cyrene. To characterize him simply as a "measurer" would not do him justice; Eratosthenes was a Renaissance man long before the Renaissance. But we focus on his measuring. He determined the inclination of the ecliptic with an error of only one-half a degree. His most memorable measurement was the difference in latitude from Syene to Alexandria. By comparing the shadow lengths at noon on the summer solstice for the two locations, he calculated that they were separated by 7.5 degrees. Knowing the distance between the two cities, he calculated the circumference of the Earth to an accuracy of 99%. Eratosthenes further collected the observations of travelers, explorers, and sailors from throughout the known world –

a very small world by today's standards – and integrated that knowledge in his *Geographica*.

Understanding of the Earth grew slowly after this Golden Age. It was not until the invention of the caravel in the 15th century, and John Harrison's chronometer in the 18th century, that man's understanding of his planet began again to grow. With ships capable of long ocean voyages, precisely navigated in latitude and longitude, maps of the oceans, continents, and islands became increasingly comprehensive and reliable.

Despite this great increase in knowledge, our planet remained in many ways almost as mysterious as it had been in Homer's time. The forces causing volcanic eruptions, earthquakes, and hurricanes remained enigmatic. The interior of the Earth, the topography of the ocean floor, and the dynamic nature of the atmosphere, the ocean currents, and the global magnetic field eluded understanding well into the 20th century. Two world wars stimulated impressive improvements in instruments and methods.

The late 20th century also brought new caravels, ships that could sail the oceans of space. The fortuitous development of the liquid-fueled rocket and, at about the same time, the digital computer made flight through space a reality. Space was the new high ground, the place for a new perspective, from which the "measurers" could acquire information never before available. In just a few decades, knowledge of the Earth's secrets has increased beyond imagining.

Exploring Space, Exploring Earth describes this increase in knowledge of the solid earth – geology and geophysics. Paul Lowman is a geologist who has been involved in space research since 1959 at Goddard Space Flight Center, which has taken a leading part in space geodesy, remote sensing, lunar geology, and satellite meteorology and oceanography. He is thus one of the new "measurers" and summarizes their accomplishments since the launch of *Sputnik 1* in 1957. The book is dedicated to John O'Keefe, who in the tradition of Parmenides and Eratosthenes made a fundamental discovery about the shape of the Earth from the orbit of only the second American satellite launched, *Vanguard 1*, in 1959.

The 20th century brought remarkable changes in our understanding of the Earth, the Moon, and the universe. Let us hope that the present century is equally productive.

Neil A. Armstrong

PREFACE

Mine was the first generation in humanity's million-year history to have seen the Earth as a globe, hanging in the blackness of space.

When the spacecraft *Eagle* landed on Mare Tranquillitatis in 1969, there were still people alive who had seen the Wright brothers flying. A century of progress has permitted us to see almost the entire surface of the Earth, even the ocean floor thanks to satellite altimetry. It has also given us the first opportunity to compare the Earth with other planets, starting with the Moon (geologically a "planet").

My purpose in this book is to explore the impact of space flight on geology and its subsurface counterpart, geophysics, an impact largely unappreciated in the earth science community. We geologists tend to be conservative, perhaps more so than other scientists. One reason for this may be the nature of our subject: the solid earth, ideally the part of it we can see, feel, and hammer. Another may be the nature of our work, often involving field work in harsh and remote terrains. Whatever the cause, geologists are demonstrably conservative, which is basically why this book is needed.

Geology at the end of the 20th century had reached what appears to be a certain maturity, as I described in a 1996 review, whose abstract is reproduced here. It seems true that we really have settled some questions that in 1900 were not only unanswered but even unasked. The age of the Earth is no longer debated, except in the third significant figure. The mechanism responsible for most earthquakes is now well understood, to the point that seismologists can often tell us which fault slipped, in what direction, and by how much. The origin of granite, intensely controversial as late as 1960, is, in general, understood for most granites – caveats inserted because nature has fooled us before.

It was a magnificent century for science in general, and for the

Twelve Key 20th-Century Discoveries in the Geosciences

Paul D. Lowman Jr.
Goddard Space Flight Center, Code 921
Greenbelt, Maryland 20771

ABSTRACT

This paper reviews 12 major discoveries in geology and geophysics during the 20th century, reasonably mature and discrete with respect to subject, time, or discoverer. The *Textbook of Geology* (Geikie, 1903) and *Understanding the Earth* (Brown and others, 1992) are used as benchmarks, supplemented by Yoder's 0007-1992 "Timetable of Petrology" (1993). The discovery of **the radioactive decay law** by Rutherford and Soddy was fundamental, bearing on the age of the Earth, internal energy, nature of the crust, and a quantitative geologic time scale. Discoveries of the **internal structure of the Earth,** investigated seismically, included definition of the core boundary (Gutenberg and Weichert), the inner core (Lehmann), the mantle/crust boundary (Mohorovic), the lithosphere, and the asthenosphere. **Deep structure of the continental crust,** revealed since 1975 by reflection profiling and study of exposed sections, has been found to be drastically different from previous concepts. **The magnetic reversal time scale** was developed by correlating reversed magnetization in terrestrial and marine rocks with radiometric and stratigraphic data, interpreted in light of sea-floor spreading. **Sea-floor spreading,** the key element of plate-tectonic theory, discovered by several independent lines of inquiry, has been directly confirmed by space-geodesy measurements of plate motion and intraplate rigidity in the Pacific Basin. **Elastic rebound,** the cause of shallow-focus earthquakes, was proposed after the 1906 California earthquake by Reid, using geodetic evidence; deep-focus events are still not understood. The **mechanism of overthrust faulting,** a major problem as late as the 1950s, was discovered by Hubbert and Rubey, who demonstrated that high fluid pressure could reduce normal stress to the point that gravity sliding would be effective. **X-ray diffraction** by the crystal lattice was discovered in 1912, and its applications revolutionized mineralogy, confirming solid solution and clarifying many crystallographic phenomena. The **origins of basaltic and granitic magmas,** completely unknown in 1900, were definitively explained for basalts by Bowen as partial melting of a peridotitic mantle, and for most granitic magmas as partial melting of crustal rocks by Tuttle and Bowen. Space exploration has revealed common patterns of **planetary differentiation,** including a first (or felsic) differentiation forming global primitive crusts, followed by a second (or basaltic) differentiation, including continued basaltic magmatism. Another major result of space exploration and related research has been discovery of the importance of **impact cratering,** proposed as being responsible for formation of continental nuclei, the first ocean basins, and many mass extinctions in the geologic record. The most important scientific achievement of the 20th century is suggested to be the discovery of **the DNA structure,** revealing not only the molecular basis for all life on Earth but a critical line of investigation, formation of RNA, for study of the origin of life.

Keywords: Extraterrestrial geology: geochemistry; geochronology; geology – general; geophysics – general; history of geology; history of science; mineralogy and crystallography; petrology – general; plate tectonics; reviews – articles; structural geology; volcanoes and volcanism.

Introduction

The 20th century opened with a surge of technological progress never equaled before or since. A mere two decades, from 1890 to 1910, saw the development of radio, aviation, electronics, the automobile industry, the steel-frame skyscraper, the Diesel engine, the steam turbine, and many other features of today's world existing, if at all, only in rudimentary form in 1890. In combination with the well known revolution in physics, these developments laid the infrastructure for a century of extraordinary scientific progress. The purpose of this review is to focus on advances in the solid-earth sciences, geology and geophysics, during the 20th century.

The approach will be to outline 12 discrete and reasonably mature discoveries of features or phenomena, and their consequences. Stress will be on the discoveries rather than on the people who made them, primarily to keep the paper within practical length. Two textbooks provide excellent benchmarks of progress since 1900: Geikie's (1903) *Textbook of Geology* and *Understanding the Earth* (Brown and others, 1992). Sullivan's (1991) *Continents in Motion*, though focussed on plate tectonics, is a well documented history of the earth sciences in general from the mid-19th century on. Yoder's (1993) "Timetable of Petrology" lists and documents specific developments from 0007 to 1992. Citations given will be largely of original sources.

Journal of Geoscience Education, 1996, v. 44, p. 485

solid-earth sciences. However, the "certain maturity" just described may be instead a certain stagnation, a feeling that the big problems in geology have now been solved. Most scientists will recognize this situation; it describes the prevailing view in physics around 1890.

I think that geology is on the verge of a major paradigm shift, to use the fashionable term, comparable to that in physics between 1895 and 1905. Geologists today subscribe almost unanimously to what W. K. Hamblin has called a "master plan," the theory of plate tectonics. The three essential mechanisms of plate tectonics – sea-floor spreading, subduction, and transform faulting – have in fact been confirmed by so many independent lines of evidence that we can consider them observed phenomena, at least in and around the ocean basins. Plate tectonic theory is called upon to explain, directly or indirectly, almost all aspects of terrestrial geology above the level of the crystal lattice. Even metamorphic petrology, in particular the new field of ultra-high pressure metamorphism, invokes phenomena such as continental collision to explain how rocks recrystallized 150 kilometers down are brought to the surface.

I think this is a mistake. We now know, from space exploration, that bodies essentially similar to the Earth in composition and structure have developed differentiated crusts, mountain belts, rift valleys, and volcanos without plate tectonics, in fact without plates. Furthermore, we now know, thanks partly to remote sensing from space, that the Earth's crust can not realistically be considered a mosaic of 12 discrete rigid plates. For these and other reasons, I disagree with certain aspects of plate tectonic theory, as will be explained in the text.

Is the Earth fundamentally unique? Most geologists think it is, and consider the discoveries of space exploration to be interesting but irrelevant to terrestrial geology. The basic objective of this book is to show the contrary: that **the exploration of space has also been the exploration of the Earth**, and that real understanding of its geology is just beginning now that we can see our own planet in almost its entirety, and can compare it with others.

Exploring Space, Exploring Earth is aimed primarily at geologists and geophysicists. However, it has been written to be understood even by readers without a single geology course. Such readers may in fact have the advantage of freedom from preconceived notions. Jargon is unavoidable, but has been kept to the minimum possible. Technical terms are explained either in context or in the glossary. Petrologic topics are presented without the use of phase diagrams, with one exception. The book is quantitative but non-mathematical, without a single equation. Readers who may be uneasy about this are invited to read the appropriate references, where they will find equations reaching to the horizon.

A serious word about mathematics: students should not be misled by the absence of equations in this book. Most of the topics covered here, in particular space geodesy and geomagnetism, involve enormous amounts of mathematical analysis and computer data handling. Any student considering a career in geophysics or geology in the 21st century must have a fundamental grasp of higher mathematics, such as algebra, calculus, and numerical analysis, and computer science (such as programming).

A minor stylistic point: the expressions "manned space flight" and "manned missions" will be used without apology in this book. Women have for years flown on space missions, manned warships, piloted eight-engined bombers. They no longer need condescending euphemisms.

A suggestion to the reader: geology is very much a visual science. If you have never had a geology course, I urge you to buy one of the many popular geology guides, get out of town, and *look* at your local geology. (If you live in San Francisco, you don't have to leave town.) Once you have learned even the most elementary facts about geology, you will be amazed at what you can see – and understand – in even a road cut. When you travel by air, especially over the western US, try to get a window seat (on the shady side: left going east, right going west), ignore the movie, and enjoy the panorama from 35,000 feet. The strangest planet in the solar system is our own; explore it yourself.

Space exploration begins on the ground. If you have never had an astronomy course, I suggest that you buy a $40 pair of binoculars (not a telescope), and *look* at the sky, in particular the Moon. Even 7 × 50 binoculars reveal a surprising amount of detail along the terminator, and may show the Galilean satellites of Jupiter, which were invisible until the invention of the telescope. Even without binoculars, simply learning your way around the night sky with a cardboard star finder will be fun. The constellations are a compass, a map, a clock, and a calendar; learn why.

Paul D. Lowman Jr.

ACKNOWLEDGEMENTS

This book summarizes the efforts of thousands of people, and the reference list is implicitly a partial acknowledgement of their contribution. I am particularly indebted to my colleagues at Goddard Space Flight Center, far too many after 41 years to name individually, with a few outstanding exceptions.

The first of these is the late John A. O'Keefe, without whom this book would never have been written. He and Robert Jastrow, in 1959 head of the Theoretical Division, took a chance on me when I was a graduate student on the "Korean G.I. Bill," with no academic credentials beyond a B.S. from a state university. I owe my career to them. However, beyond that, John O'Keefe was the founder of space geodesy, as the dedication indicates. The late Eugene Shoemaker also credited him with being "the godfather of astrogeology." He was instrumental in bringing the US Geological Survey into NASA programs in 1960, and helped draw up the first scientific plans for the *Apollo* landings. His tektite field work took him to North Africa, Australia, and many parts of North America. He also played a little-known but critical role in the *Landsat* program, by getting terrain photography onto the experiment list for the later *Mercury* flights, as documented in the remote sensing chapter. John O'Keefe was, in summary, one of the great figures of the heroic age of space exploration, ranking with Goddard, von Braun, Shoemaker, Van Allen, and von Karman. He died in September 2000, but I am gratified that he saw the manuscript of the book and its dedication well before his death.

I was originally invited to write this book in 1993 by Dr Catherine Flack, then with Cambridge University Press. I accepted the invitation at once, one reason being that Neil Armstrong and I had co-authored a paper "New Knowledge of the Earth from Space Exploration" in 1984, which Neil presented at the Royal Academy of Sciences in Morocco. We covered the same major topics presented in this book, now supplemented by 15 years of progress.

Exploring Space, Exploring Earth was written with the encouragement and support of the Goddard Space Flight Center management. However, I wish to specifically acknowledge the support of two Director's Discretionary Fund (DDF) grants. The first of these, from Tom Young in 1983, was for a lineament study by Nick Short and me on the Canadian Shield. It led to a successful *Shuttle* Imaging Radar experiment and gave me access to both the geology and the geologic community of Canada, a world leader in the field. The second DDF grant, from Joe Rothenberg in 1997, provided support for the Digital Tectonic Activity Map that opens the book.

In a long career I have had several branch chiefs, all of whom were friends as well as supervisors. I must specifically acknowledge the decision of the late Lou Walter, first chief of the Planetology Branch, to transfer me to the Geophysics Branch during a major reorganization in 1973. By forcing me to broaden my scientific scope, Lou helped lay the foundation for this book. My present branch chief, Herb Frey, has given me continual encouragement and support through the years the book has been in preparation. Herb's own scientific work on the tectonic effects of large impacts, and on crustal magnetism, cited in the text, has been important in my studies of crustal evolution.

I wish to thank the many Canadian geologists who have helped me, often by vigorous arguments, in particular Lorne Ayres, Hayden Butler, Ken Card, Tony Davidson, Mike Dence, Dave Graham, Jeff Harris, Paul Hoffman, Darrel Long, Wooil Moon, John Murray, John Percival, Walter Peredery, Don Rousell, Vern Singhroy, John Spray, Hank Williams, and Susan Yatabe. My wife, Karen, has accompanied me on many Canadian trips, making them much broader experiences by persuading me to look at something beside outcrops.

The late Eugene Shoemaker was both a friend and a continual inspiration as a pioneer in comparative planetology. His sudden death in a 1997 auto accident in the Australian outback was felt by thousands of geologists, as a personal and professional loss.

The Library of Congress Geography and Map Division has been an invaluable resource in writing this book. Ruth Freitag and Barbara Christy in particular have been generous with critical reviews, and in providing information for compilation of the tectonic activity maps. At Goddard Space Flight Center, the staff of the Homer E. Newell Library has been continually cooperative and helpful for many years.

Technical reviews of specific chapters have been provided by Barbara Christy, Steve Cohen, John O'Keefe, Mike Purucker, Tikku

Ravat, Dave Rubincam, Terry Sabaka, Pat Taylor, and Jacob Yates, to whom I am indebted.

An eminent French scientist, commenting on the value of the *International Space Station*, said, in 1998, that he was unaware of any scientific discovery made by an astronaut. Taking a narrow view, neither am I, but this book would not have been possible without the enormous accomplishments of the American manned space flight program.

Since 1962 I have worked frequently in Houston with the staff of Johnson Space Center, originally the Manned Spacecraft Center, and thank them collectively for their invaluable support in efforts ranging from terrain photography to analysis of lunar samples.

A specific acknowledgement, also collective, is to the astronauts of the *Mercury*, *Gemini*, *Apollo*, and *Skylab* programs for their outstanding accomplishments in hand-held terrain photography. It is not generally realized that the *Landsat* program was originally stimulated by the spectacular 70 mm color pictures they returned. I had the privilege of working with these men – the best of the best – and have tried to document their fundamental contributions in the remote sensing chapter and in the referenced article on the *Apollo* program.

This book is in a sense the final report on two radar experiments carried on the 41-G *Shuttle* mission of 1984, focussed on the origin of the Grenville and Nelson Fronts, east and west boundaries respectively of the Superior Province of the Canadian Shield. I thank the crew of the 41-G mission for their performance in overcoming in-flight problems, my co-investigators in the US and Canada for their contributions, and Dr R. E. Murphy for his support of the 41-G investigation.

The contribution of astronauts to space research, returning to the comment cited above, is not generally appreciated. The crew of HMS Beagle made no scientific discoveries. But their passenger did; he was Charles Darwin. My final acknowledgement is therefore to those whose day-to-day work laid the foundation for *Exploring Space, Exploring Earth*: astronauts, of course; deck hands on oceanographic expeditions; field geologists the world over; jug-hustlers in seismic reflection profiling; darkroom technicians; computer programmers; motor pool dispatchers; librarians; secretaries; and many more, too many to thank specifically. This is their book.

Paul D. Lowman Jr.

CHAPTER 1

Preview of the orbital perspective: the million-year day

1.1 Introduction

"The present is the key to the past." This axiom is familiar to all geologists, expressing the belief that the Earth has evolved over billions of years by processes still going on today. In recent years there has been something of a revival in catastrophism, with the realization that rare events such as meteoritic impacts and sudden glacial floods do occasionally happen. These events – and better known sporadic ones, such as volcanic eruptions and great earthquakes – require us to reconsider our concept of the geologic "present." How long must we watch the Earth to get a realistic picture – a day in the life of the Earth, so to speak – of geologic processes?

The geologic events of a month or even a decade clearly do not represent the full spectrum of geologic activity. The map (see Fig. 1.1) that opens this book, derived largely from space-acquired data, was designed to illustrate the tectonic and volcanic activity of the last one million years – a "million-year day." Mathematically-inclined readers may think of it as a first derivative – the instantaneous rate of change – of a conventional tectonic map. The tectonic activity map and related ones are presented at this point as a preview, to demonstrate at once the fundamental impact of orbital data on geology and geophysics.

1.2 A digital tectonic activity map of the Earth

It is only in the last 40 years, since about 1965, that we have finally arrived at something like a true understanding of "the way the Earth works" in the title of Wyllie's (1976) geology text. As late as 1962, for example, it could still be reasonably proposed in a leading journal (Chenoweth, 1962) that the deep ocean floor, if uncovered, would be a primordial cratered surface resembling the face of the Moon. As it happened, the same year saw the publication of Hess's (1962) classic "essay in geopoetry" (his words), which laid the foundation for the new global tectonics, soon to be termed "plate tectonics," showing that the oceanic crust is geologically young and mobile.

Plate tectonic theory holds that roughly two-thirds of the Earth's crust, the ocean basins, is both active and, in a geologic sense, ephemeral. Oceanic crust, chiefly basaltic, or "mafic," is continually generated by volcanic eruptions along features known but little understood for years: the mid-ocean ridges such as that bisecting the Atlantic Ocean. Its Pacific counterpart is the East Pacific Rise. These ridges are seismically active, an expression of processes in the deep interior generating basaltic magma that is erupted along the ridges (generally under water with the notable exception of Iceland, sitting astride the Mid-Atlantic Ridge). This newly generated crust moves away from the ridges at a few centimeters a year in the now-familiar process of "sea-floor spreading," a term coined by Dietz (1961), who credited Hess with the concept although an earlier version had been proposed by Arthur Holmes (1931).

The Earth is tectonically a closed system, and newly generated crust in most areas is eventually destroyed by "subduction," in zones several hundred kilometers deep in which the oceanic crust descends, to be re-absorbed in the mantle by processes still not fully understood. (Most active volcanos, notably those of the Pacific rim, occur over these subduction zones.) These phenomena collectively account for the term "ephemeral" used to describe oceanic crust, in that sea-floor spreading and subduction recycle this crust in a few hundred million years. Far from being primordial as suggested by Chenoweth, very little of the basaltic ocean crust is more than about 200 million years old. The continental crust in contrast is at least four billion years old and as will be argued later – from comparative planetology – may be fundamentally "primordial."

Since the emergence of the theory just outlined, many global tectonic maps have been published. However, regardless of the validity of plate tectonic theory itself, these maps have all been unconstrained by time limits, showing all mappable features of whatever age. The word "features" is critical to understanding the map presented here, which is focussed on phenomena rather than features, the phenomena being tectonic and volcanic activity of the "million-year day."

Two versions of the digital tectonic activity map (DTAM henceforth) are presented. The first (Fig. 1.1) is a composite of shaded relief with superimposed tectonic and volcanic features; the second (Fig. 1.2) is a schematic map showing the same features in purely symbolic form, with the addition of continental/oceanic crust boundaries. The following discussion of the DTAM will be focussed

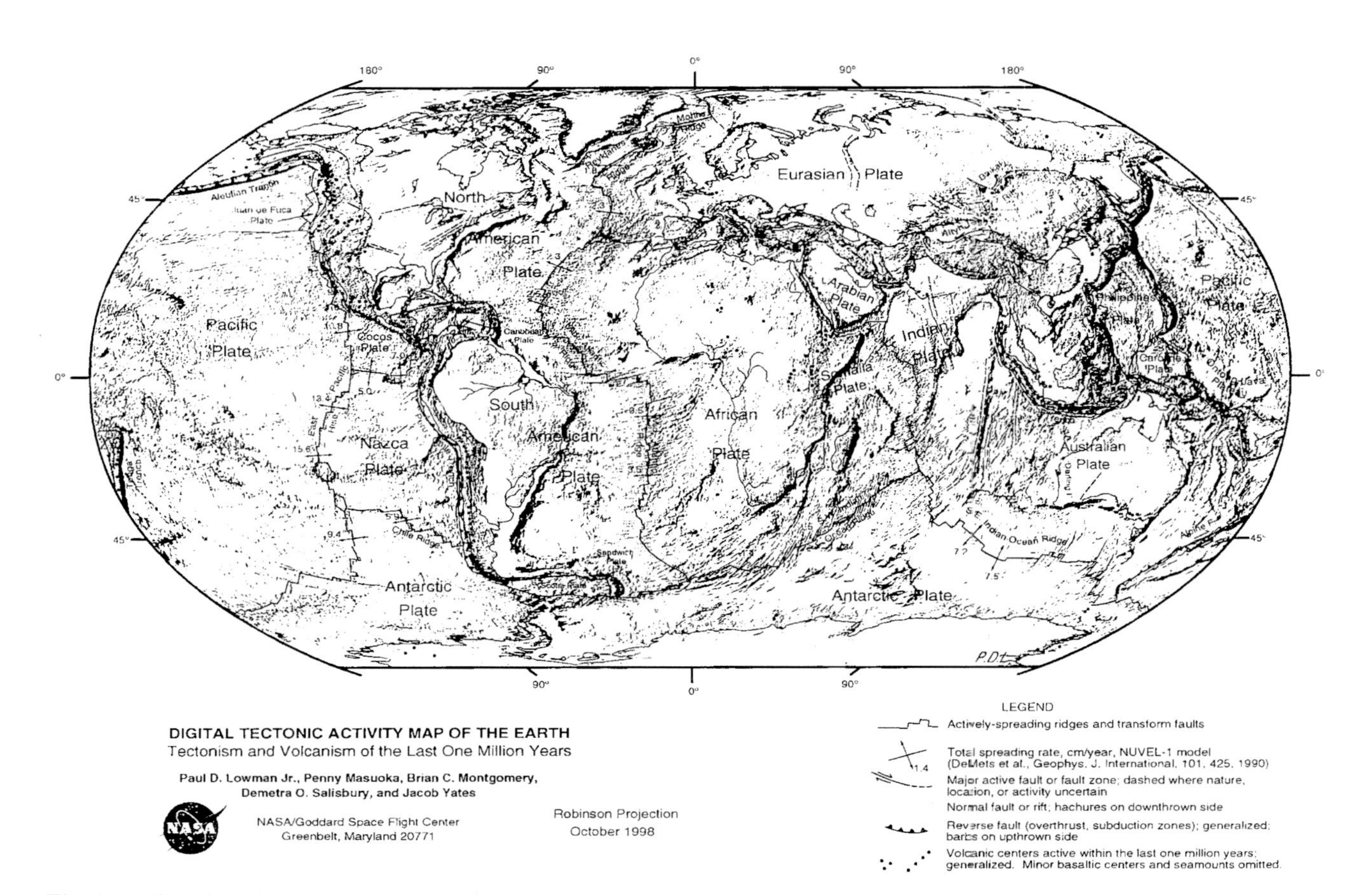

Fig. 1.1 (See also Plate I) Digital tectonic activity map (DTAM) of the Earth, based on shaded relief map largely generated from satellite altimetry.

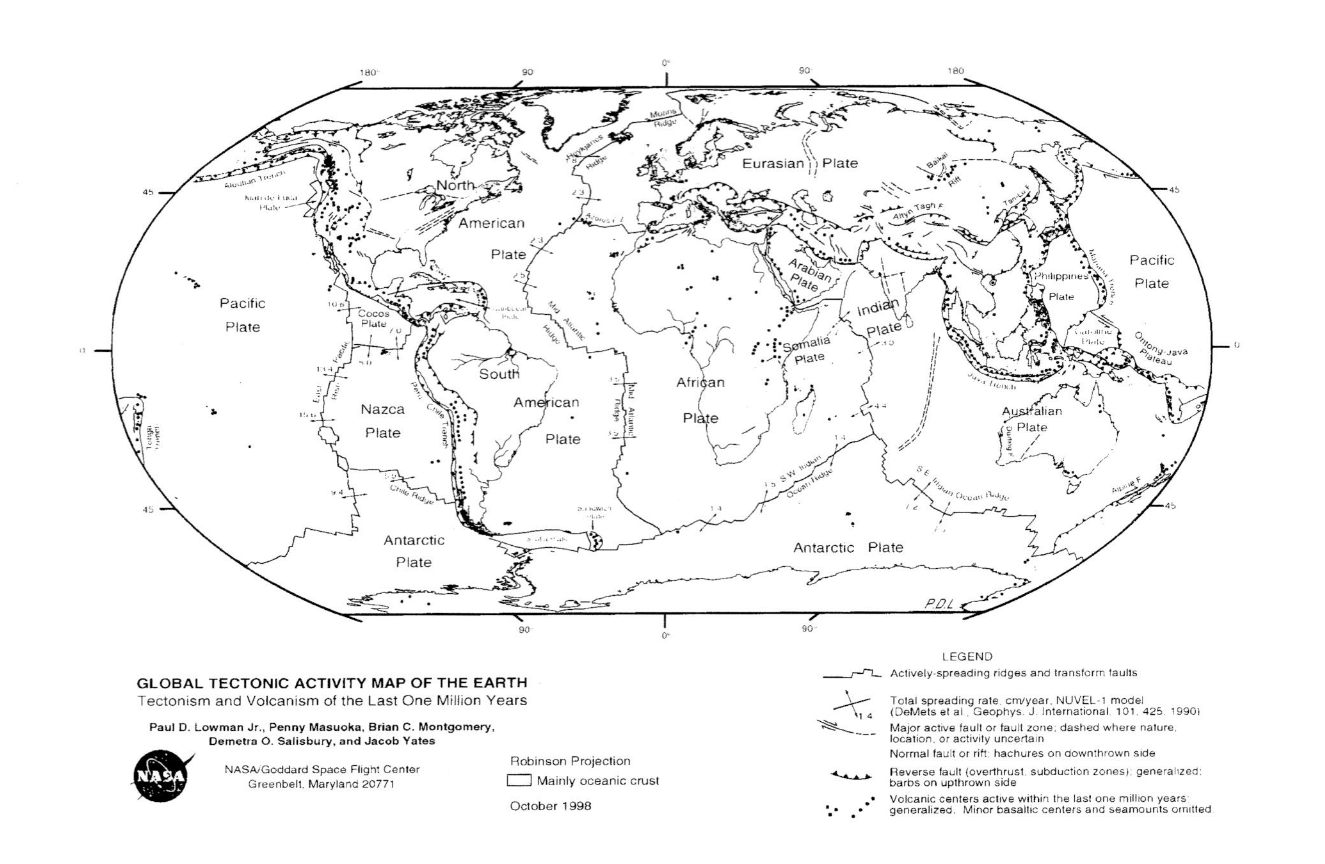

Fig. 1.2 (See also Plate II) Schematic global tectonic activity map (GTAM), from Fig. 1.

primarily on the relief version, with the objective of summarizing the contributions of space data to its compilation. A series of seismicity maps, computer-drawn with the same scale and projection, was essential for the DTAM and one is accordingly included (Fig. 1.3). Software and major data sources are given by Lowman *et al.* (1999) and Yates *et al.* (1999).

1.3 Sea-surface satellite altimetry

The DTAM is derived from an enormous data base of surface and space-related surveys and studies. The contribution of space data begins with the delineation of the topography of the ocean basins (that is, bathymetry) by satellite altimetry. Comparable maps became available in the 1960s with publication of the now-classic hand-drawn maps of Bruce Heezen and Marie Tharp, for several decades familiar features of most introductory geology books. The Heezen–Tharp maps were based on conventional marine surveys, chiefly echo-sounding. But, since such surveys produced depth data along single survey lines they could not begin to show the topography of large uninterrupted areas of the ocean basins. Consequently, many features had to be drawn by extrapolation in unsurveyed areas.

This situation has now been remedied by sea-surface satellite altimetry (Sandwell, 1991; Smith and Sandwell, 1997). This method (discussed in detail in Chapter 2) depends on the fact, first demonstrated in 1973 by a radar altimeter carried on *Skylab*, that the mean sea surface (after correction for tides, currents, and the like) forms a very subdued replica of the underlying ocean-floor topography. The effect is suggestive of snow-covered ground. It is caused by the lateral gravitational attraction of ocean-floor relief features. A submerged volcano, for example, will pull the surrounding ocean toward itself, forming a very slight hump (generally a few meters at most) on the overlying ocean. The ocean floor adjacent to a trench will similarly pull the water very slightly away from the trench, which will thus be mirrored in the overlying sea surface.

In the decades since the phenomenon was first demonstrated, satellite altimetry has become an essential tool for mapping the ocean floor. Hundreds of thousands of satellite altimetry passes have been combined to produce a digital elevation model, available from the National Geophysical Data Center, of almost the entire ocean floor. It is that model on which the computer-drawn shaded relief map of Fig. 1.1 is based. The software used for its construction exaggerates the relief, emphasizing it with pseudo-illumination from the northwest. However, this map is fundamentally different

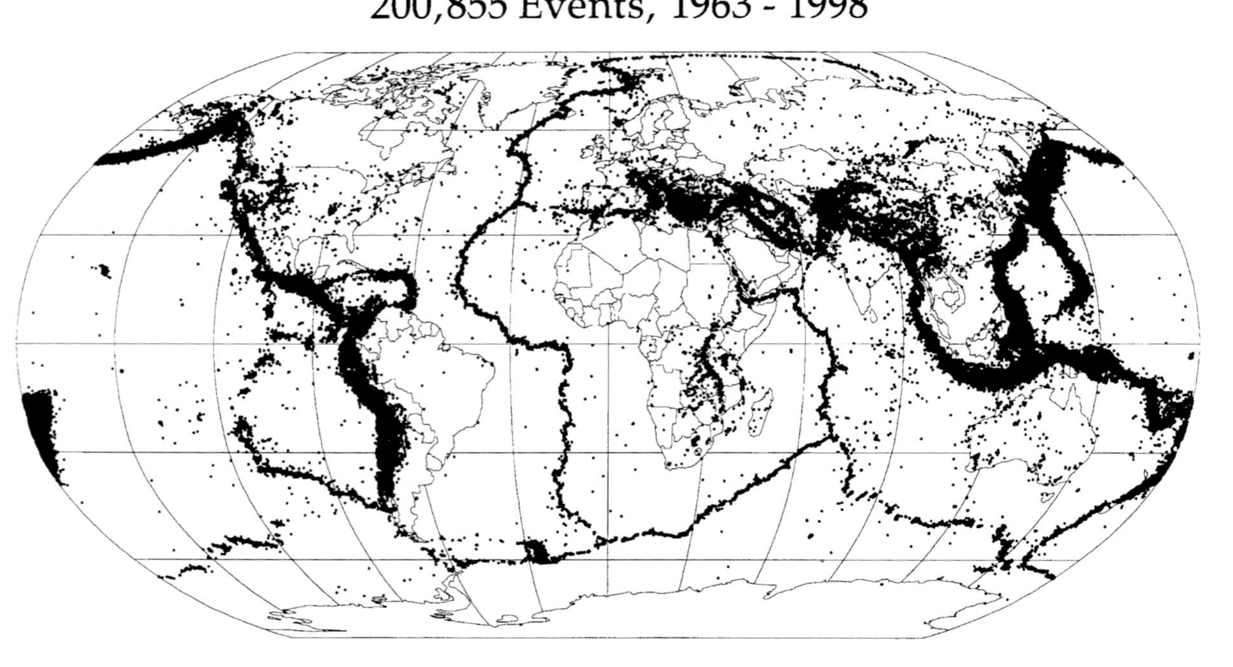

Fig. 1.3 Earthquake epicenter map, used to delineate zones of tectonic activity.

from previous maps in being an objective one, not an artist's rendition. Subject to scale and resolution limits, satellite altimetry has given us a view of some ⅔ of the Earth's surface that was until recently much more poorly mapped than the near side of the Moon. The active tectonic features shown in the schematic DTAM (Fig. 1.2) are thus offered as reasonably objective representations.

1.4 Satellite measurement of plate motion and deformation

The digital tectonic activity map shows recent estimates of total sea-floor spreading rates from the mid-ocean ridges. These estimates owe nothing to space techniques, being based on the distances of dated magnetic anomalies from spreading centers, as will be explained in Chapter 3. However, space geodetic techniques (Robbins *et al.*, 1993), specifically satellite laser ranging (SLR) and very long baseline interferometry (VLBI) have made it possible to measure intercontinental distances with precisions on the order of one centimeter. The *Global Positioning System* (*GPS*) is now filling in areas with denser measurement nets. It is hardly necessary to point out that such an achievement was unimaginable before satellite methods and radio astronomy were available.

The contribution of satellite distance measurements to the DTAM lies in the direct demonstration (Fig. 1.4) independently by SLR and VLBI, that several islands in the Pacific Ocean, such as Kauai, are moving northwest toward Japan at rates of roughly 6–7 centimeters per year. These rates are very close to those implied by the NUVEL-1 sea-floor spreading estimates, which are averages for about the last 3 million years. To extrapolate these spreading rates to the motion of entire plates of course requires the assumption that the plates are internally rigid. But here SLR and VLBI are again useful, by showing that the inter-island distances are essentially constant, that the Pacific Plate is indeed rigid to a close approximation.

In summary, the dynamic behavior of the oceanic crust as outlined by plate tectonic theory has been directly observed by space geodetic methods, confirming estimates made by totally independent measurements along the mid-ocean ridges.

1.5 Satellite remote sensing

The tectonic activity of large areas, especially in southern Asia, as shown on the DTAM is derived to a major degree from satellite remote sensing data, especially *Landsat* imagery. As will be

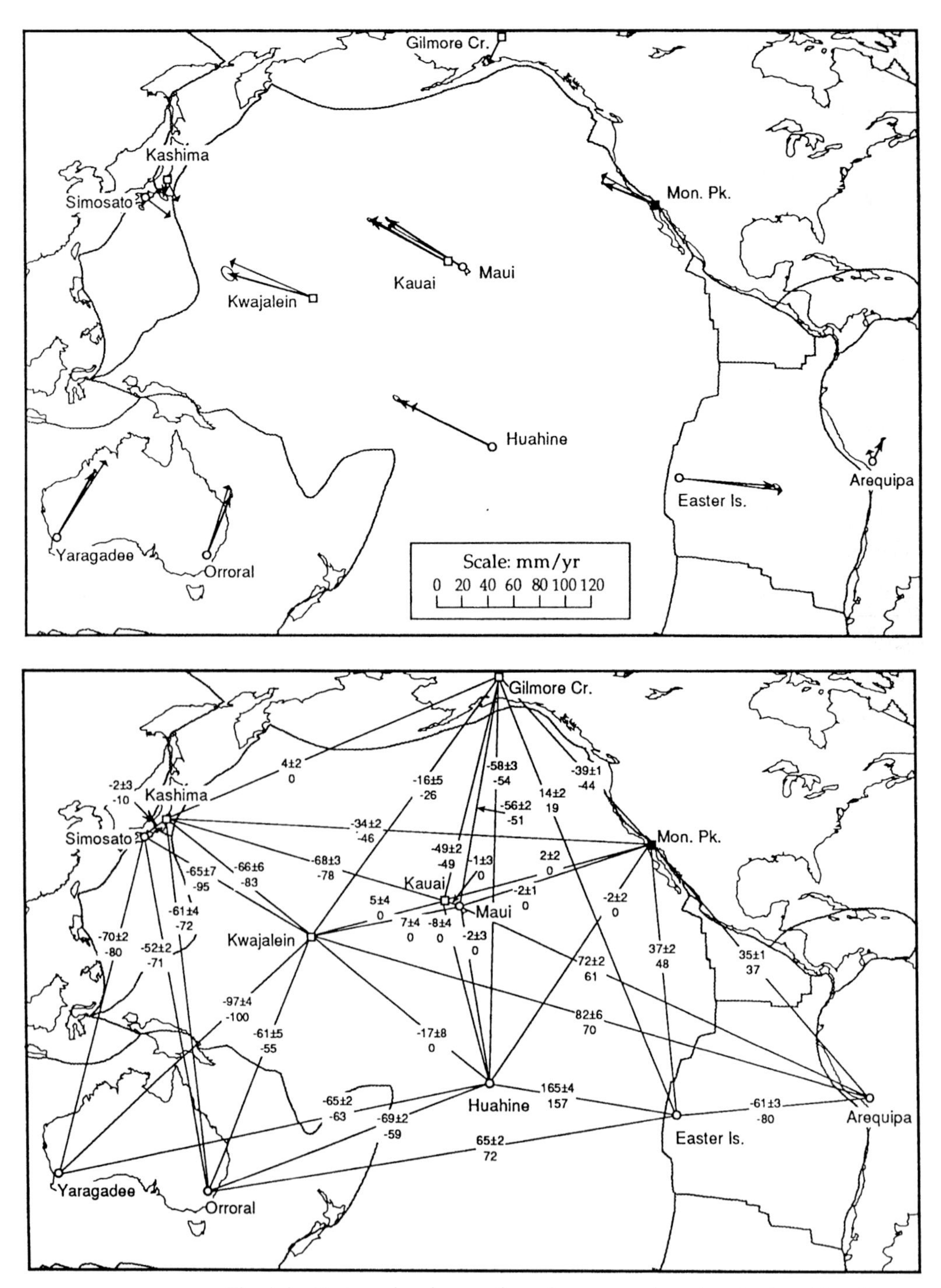

Fig. 1.4 Vectors showing motion of space geodesy sites (*top*) and baseline lengths (*bottom*). Values in lower diagram are observed length changes (combined SLR and VLBI solutions) above and length changes predicted by NUVEL-1 model. From Robbins *et al.* (1993).

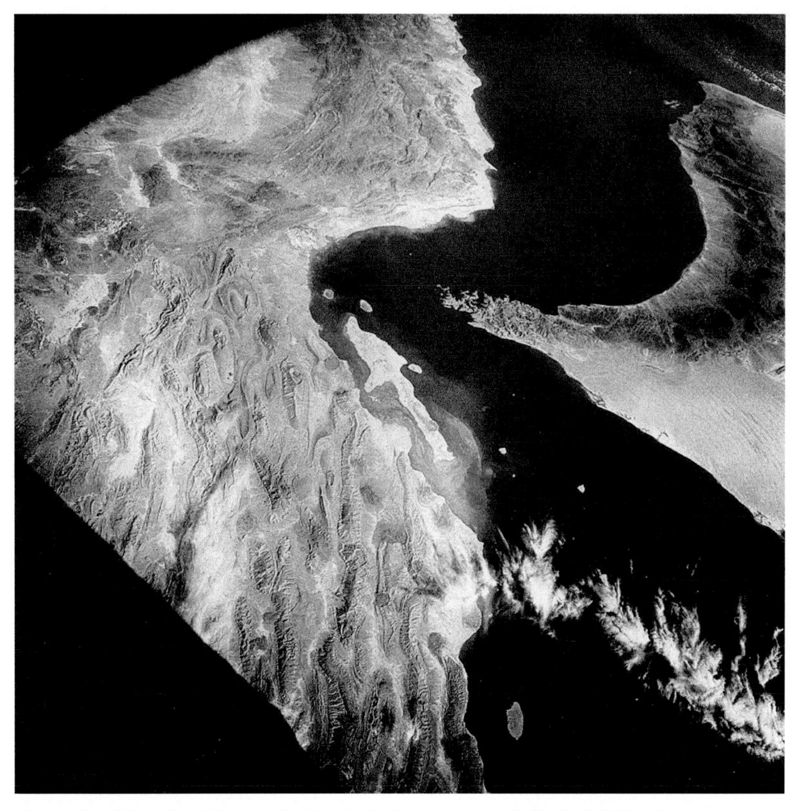

Fig. 1.5 (See also Plate III) Gemini 12 photograph S66-63082; view to east over the Zagros Mountains (left), Strait of Hormuz, and Makran Range. Width of view 600 km. From Lowman (1999).

described in Chapter 4, the tectonic structure of Asia, especially north of the Himalayas, was almost completely unknown until the availability of satellite imagery, starting with hand-held 70 mm photographs (Fig. 1.5) taken by *Mercury* and *Gemini* astronauts (Lowman, 1999). The Tibetan Plateau, for example, was essentially inaccessible for many years because of its remoteness and, after World War II, political barriers. Satellites surmounted these barriers, providing superb synoptic views of hundreds of thousands of square kilometers. *Landsat* images were first used to produce tectonic maps of Tibet. Chinese geologists were among the first to use *Landsat* for similar maps, and in the years since, satellite imagery has become widely used to map the tectonic structure of not only southern Asia but other parts of the Alpine fold belt and even supposedly well-mapped areas such as northern Norway.

Satellite imagery not only reveals the existence of large faults in remote areas, but often gives a good idea of their current activity by showing features such as fault scarps and offset streams. This is only the beginning of a rapidly expanding field, since satellite radar interferometry and space geodesy by means of the *Global Positioning System*, to be described in Chapter 2, are rapidly complementing satellite imagery.

The DTAM is explicitly a very generalized map, especially in its representation of continental volcanism. The contribution of space data here also comes from satellite imagery. As will be described, many previously-unmapped volcanos have been found with *Landsat* imagery and astronaut 70 mm photography starting with the *Gemini* missions of the mid-1960s. However, the real value of satellite imagery for showing volcanism is simply in making the map's compiler aware of geomorphically fresh, and hence young, volcanos and lava flows in previously-mapped but remote areas. The best available compilations of active volcanos are restricted to historical records, covering the last 10,000 years. By revealing many other young but historically inactive volcanos and lava flows, satellite imagery has made it possible to produce the first global representation of volcanic activity extending back about one million years.

1.6 Satellite magnetic surveys

The existence of the Earth's magnetic field has been known for centuries, and scientific study goes back to the time of Queen Elizabeth I, as we will see in Chapter 3. However, it has been an extremely difficult feature to study. To begin with, it is highly variable, from minute to minute during solar storms, or over hundreds of thousands of years during reversals of the Earth's main magnetic field. Additionally, the magnetic poles are constantly moving. A recent Canadian geophysical expedition on Ellesmere Island found that the north magnetic pole actually passed under them while they were at one camp site. (Canada has been a traditional leader in geomagnetic studies; it may help to own one of the magnetic poles.) Moreover, magnetism in general was not understood until the 19th century, in particular until the work of James Clerk Maxwell.

Study of the Earth's crustal magnetism was given a jump start, so to speak, during World War II, when greatly improved magnetometers for submarine detection were developed. Aeromagnetic studies in the early post-war years discovered many valuable ore deposits, in Canada and elsewhere. However, such studies are time-consuming and expensive. Mapping of crustal magnetism in the oceans is even more time-consuming when done by ships. The next big step forward in the study of geomagnetism came with the launching of artificial satellites.

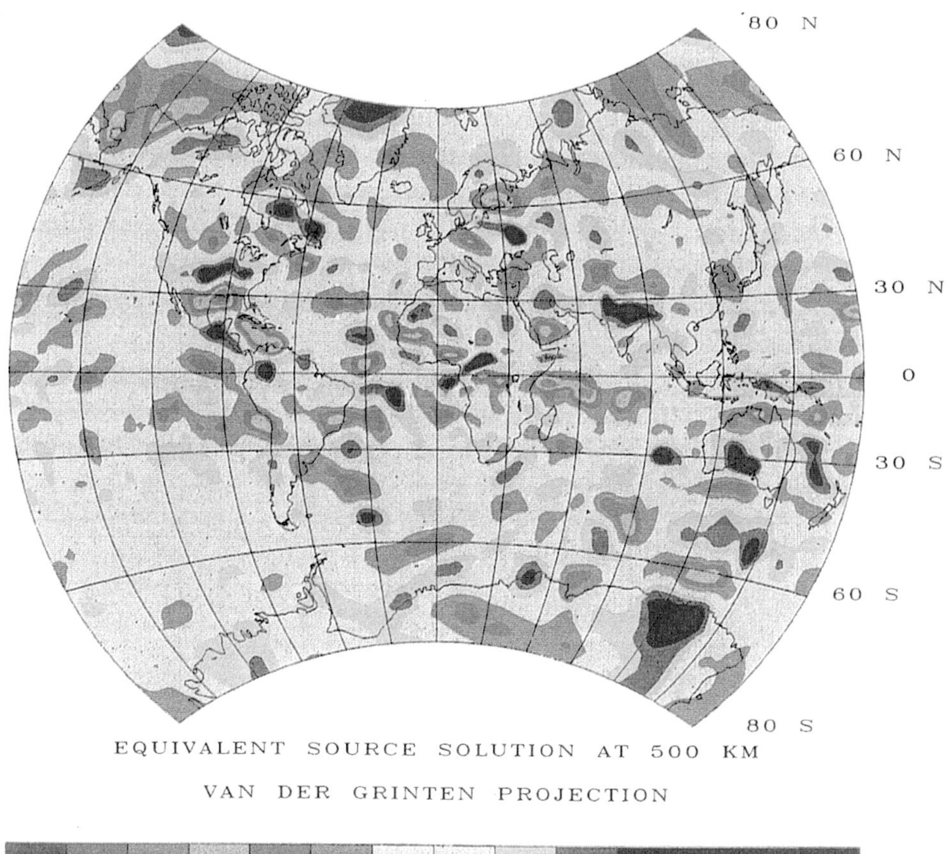

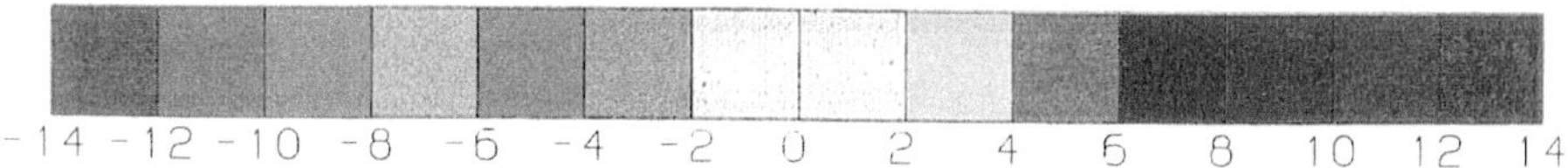

Fig. 1.6 (See also Plate IV) Scalar (non-directional) magnetic anomaly map from *POGO* data, equivalent sources at 500 km altitude, reduced to pole. Values in nanoteslas (nT). From Langel (1990).

A major contribution of satellite data has come from measurements of crustal magnetism by satellites, notably the *Polar Orbiting Geophysical Observatory* series (*POGO*) and *Magsat*. They are illustrated (Fig. 1.6) by an early map from *POGO* data, compiled by the late R. A. Langel (1990), *Magsat* Project Scientist and a pioneer in the use of satellite magnetic data. Principles and details of such maps will be covered in Chapter 3. At this point, it will just be mentioned that this *POGO* map, actually using data from three different satellites, illustrates the promise of satellite data as well as the difficulties in using them.

The "promise" is demonstrated simply by the global coverage of this map (accompanied by polar projections, not reproduced here). Entire ocean basins, and large unsurveyed continental areas, are shown, admittedly in relatively little detail. One previously unknown feature, the Bangui anomaly in central Africa, was essentially discovered in satellite data, as will be discussed in Chapter 3. However, the compiler of this map had to allow for different satellite altitudes, geomagnetic latitudes, and short-term variations in the magnetic field caused by external influences. The term "reduced to pole" refers to the problem of the inclination of the Earth's main magnetic field, assumed to produce the observed features by induction. If one were at the north magnetic pole, the field would be pointing essentially straight down, in contrast to the magnetic equator, where it would be horizontal. In this map, Langel reduced all data to a common inclination, vertical, as if the features were at the north magnetic pole.

This and other magnetism maps typify the advantages of satellites for earth observations: global coverage, rapid and systematic repetition of coverage, and broad field of view. These advantages can be discussed further by returning to the digital tectonic activity map.

1.7 Origin and significance of the digital tectonic activity map

The post-*Apollo* realignment of NASA roles and missions, in the early-1970s, led to increased emphasis at Goddard Space Flight Center on solid-earth geophysics and space geodesy. Focussed on the Earth, these fields required realistic global maps of tectonic activity. However, it was found that even the best small-scale tectonic maps were highly interpretive representations intended to illustrate plate tectonic theory. Furthermore, they were generally not time-constrained, illustrating many features as old as Mesozoic (some 200 million years ago).

Fortunately, in 1975 the National Geographic Society published "The Physical World," a hand-drawn Van der Grinten projection color representation of the Earth. Far outclassing all previous maps of this type, the National Geographic map was used (Lowman, 1982) to compile "Global Tectonic and Volcanic Activity of the Last One Million Years," (an updated version of which appears elsewhere in Chapter 4). This global tectonic activity map has proven useful in many applications. It served as the base map for laying out baselines for SLR and VLBI measurements in the NASA Crustal Dynamics Program, and was used for interpretation of satellite gravity and magnetic measurements as illustrated in Chapters 2 and 3. However, its widest application was in geologic education, appearing in more than a dozen textbooks and many general articles.

The flood of new information about the Earth's topography, from satellite remote sensing and sea-surface altimetry among other things, combined with new computer-based cartographic techniques, made a modernized version of the 1982 tectonic activity map desirable and possible. The DTAM (Lowman *et al.*, 1999) was the result. The point of this historical account is that the very origin of the DTAM can be directly traced to space research, which provided not only the means to compile the DTAM but the necessity to produce it and its predecessor.

This chapter is essentially an illustrated essay, leaving out most references and postponing detailed discussion of the DTAM to later chapters. However, a few of the most fundamental implications of this map should be briefly mentioned.

The most significant aspect of the DTAM concerns the question: *How well does plate tectonic theory describe and explain the structure and activity of the solid earth?* The theory has been eloquently described by Hamblin (1978) as "a master plan into which everything we know about the Earth seems to fit." A more recent text (Press and Siever, 2001) describes it as an "all-encompassing theory," treating most physical geology topics in a plate tectonic context. Plate maps are familiar features of every geology text and most popular articles about geology, and any good student can point out the major plates. However, the DTAM shows that such maps, and the theory they illustrate, are at best oversimplifications of a complex and active planet. For example, the Eurasian Plate is shown on most "plate maps" as including the entire area from Iceland to Indonesia. If a plate is defined, loosely, as a torsionally rigid segment of lithosphere bounded by some combination of ridges, trenches, and transform faults, the artificiality of an Iceland–Indonesia "plate" is obvious.

The California part of the boundary between the North American and Pacific Plates is generally shown, and described, as the San Andreas fault, whereas the DTAM shows tectonic activity reaching to at least the Wasatch Mountains and probably beyond. (As described in Chapter 2, space geodesy has verified this.) The boundary between the African and Eurasian Plates is not a line but a broad zone including the intensely active Mediterranean Sea and the Alpine fold belt. It is clear that there are large areas of the Earth's surface, notably on the continents, that simply can not be assigned to any particular plate. A frequent approach to this problem is to postulate "microplates." However, as discussed by Thatcher (1995), the crust in many areas seems to behave as a continuum. Which description – continuum or microplate – is correct remains to be seen. This problem will be discussed in more detail in Chapter 2. At this point, it will simply be suggested that plate tectonic theory is at best an incomplete description of the Earth.

A related aspect of plate tectonic theory is its explanation of the Earth's geologic behavior, i.e., its tectonic and volcanic activity. The theory as commonly presented in today's textbooks often explains features such as folded mountain belts as the result of "continental collisions," or more accurately "plate convergence." The DTAM and its associated seismicity map (Fig. 1.3) should raise questions at once about this facile explanation. For example, the western part of the Alpine fold belt, from Pakistan to the Atlantic Ocean, is intensely active. Can this activity be explained by continental collisions, or plate convergence? Plate tectonic models, such as the one (NUVEL-1) used for the DTAM spreading rates, all show even the schematic Eurasian and African Plates as moving extremely slowly. Two leading proponents of plate tectonic theory, K. C. Burke and J. T. Wilson (1972) have pointed out that the absence of age-progressive volcano chains on Africa ("hot-spot trails" in plate tectonic terms) indicates in fact that Africa has been *stationary* with respect to the mantle for at least 25 million years (Burke, 1996). The DTAM thus brings out the paradox that some of the most intense tectonic activity on Earth occurs in a broad zone between two plates that hardly appear to be moving at all.

Perhaps the most controversial and speculative question raised by the DTAM is whether plates actually carry continental crust along with them, as implied by the common phrase "plate tectonics and continental drift." Here it will simply be pointed out, with reference to Fig. 1.2, that movement of the Eurasian Plate involves the rotation of a block of lithosphere extending from Iceland to Siberia. Can such a block behave as a rigid unit? The tectonic activity of western Europe, and even of the Urals (commonly thought to have been formed in the late-Paleozoic), throws strong doubt on the rigidity of this "plate." Furthermore, the DTAM illustrates a problem brought out by the author, discussed in Chapter 3: the driving force for such a plate. Can sea-floor spreading, or ridge push, in the North Atlantic actually move a segment of lithosphere covering some 120 degrees in longitude? The answer will at this point be left as an exercise for the reader. (The author's views are given in Skinner and

Porter, 1995.) But it can be suggested that among the most important implications of the digital tectonic activity map are the uncertainties, anomalies, and questions it illustrates in the supposed "master plan," plate tectonic theory. Science progresses by recognition of such anomalies: the supposedly unsuccessful Michelson–Morley experiment, Fleming's spoiled Petri-dish culture, Plunkett's clogged Freon cylinder being just three examples. Consensus is satisfying, comfortable, and easy to teach, but it can turn into stagnation. The DTAM, and the findings of space research in general, may help prevent such stagnation.

CHAPTER 2

Space geodesy

2.1 Introduction

Space geodesy is a new term for a new science, one in which satellite tracking and related techniques are used to study the shape, deformation, and motions of the Earth. However, this "new" science has a history going back millennia. As Neil Armstrong's Foreword makes clear, scientists of today using satellites and radio telescopes are continuing the work begun by Eratosthenes, whose value for the circumference of the Earth was within 1% of the true value. Even the very first artificial satellites produced new knowledge of the Earth's size and shape. Four decades of successive satellite and manned spacecraft missions have added enormously to this knowledge, and the field of space geodesy has turned out to have great scientific and applied value, value only now beginning to be realized.

The first formal proposal to use satellites for what is now called space geodesy was by J. A. O'Keefe (1955), as part of a proposal to the National Science Foundation for "an artificial unmanned earth satellite" (Haley and Rosen, 1955). O'Keefe proposed a reflective inflated satellite, to be illuminated by searchlights or radar. Tracking of such a satellite would permit (1) precise measurement of intercontinental distances, (2) mapping of absolute gravity values, and (3) calculation of the Earth's semimajor axis. This was the first and for several years the only forecast of the applications of satellites to the solid-earth sciences. O'Keefe's proposal was actually carried out in 1966 when the *PAGEOS* (*PAssive GEOdetic Satellite*) was launched (Newell, 1980), although by that time space geodesy was well under way with other satellites and techniques, as we shall see.

Space geodesy is the application of various space-related techniques to the problems of what can be called *geophysical (*or *physical) geodesy* and *geometrical geodesy*. Geophysical geodesy is the study of the interior of the Earth, by precise measurement of features such as the gravity field. An alternative and equally valid use of "geophysical geodesy" is the study of the "slow deformations of the Earth" (Lambeck, 1988). This usage reflects the fact that the apparently solid crust and mantle are in fact dynamic features, as demonstrated by earthquakes, volcanic eruptions, and tilting of the land such as that which is slowly lowering London below sea level. Geometrical geodesy is closer to the traditional definition, being concerned with the precise measurement of positions on the surface of the Earth. The term "precise" has taken on new meaning with the advent of space geodesy, with trans-oceanic distances now being measured routinely with a precision close to the length of a well-trimmed fingernail. "Geometrical geodesy" now includes topographic mapping from space by means of satellite altimetry of the mean sea surface and of land elevations. A third subdivision of space geodesy is becoming distinctive enough to merit a new name, *earth dynamics*, referring to variations in the Earth's rotation rate, its orientation in space, and similar phenomena.

Space geodesy is a difficult subject to treat concisely and clearly, for several reasons. The field is a large and rapidly growing one. Geophysical and geometrical geodesy and earth dynamics are closely interrelated, divisions among them being somewhat artificial. Data from any one technique, such as satellite laser ranging, can be applied to the study of a wide range of geophysical parameters, from sea-floor topography to rotation of the Earth. Furthermore, interpretation of geophysical data in general is beset by inherent ambiguities. Regional gravity anomalies, for example, can be interpreted in terms of variations in rock type, thickness of the crust, temperature of the lithosphere, or convection in the upper mantle. A final difficulty is the rapid pace of space geodesy technique development, a topic already filling books by itself. This new field has involved remarkable technological and mathematical progress in areas far removed from satellite launching. It is safe to say that had an artificial satellite been launched in 1940, only the most elementary geodetic use could have been made of it, lacking laser and microwave techniques and especially today's massive computing capability (Cohen and Smith, 1985). However, let us begin with a brief summary of the main techniques used in space geodesy.

2.2 Space geodesy methods

The oldest technique of space geodesy is satellite tracking, i.e., precise monitoring of a satellite's orbit. The satellite need not be artificial, for "tracking" of the Moon by astronomers long before *Sputnik* provided fundamental, if imprecise, information on the Earth's rotation and structure (O'Keefe, 1972; Rubincam, 1975).

This technique was in effect modernized when laser retroreflectors were put on the lunar surface. However, most space geodesy involves tracking of artificial satellites in low earth orbit, "low" meaning sub-synchronous (<36,000 km) altitudes. Good historical summaries of satellite-tracking techniques have been presented by O'Keefe (1958), Murray (1961), and King-Hele (1983, 1992).

Optical tracking is in principle the simplest method, requiring nothing from the satellite but its visibility. Instruments used profitably can be as simple as binoculars and stopwatch. Early satellites were tracked by the "Moonwatch" program organized by F. L. Whipple, in which arrays of ground observers with simple home-built telescopes set up optical fences for satellite observation. Most professional satellite tracking is done with wide-angle telescopes, such as the Baker–Nunn cameras operated in the early years by the Smithsonian Astrophysical Observatory, or with their modern electronic equivalents using charge-coupled device (CCD) arrays. The first American satellite launched for geodetic purposes, *ANNA 1-B*, carried a flashing light; simultaneous photographs from several ground stations permitted precise triangulation among the stations (Schmid, 1974). Optical tracking is still producing useful data, as shown by King-Hele (1983, 1992), and though being superseded by other geodetic techniques, is still needed for study of atmospheric density at orbital altitudes because this changes with solar cycle and other phenomena.

Radio tracking has also been widely used since the first satellites were orbited. The two main methods are Doppler tracking and interferometry. Doppler tracking depends on the frequency shift of an orbiting transmitter as it approaches and then recedes from the tracking antenna. Usually the antenna is on the Earth, but the new French system, DORIS, (Cazenave *et al.*, 1993) uses a network of radio beacons on the ground whose signals are received by suitable receivers on satellites. Satellites can track other satellites using the Doppler technique. Satellite-to-satellite tracking, as described by Kahn and Bryan (1972), has the advantage of permitting continuous tracking throughout an entire revolution, hard to achieve with ground stations for low satellites because of the number of stations required. Satellite-to-satellite tracking data are becoming widely used (Lemoine *et al.*, 1998b), involving the *Global Positioning System* (*GPS*, to be described) and the *Tracking and Data Relay System Satellites* (*TDRSS*).

Interferometry uses arrays of orthogonally oriented antennas and measurement of the phase difference of the satellite signals received by each antenna; a good description can be found in

Glasstone (1965). The NASA Minitrack system (now part of the Space Tracking and Data Acquisition Network) uses interferometry, especially for low earth orbit satellites.

Radar tracking is an active technique, in contrast to the optical and radio methods just described. Its big advantage is independence of satellite visibility and weather. Radar space tracking installations are of course much more expensive to operate than the those using passive methods, and they are less accurate than optical methods because of the longer wavelengths used (generally in the centimeter range). Radar and interferometry are combined in the US *Navy Naval Space Surveillance* system, using an array of six interferometer stations extending from Georgia to California in an east–west direction.

The most accurate tracking method of all uses satellite laser ranging (SLR), in effect an optical radar, now capable of range accuracies of 1–3 millimeters (Degnan, 1993). First used in the 1960s for precise orbit determination, SLR has become a primary geodetic tool. The technique uses lasers mounted in combination with telescopes, sending pulses every few seconds to cube corner reflectors mounted on satellites. Cube corners have the property of reflecting incoming radiation back in the same direction it came, and are widely used for SLR. The main disadvantage of lasers for satellite tracking is the narrowness of the beam; the satellite orbit must be determined precisely so that the laser beam will hit the satellite. In addition, the technique depends on good weather. More than a dozen reflector-carrying satellites have been launched at this writing, the flagships being *Starlette* (France), *LAGEOS I* (US), and *LAGEOS II* (Italy) (Fig. 2.1). These are high-density satellites in very stable, high-altitude orbits, planned so as to minimize the effects of atmospheric drag (Marsh *et al.* 1985; Christodoulidis *et al.* 1985; Tapley *et al.*, 1993) and the high-frequency harmonics of the gravitational field, i.e., the smaller irregularities. International efforts in both earth satellite ranging and lunar laser ranging, to be discussed, are coordinated through the International Laser Ranging Service.

There are two general types of techniques for using satellite-tracking data (Smith *et al.*, 1991): dynamic and geometric. The simplest method is geometric geodesy, in which the satellite is observed simultaneously from more than one ground station and the observations then used for triangulation or trilateration. This is essentially the method originally proposed by O'Keefe (1955), in which the satellite serves as a passive target, whose orbit must be known only well enough to find it. The method is limited by several factors, such as the location of ground stations, weather conditions for all ground

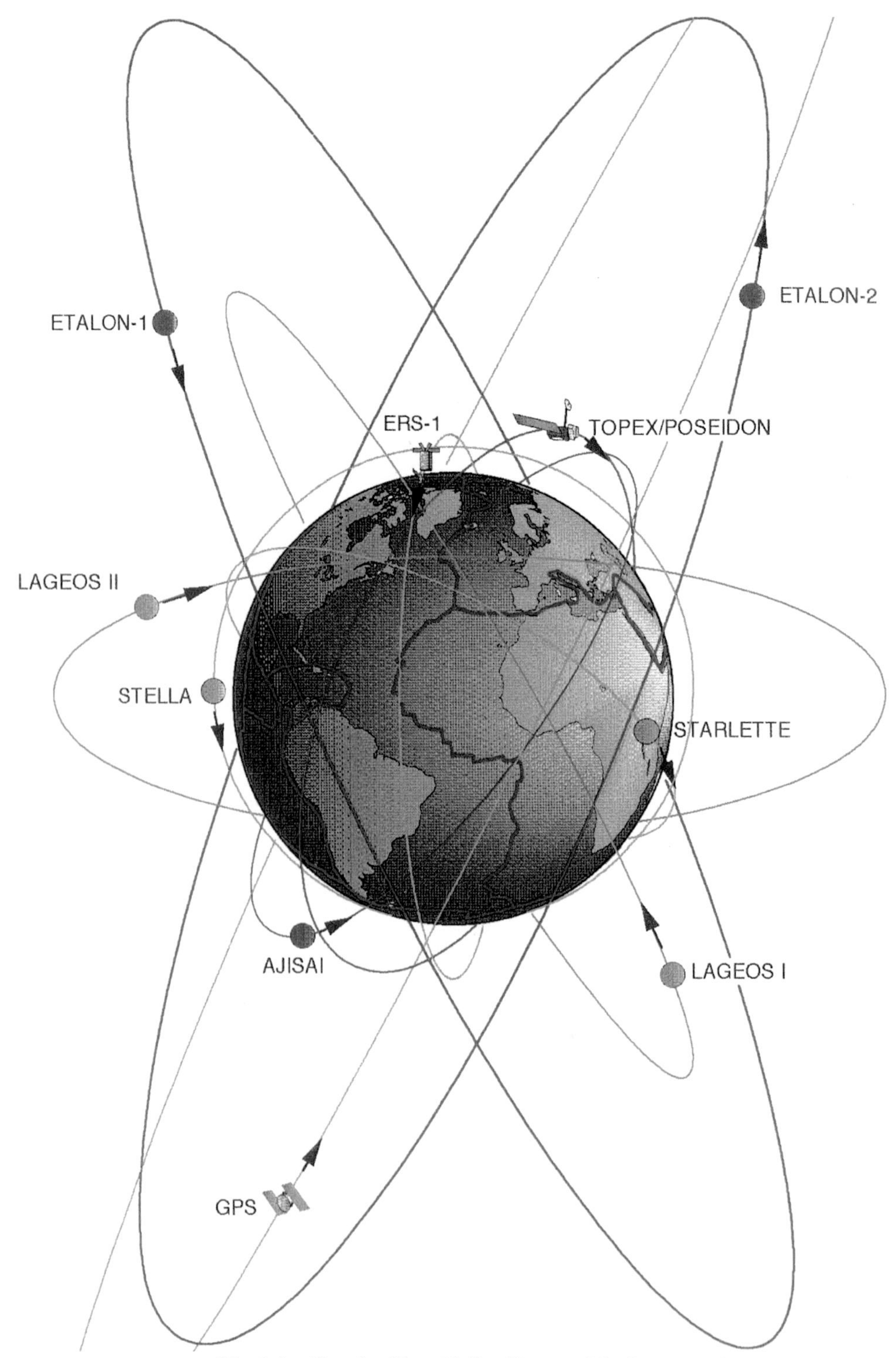

Fig. 2.1 (See also Plate V) Satellites used for laser ranging.

stations at any given time, and the accuracy of data from each station. Consequently, the increased number of retroreflector-bearing satellites (Tapley *et al.*, 1993) has greatly improved the effectiveness of geometric space geodesy. Dynamic satellite geodesy, in contrast, uses the satellite in effect as a sensor passing through the Earth's gravity field. It is far more complicated, requiring accurate modeling of the satellite orbits and great computing power. In addition to the gravitational forces, the main factor controlling the orbits, a wide range of other disturbing forces such as atmospheric drag, solar radiation, and terrestrial radiation must be allowed for. For meaningful geodetic use of the tracking data, the motions of the ground stations, such as earth tides, must be known. Tectonic setting and local ground conditions must also be considered (Allenby, 1983). However, if all these complexities can be overcome, dynamic satellite geodesy as carried out with *LAGEOS* can produce valuable data in addition to the precise distance between ground stations: the broad features of the gravity field (i.e., the low harmonics), polar motion, earth rotation, and earth tides. Frequent intercomparison among SLR, VLBI, and terrestrial geodesy show that the dynamic approach has been highly successful (Sauber, 1986; Ryan *et al.*, 1993). To show the wide range of data types used in mapping the global gravity field, a table of satellites used for earlier models, from Lemoine *et al.* (1998b), is presented (Table 2.1).

The general process of geodetic satellite tracking is an iterative one, so to speak, in that tracking data can be repeatedly fed back into the determination of satellite orbits, increasing the accuracy of the final result. This is especially important for the *Global Positioning System*, as will be discussed below.

The ultimate SLR experiment uses earth-based lasers and cube corner reflector arrays on the surface of the Moon (Fig. 2.2) (Bender *et al.*, 1973; Williams, *et al.*, 1993). A diffuse laser return had been received in 1962 from the surface by L. D. Smullin and G. Fiocco (Lambeck, 1988), but could not be used for precise ranging. However, beginning with *Apollo 11*, three retroflector arrays were emplaced by *Apollo* astronauts, and two French arrays were carried on the Soviet *Lunokhod* remotely-driven vehicles. The simplicity of the technique may obscure the tremendous technological difficulty of lunar ranging. Despite their coherence, laser beams obey the inverse square law (both ways, so that the beam loses strength in proportion to the *fourth* power of the Earth–Moon distance). The telescopes that detect reflections from the lunar arrays are sometimes detecting only one photon of the 10^{18} transmitted by each pulse. Laser-ranging measurements with a precision of a few centimeters

Table 2.1 *Satellites and orbital characteristics used for JGM-1 and -2 (Joint Gravity Models, also in honor of late James G. Marsh, GSFC). From Lemoine* et al. *(1998b).*

Satellite	*a* (km)	*e*	Inclination (°)	Perigee height (km)	Mean motion (rev/d)	Primary resonant period (d)	Data type
ATS-6	41867	0.0010	0.9	35781	1.01	92.8	SST
Peole	7006	0.0162	15.0	515	14.82	2.1	L
Courier-1B	7469	0.0174	28.3	989	13.46	3.8	O
Vanguard-2	8298	0.1648	32.9	562	11.49	2.7	O
Vanguard-2RB	8496	0.1832	32.9	562	11.09	294.3	O
DI-D	7622	0.0842	39.5	589	13.05	8.4	O,L
DI-C	7341	0.0526	40.0	587	13.81	2.5	O,L
BE-C	7507	0.0252	41.2	902	13.35	5.6	O,L
Telstar-1	9669	0.2421	44.8	951	9.13	14.9	O
Echo-1RB	9766	0.0121	47.2	1501	12.21	11.9	O
Starlette	7331	0.0200	49.8	785	13.83	2.8	L
Ajisai	7870	0.0010	50.0	1487	12.43	3.2	L
Ana-1B	7501	0.0070	51.5	1076	13.37	4.8	O
GEOS-1	8075	0.0725	59.3	1108	11.96	7.0	O,L
ETALON-1	2550	0.0007	64.9	19121	2.13	7.9	L
TOPEX/POSEIDON	7716	0.0004	66.0	1342	12.80	3.2	L,D
Transit-4A	7322	0.0079	66.8	806	13.85	3.5	O
Injun-1	7316	0.0076	66.8	895	13.87	3.8	O
Secor-5	8151	0.0801	69.2	1140	11.79	3.4	O
BE-B	7354	0.0143	79.7	902	13.76	3.0	O
OGO-2	7341	0.0739	87.4	425	13.79	3.8	O
OSCAR-14	7448	0.0030	89.2	1042	13.50	2.2	Dp
OSCAR-7	7411	0.0242	89.7	848	13.60	3.2	Dp
5BN-2	7462	0.0058	90.0	1063	13.46	2.4	O
NOVA	7559	0.0010	90.0	1123	13.20	6.3	Dp
Midas-4	9995	0.0121	95.8	1505	8.69	3.0	O
SPOT-2	7208	0.0015	98.7	840	14.17	6.2	Dp
GEOS-2	7711	0.0308	105.8	1114	12.82	5.7	O,L
Seasat	7171	0.0010	108.0	812	14.29	3.1	O,L,R,A
Geosat	7169	0.0010	108.0	754	14.30	3.0	Dp,A
LAGEOS	12273	0.0010	109.9	5827	6.39	2.7	L
GEOS-3	7226	0.0010	114.9	841	14.13	4.5	L,A,SST
OVI-2	8317	0.1835	144.3	415	11.45	2.2	O

Key: L = Laser, Dp = TRANET/OPNET Doppler, O = Optical, D = DORIS, R = Radar, A = Altimetry, SST = sat.-to-sat. range rate

Fig. 2.2 Laser retroreflector placed on the Moon during *Apollo 11* mission to Mare Tranquillitatis. Photograph by N. A. Armstrong.

to these arrays are being carried out by two observatories at this writing (in the US and France), and continue to produce useful data. (The fact that these unprotected optical surfaces continue to reflect normally more than twenty-five years after emplacement is an important demonstration that the lunar surface would be suitable for emplacement of astronomical instruments (Lowman, 1996).)

Some of the most productive space geodesy techniques have been termed "satellite positioning." These invert the procedure described for satellite tracking, in that the satellites are used to, in effect, track points on the surface of the Earth. Early navigation satellites, such as the US *Transit* system, transmitted radio signals which, when received on the surface, could locate the receiver to within roughly a kilometer by the Doppler frequency shift. Since then, there have been enormous increases in accuracy and coverage of navigation satellites, with the 24-satellite American *Global*

Positioning System (*GPS*) (Hager *et al.*, 1991) and a corresponding Soviet one, the *GLONASS* system (Daly, 1993). The *GPS* was started by the Department of Defense in 1978, but by the early-1990s, as the constellation was completed, it became very widely used for geodesy in addition to its primary function of real-time navigation. Accordingly, it will be described in some detail.

The *GPS* uses a constellation of 24 active *NAVSTAR* satellites (including spares) in 20,000-km high 12-hour orbits, arranged so that at least four satellites are visible to ground receivers at any one time (UNAVCO, 1998). The satellites, whose orbits are continually updated, transmit coded radio signals giving their positions and the time of transmission (Yunck *et al.*, 1985). Ground receivers compute the transit time, thus finding the range to each satellite. Ranges from three satellites gives the position of two points (intersection of three spheres), one of which is known to be on the surface of the Earth. Range to a fourth satellite allows correction for clock errors. The *GPS* is intended to give real-time absolute locations to within 10 m accuracy, which it does very well. However, it was found rapidly that relative locations, horizontal and vertical, could be measured to accuracies of a few centimeters despite the necessary security measures (e.g., selective availability) necessary for this military system. Differential *GPS*, using fixed reference receivers in combination with mobile ones, is coming into very wide use for precision location. Small hand-held units can achieve accuracies of 2–10 meters, and survey-grade ones better than 1 meter (UNAVCO, 1998). The *GPS* has opened a "new era" (Hager *et al.*, 1991) for the study of tectonic activity and many other scientific applications, as will be described later in this chapter. An interesting synergism is between geodetic satellites carrying laser retroreflectors and *GPS*, in that new satellites are equipped with *GPS* receivers.

Most satellite geodesy techniques might have been understood by Eratosthenes, but one of the most productive – satellite radar altimetry – is genuinely novel. Radar altimetry from satellites measures the instantaneous distance above the ocean surface (Kahn *et al.*, 1979). This technique was originally intended to monitor the oceans themselves – currents, eddies, and fronts. However, as shown when first tried from *Skylab*, radar altimetry also reveals the topography of the ocean floor, i.e., the bathymetry. Depressions such as trenches produce geometrically-similar depressions in the mean sea surface; elevations, such as seamounts, produce slight elevations of the sea surface. The principle behind this is roughly that positive topography tends to pull the ocean horizontally toward it; for depressions, the surrounding higher topography pulls the ocean away. A useful

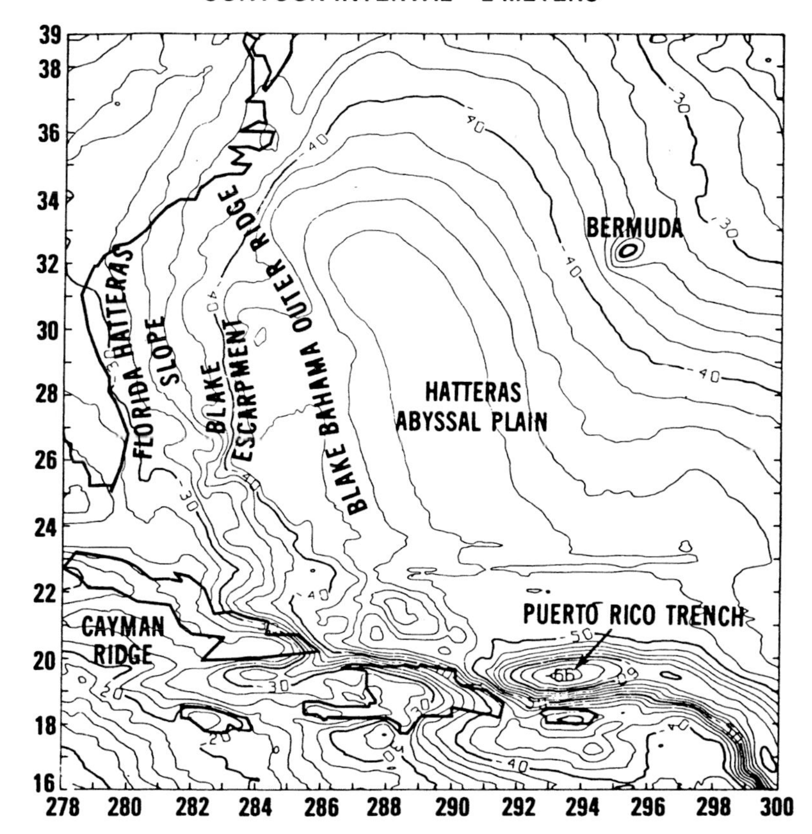

Fig. 2.3 Map of marine geoid north of Puerto Rico, from *GEOS-C* radar altimetry. From Marsh and Chang (1985).

example is from an early study by Marsh and Chang (1985), using *GEOS-C* altimeter data over the Atlantic Ocean (Fig. 2.3). The effects on the mean sea surface are surprisingly great. Bermuda, for example, has pulled the surrounding ocean up roughly 3 meters; the Puerto Rico Trench is reflected in a 22 meter depression relative to the coast.

The ratio between the sea-surface deflection and the underlying topography, termed the "admittance" by Cazenave and Dominh (1987), is generally a few meters per kilometer, and is by itself a potentially valuable indicator of crust and mantle conditions, such as the isostatic compensation mechanism (Rapp, 1989). Satellite altimetry, although originally intended to monitor oceanographic conditions, has proven a remarkably effective way to map the ocean-floor topography (Sandwell, 1991; Smith and Sandwell, 1997). The base map for the digital tectonic activity map (see Fig. 1.1) is the latest version of a series of such maps, earlier versions of which have been produced by Haxby (1987) and others. A popular account of Haxby's work has been published by Hall (1992).

The map based on the Sandwell and Smith data is presented here (Fig. 2.4), this time without tectonic features overlaid, to illustrate the resolution of their techniques, combining surface and space data. Their map is properly termed a gravity map, but because of the effects just described is also a bathymetric map. The great advantage of satellite altimetry over satellite tracking is that because of the much smaller footprint of the radar (usually around 10 km), the spatial resolution is far better than that of purely orbital methods, as shown by the shaded relief map.

The most comprehensive map of the Earth's gravity field at this writing is the Earth Gravitational Model 1996 (EGM96) (Lemoine *et al.*, 1998a, b). To illustrate both the techniques used and their relative precisions, a table of data types is presented (Table 2.2) To show two ways in which the gravity field can be represented, earlier maps by Marsh (1979), based on satellite tracking, are presented. The free-air gravity anomaly map (Fig. 2.5) shows the Earth's gravity field, in milligals. This is primarily the Earth's topography, since a free-air anomaly is that remaining after correction for height (but not intervening crust) has been made. The gravimetric geoid map (Fig. 2.6) shows the broad features of the equipotential surface, equivalent to an undisturbed global ocean. The geoid expresses deeper features, chiefly in the mantle. A tectonic and volcanic activity map (Fig. 2.7), similar to that in Fig. 1.2 but with the same projection as the gravity maps, is presented for comparison with them.

Another important technique, Very Long Baseline Interferometry (VLBI), can be considered either not true space geodesy or the ultimate space geodesy, since it depends on ground-based radio telescope interferometry of the radiation emitted by quasars several billion light-years distant (Fig. 2.8). The technique uses fixed or portable radio telescopes that can be separated by any distance up to the diameter of the Earth. (The technique could be used for telescopes on the Earth and Moon, the Ultra Long Baseline Interferometry proposed by J. O. Burns, 1988.) Originally developed as an astronomical technique to produce very high resolution radio images of distant objects, it was found that the distance between widely separated radio telescopes could be determined to precisions of a few centimeters. VLBI has developed more or less in parallel with SLR, permitting frequent intercomparison of the results from each method (Kolenkiewicz *et al.*, 1985).

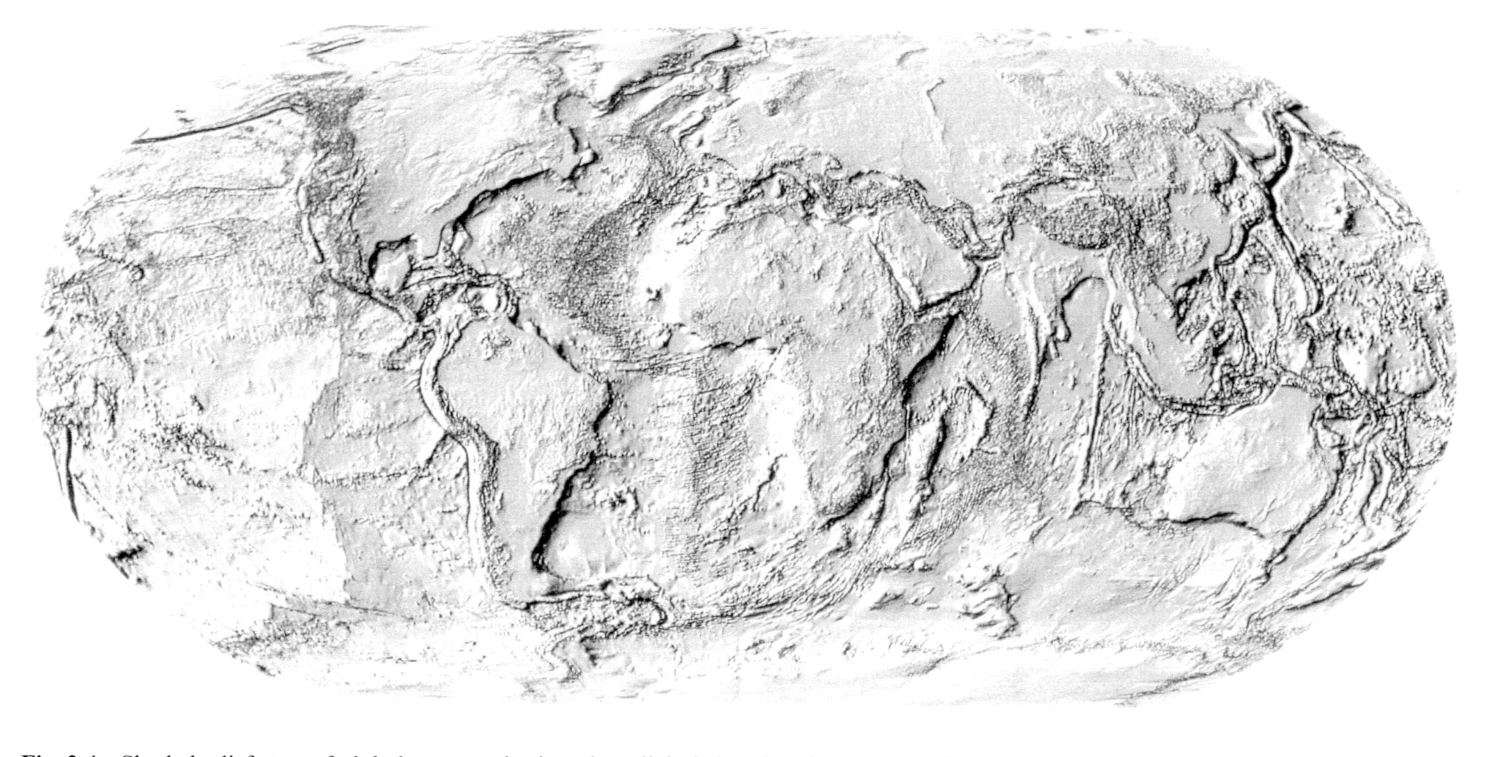

Fig. 2.4 Shaded relief map of global topography, based on digital elevation data from National Geophysical Data Center; marine gravity from Smith and Sandwell (1997). Robinson Projection; prepared by Penny M. Masuoka, GSFC.

Table 2.2 *Satellite-tracking methods. From Lemoine* et al. (*1998b*).

Technology	Configuration Observable types	Precision	Typical orbit fit	Strengths	Weakness	Period of use
Camera: Baker–Nunn MOTS SPEOPT	satellite image against stars, right ascension and declination; passive and/or active (i.e., spaceborne flashing lamp)	1–2 arcsec (10–20 m)	1–2 arcsec	first precision tracking systems	atmospheric shimmer star catalog errors passive data tracking limited to dawn/dusk geometry	1960–1974
Satellite laser ranging	2-way range, use restricted to satellites carrying retroreflectors	0.5 cm	2 cm (LAGEOS) 5 cm (Starlette)	most precise absolute range unbiased excellent optical refraction modeling	clouds obstruct observations only 40–60% of passes acquired early network limited in distribution	1968+
Radiometric ground-based	2-way range 2-way range-rate S band-> NASA-> active C-band-> DoD-> passive	1 m 0.3 cm/s	5 m 1 cm/s	first all-weather precision tracking system	single-frequency results in large ionospheric error measures biases	1972+
TDRSS (NASA)	1-/2-way ground–TDRSS–sat. range/range-rate single-frequency S and K band links	1 m 0.4 mm/s	1.5 m 0.8 mm/s	excellent global coverage of user sats. high precision	single-frequency transponder delay (range biases) TDRSs orbit errors	1983+

OPNET/ TRANET (USN)	1-way sat.–ground range-rate dual frequency (150 and 400 MHz)	0.2 cm/s	0.7 cm/s	good global network distribution	poor clocks large third-order ionospheric refraction errors 40% of data rejected	1965–1995, TRANET phasing out
DORIS (CNES)	1-way ground–sat. range-rate dual frequency (401.25 and 2036.25 MHz)	0.4 mm/s	0.5 mm/s	high-precision, all weather excellent global coverage	sat. tracks only one ground station at a time Note: the new DORIS system envisioned for the JASON mission will track two stations simultaneously. Additionally, the noise floor should be reduced to 0.1 mm/s	1992+
GPS (DoD)	pseudo-range/carrier phase (sat.-to-sat.)/(sat.-to-ground)	1–2 cm	1–2 cm	3-D navigation of low satellites unsurpassed coverage	controlled by DoD some on-orbit receivers cannot cope with antispoofing future receivers will use codeless technology and track 12+ sats.	1992+
Altimetry	2-way range (sat.–ocean) both single- and dual-freq. altimeters flown	1–2 cm	7 cm	precise range to directly map ocean-surface topography	limited by modeling of complex ocean-surface signals	1975

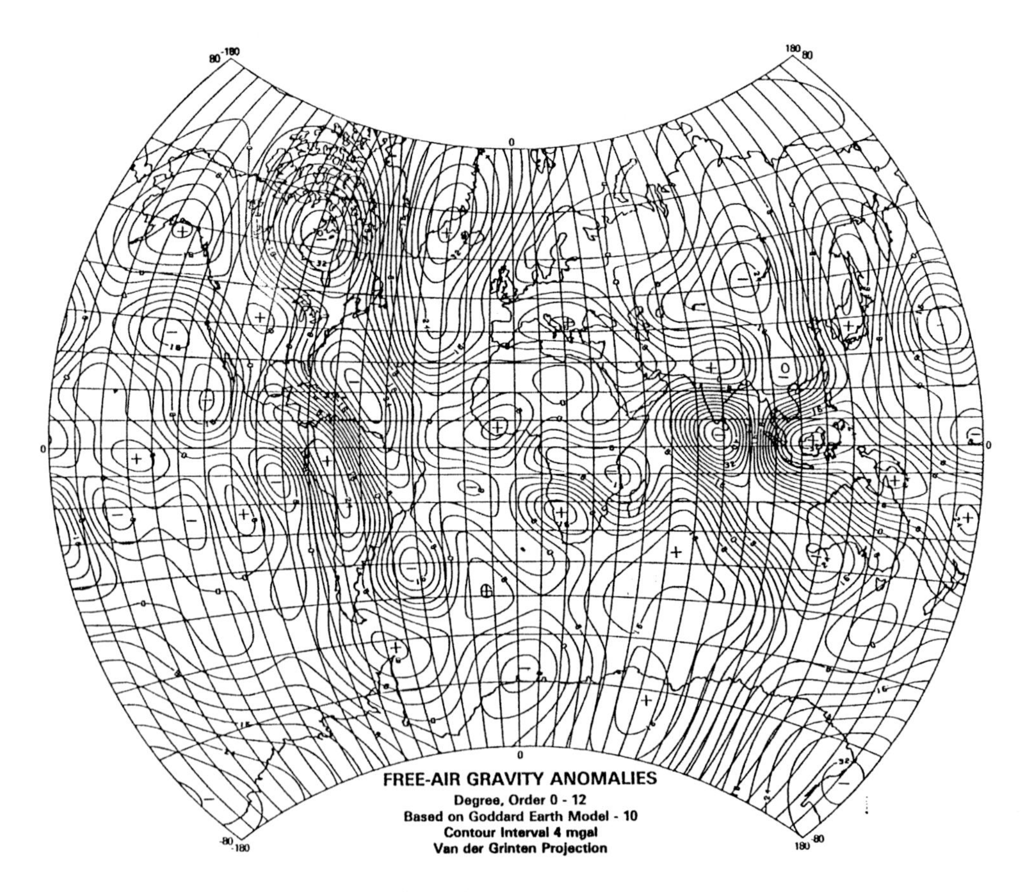

Fig. 2.5 Free-air gravity anomalies. From Marsh (1979).

In addition to its use as a terrestrial survey device, VLBI permits measurement of the Earth's rotation rate with unprecedented accuracy (Carter and Robertson, 1986; Dickey and Eubanks, 1985). The rotation rate of the Earth, or more properly changes in this rate, has been a long-standing issue, the problem being to explain the various changes in length of day (LOD). This topic clearly belongs in the "earth dynamics" category, but full understanding of changes in LOD bears on many problems in solid-earth geophysics, oceanography, and even meteorology.

Before moving on to the results of the techniques now in use, it should be pointed out that serendipity or "spinoff," chiefly in the form of unexpected uses for military technology, has played a continuing role in their geodetic application. J. A. O'Keefe's original proposal for a geodetic satellite was stimulated by his World War II experiences with mis-matched and inaccurate topographic maps,

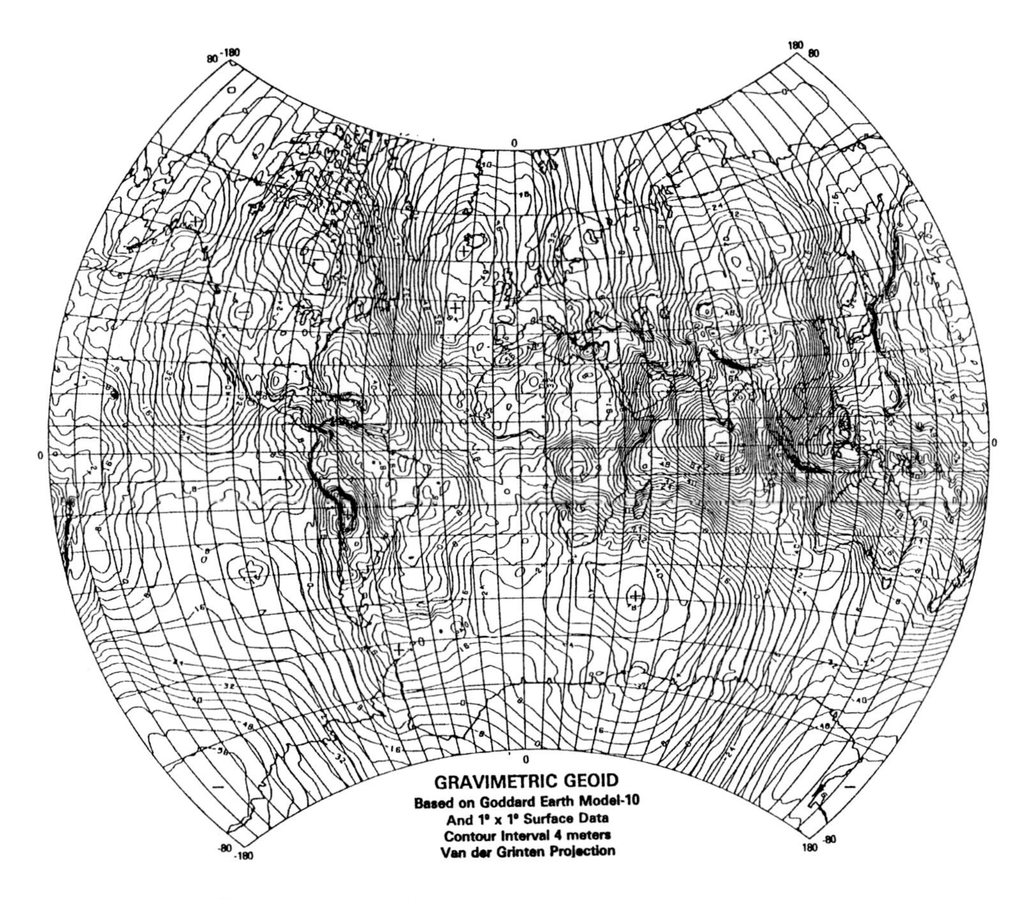

Fig. 2.6 Gravimetric geoid, from satellite and surface data. From Marsh (1979).

problems that had "serious effects" on the later conduct of the war in Europe (O'Keefe, 1966; this paper is highly recommended as an inside look at geodesists at work). Accurate global knowledge of the Earth's gravity field is a military necessity (Newell, 1980) for accurate control of navigation and photographic satellites, as well as for aiming long-range ballistic missiles. Radar altimetry was originally developed for the US Navy, for monitoring the condition of the ocean surface. The *Global Positioning System* was developed by the US Department of Defense as a military navigation tool. Satellite laser ranging was intended primarily for satellite tracking, and very long baseline interferometry was intended as an astronomical technique. The "spinoff" argument is often criticized by saying that we could get the same results by a direct approach. But the history of satellite geodesy shows that we often do not know what the direct approach should be. Furthermore, military requirements can

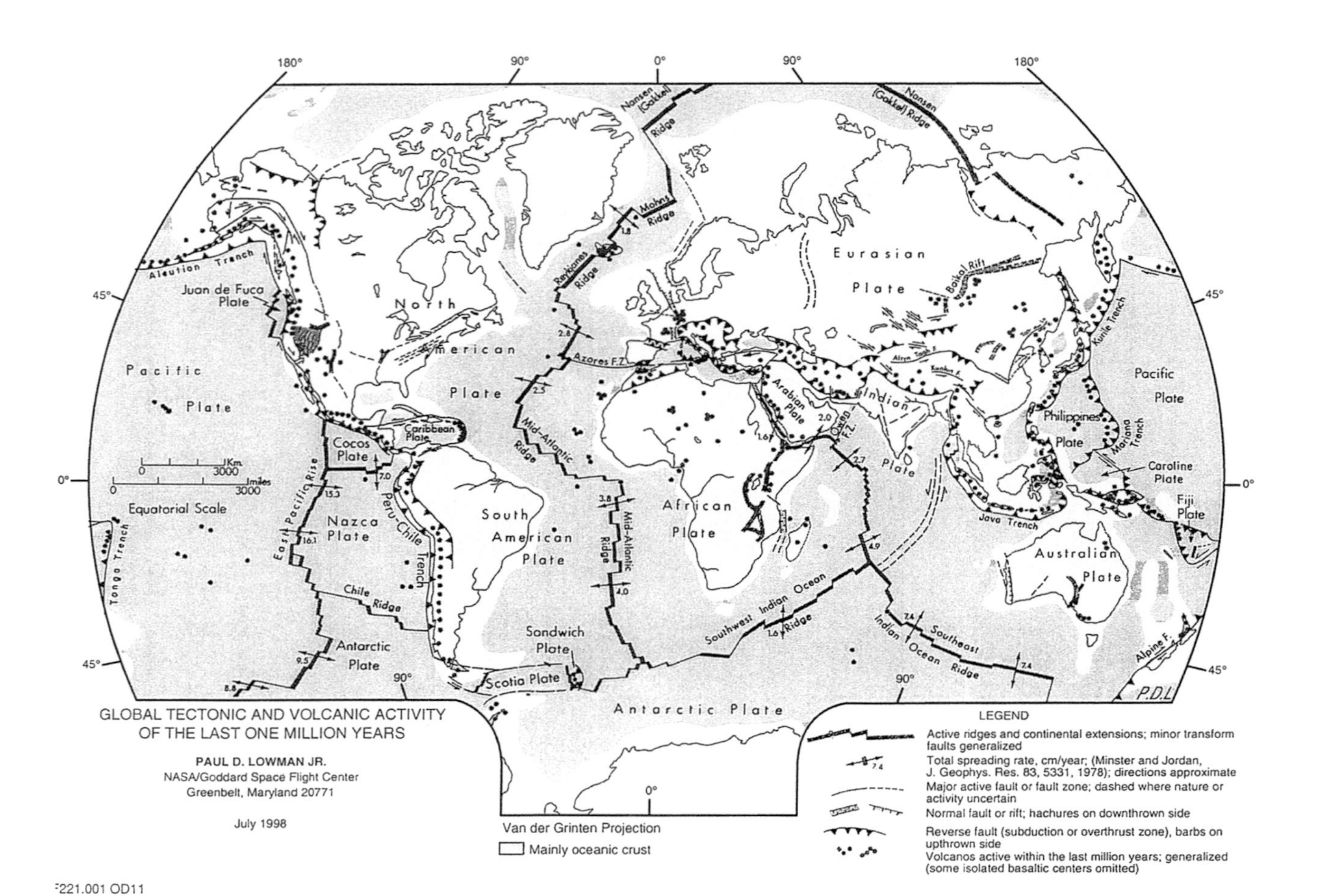

Fig. 2.7 Tectonic and volcanic activity map. See Chapter 1, this book.

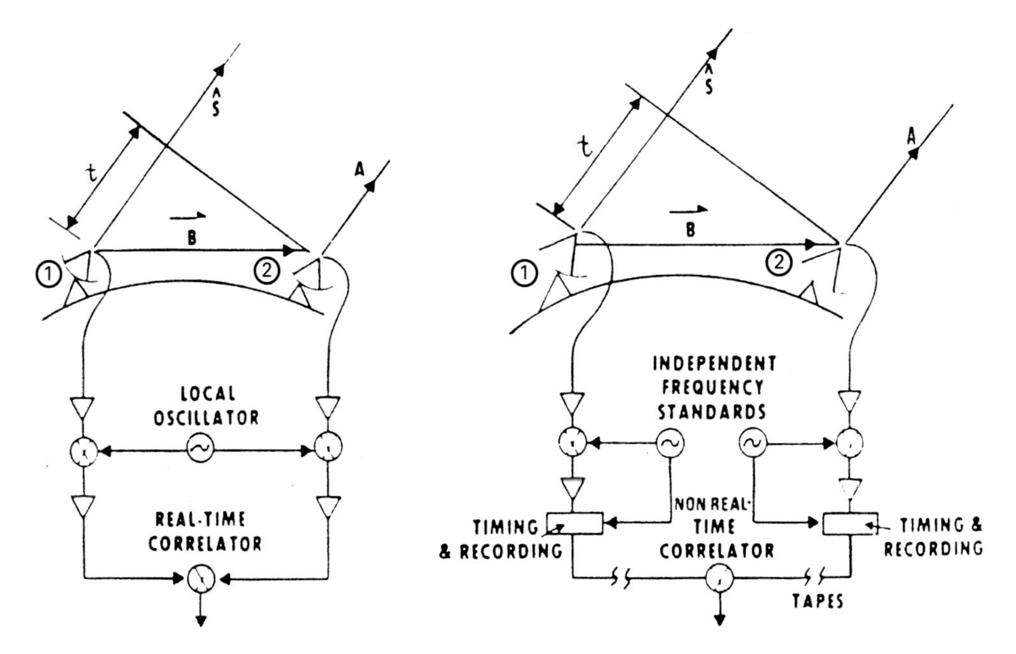

Fig. 2.8 Principles of radio telescope interferometry, for cable-connected instruments (*left*) and independent instruments (*right*). Phase differences measured on radio signals from extremely distant sources, so that wave front is effectively a plane, permit precise determination of straight-line distance B between instruments.

understandably command far bigger budgets than science. Proposed civilian space geodesy programs, such as the *Geopotential Research Mission* (two satellites), have in recent years been routinely turned down on grounds of cost. No one can seriously imagine NASA getting $10.5 billion for a constellation of 24 large active satellites, i.e., the *Global Positioning System*.

Turning from these bleak reflections, let us examine the more general scientific results of space geodesy, treating its major fields of application in very rough historical order. For background, publications by NASA (1983, 1988) and the National Research Council (1987) will be helpful.

2.3 Shape of the Earth

The first scientific discoveries of the Space Age included new knowledge of the Earth's gravity field, and from that its shape and internal structure. The very first satellite, *Sputnik 1*, caught the western world by surprise on October 4, 1957, although the Soviet Union had anounced its intentions several months earlier. However, as described by Massey and Boyd (1958), *Sputnik 1* was successfully tracked by radio and radar in Britain, producing new values for the atmospheric density about 10 times higher than the then standard USAF model. *Sputnik 1* and its final rocket stage (which is what most people actually saw) produced little information beyond this but, in addition to galvanizing the United States into action, it also gave western satellite trackers valuable practice that was applied to *Sputnik 2*, launched a month later.

A minor non-technical comment may not be out of order at this point. The many failures and misdeeds of the Soviet Union are now well known, and few people inside or outside the former USSR would wish to re-establish it. But any impartial historian must agree that the Soviet Union was in its day a true leader in space flight, regardless of its motives. Beside launching the first satellite, the Soviets for decades pursued an ambitious program of space exploration, often in the face of repeated failures. Space flight has long since become truly international, and the "Space Race" in its original sense is over. But Russia, Ukraine, and other former members of the Soviet Union can be justly proud of their role in a competition in which all sides were the ultimate winners.

Returning to space geodesy: *Sputnik 2* was much bigger than *Sputnik 1* and stayed in orbit much longer. Accordingly, it was observed many more times than its predecessor; observations bearing on the shape of the Earth, that mark the true beginning of space geodesy. As used in this context, "shape of the Earth" means the geoid, essentially the undisturbed sea-level surface if the Earth were completely covered with water (King-Hele, 1976). It is an equipotential surface, to which the acceleration of gravity is everywhere perpendicular. If the Earth were internally homogeneous, spherically symmetrical, non-rotating, and alone in the universe, the geoid would be a sphere. None of these conditions prevail, in particular the absence of rotation, and it was recognized by Isaac Newton that the Earth would be slightly flattened by its rotation. The degree of flattening he calculated, assuming a homogenous interior, was 1/230, a fraction expressing the difference between equatorial and polar diameters. (A concise mathematical treatment of this subject has been presented by Kaula, 1968.)

This value is important for regional surveys, and efforts were made to improve Newton's estimate by, for example, measuring the width of a degree of latitude in South America and then in Scandinavia. After centuries of effort, by 1957 a figure of 1/297.1 had been agreed upon (O'Keefe, 1966).

It was at this point that astronautics began to influence geophysics, for orbital observations of *Sputnik 2* were immediately used to calculate the flattening value by application of Clairaut's Theorem relating gravity to the geoid (Garland, 1965). The first value, obtained by Buchar (1958), was 1/297.4, not far from the accepted one. However, Merson and King-Hele (1958), using many additional observations, soon calculated a value of 1/298.1 (the presently-accepted value is 1/298.3). This may seem a trivial improvement, but as King-Hele (1983) has pointed out, it was not trivial for geodetic surveying. Geodesists had been working to an accuracy of 10 meters for long baselines, and the new value for flattening meant the assumed size for the Earth was 170 meters off. Thus the very shape and size of our planet were re-measured, in a few months, by the second satellite ever launched.

Further improvement in our knowledge of the Earth's shape followed rapidly, from radio tracking of *Vanguard 1* in 1958 (Siry, 1959). It had been shown earlier by O'Keefe and Batchlor (1957) how the ellipticity of the Earth might be derived from motion of the node (equator crossing point of the orbit) of a close satellite. Using this method, O'Keefe, Eckels, and Squires (1959) discovered the third zonal (latitudinal) harmonic of the gravity field, expressing the "pear-shaped" component of the geoid. The formal publication of this discovery is so short, and so elegantly written, that it will be reproduced in full here (Fig. 2.9).

The "pear-shaped Earth," as it was labeled in headlines the world over, was fascinating by itself (Fig. 2.10). But it also had major implications for the internal structure of the Earth, typifying "geophysical geodesy." A widely-accepted concept of the Earth's internal structure at the time held it to be close to that of a fluid in equilibrium, the "basic hypothesis of geodesy" of Heiskanen and Vening Meinesz (1958). Opposed to this was the view of Jeffreys (1962), that the Earth could support substantial stress differences. Discovery of the third harmonic showed that Jeffreys was more nearly correct, and that either mechanical strength or large-scale convection currents in the mantle must be supporting stress differences of the magnitude he had estimated (O'Keefe, 1959). Gravity interpretations are inherently non-unique by themselves, since any given value is the expression of mass *and* distance. Runcorn (1967) and many others have shown that the satellite data can be interpreted in terms of mantle convection. The weight of other evidence, such as glacial rebound and sea-floor spreading, has led most geophysicists to accept mantle convection as the cause of the broadest features of the gravity field. A comprehensive (and highly mathematical) discussion has been presented by Peltier (1985).

Vanguard Measurements Give Pear-Shaped Component of Earth's Figure

The determination of the orbit of the Vanguard satellite, $1958\beta_2$, has revealed the existence of periodic variations in the eccentricity of that satellite (*1*). Our calculations indicate that the periodic changes in eccentricity can be explained by the presence of a third zonal harmonic in the earth's gravitational field. The third zonal harmonic modifies the geoid toward the shape of a pear. In the present case, the stem of the pear is up—that is, at the North Pole. According to our analysis, the amplitude of the third zonal harmonic is 0.0047 cm/sec² in the surface acceleration of gravity, or 15 meters of undulation in the geoid.

Figure 1 shows the observed variation in eccentricity. The period of the variation in eccentricity is 80 days, approximately equal to the period of revolution of the lines of apsides. The eccentricity is a maximum when the perigee is in the Northern Hemisphere. The amplitude of the variation is 0.00042 ± 0.00003. Similar perturbations may exist in the angle of inclination of the orbit, although the data for them are much less accurate. No perturbations of this magnitude appear to exist in the semimajor axis.

In principle, the perturbation might be caused by both odd and even harmonics. However, the even harmonics can be excluded because the observed effect has opposite signs in the Northern and Southern hemispheres. Furthermore, we can also exclude tesseral harmonics (those which depend on longitude as well as latitude) because these also are the same in the Northern and Southern hemispheres, apart from a shift in longitude. We are left with the zonal harmonics (those which depend only on latitude) of odd degree.

Of the odd zonal harmonics, the first degree is forbidden; and those of higher degree are unlikely to have a large effect because they die out inversely as the $(n+1)$ power of the distance. The effect is therefore due mostly to the third zonal harmonic, with a possible contribution from the fifth.

Accordingly, a calculation was made of the effect of the third zonal harmonic on the orbit elements of $1958\beta_2$, by methods developed by O'Keefe and Batchlor (*2*). In the resultant expression for the eccentricity, the dominant terms were those whose argument was the mean motion of perigee. These were larger than the others by a factor of 10^3. Keeping only the large terms, we find

$$e = e_0 + \frac{3}{2} A_{3,0} \frac{(1-e^2)^{1/2}}{na^6} \frac{1}{n'} \times \sin i \left(1 - \frac{5}{4}\sin^2 i\right) \sin \omega \qquad (1)$$

where $A_{3,0}$ represents the coefficient of the third zonal harmonic in the notation of Jeffreys (*3*), n is the orbital mean motion and n' is the mean motion of the perigee, e is the eccentricity and e_0 the mean eccentricity, i is the angle of inclination, ω is the argument of perigee, and a is the semimajor axis.

Setting in the constants of the orbit and the observed amplitude of e, we find

$$A_{3,0} = (2.5 \pm 0.2) \times 10^{29} \qquad (2a)$$

in meter-second units. Utilizing the relation given by Jeffreys,

$$A_{n,s} = \frac{c^{n+2}}{n-1} g_{n,s}$$

(where $A_{n,s}$ is the coefficient of the disturbing potential, $g_{n,s}$ is the acceleration of gravity at the surface of the earth, and c is the earth's equatorial radius), we find that the third zonal harmonic of gravity at the earth's surface, in milligals, is

$$g_{3,0} = 4.7 \pm 0.4 \qquad (2b)$$

Equation 2 is relevant to what Vening Meinesz (*4*) and Heiskanen call the "basic hypothesis of geodesy." These authors assume that the earth's gravitational field is very nearly that of a fluid in equilibrium. They consider that the deviations from such an ellipsoid, in any given area, do not exceed about 30 milligal-megameter units—that is, they assume that one will not find deviations of more than 30 milligals over an area of 1000 kilometers on a side, or deviations of more than 3 milligals in an area 3000 kilometers on a side.

Our determination of the third-degree zonal harmonic shows that the hypothesis of Vening Meinesz and Heiskanen is not justified; for example, each of the polar areas has a value of about 120 milligal-megameters, and each of the equatorial belts a value more than twice as great.

The presence of a third harmonic of the amplitude (*2*) indicates a very substantial load on the surface of the earth. Following the arguments of Jeffreys, we may calculate the values of this load and the minimum stress required in the interior to support it. We find a crustal load of 2×10^7 dy/cm². We can choose between assuming that stresses of approximately this order of magnitude exist down to the core of the earth, or that stresses of about 4 times that amount exist in the uppermost 700 kilometers only (*3*, p. 199). These stresses must be supported either by a mechanical strength larger than that usually assumed for the interior of the earth or by large-scale convection currents in the mantle (*5*).

J. A. O'Keefe
Ann Eckels
R. K. Squires
Theoretical Division, National Aeronautics and Space Administration, Washington, D.C.

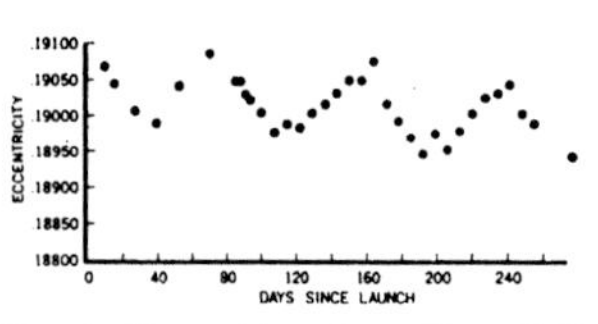

Fig. 1. Eccentricity of satellite $1958\beta_2$ (Vanguard).

References and Notes

1. J. W. Siry, distribution to orbit-computing centers, 1958.
2. J. A. O'Keefe and C. D. Batchlor, *Astron. J.* 62, 183 (1957).
3. H. Jeffreys, *The Earth* (Cambridge Univ. Press, Cambridge, 1952).
4. W. A. Heiskanen and F. A. Vening Meinesz, *The Earth and Its Gravity Field* (McGraw-Hill, New York, 1958), pp. 72, 73.
5. We would like to thank the Vanguard Minitrack Branch, the IBM Vanguard Computing Center, and Dr. Paul Herget, whose work in obtaining and processing the data made this study possible.

27 January 1959

Fig. 2.9 Complete report of discovery of "pear-shaped Earth." From O'Keefe, Eckels, and Squires (1959).

Further refinements of the gravity field followed rapidly. Following in Newton's tradition, British scientists took a leading role in this area. By 1961 the second-, fourth-, and sixth-degree harmonics had been obtained from satellite tracking (Smith, 1961). The areal extent of these "harmonics" decreases with increasing degree. For zonal, or latitudinal harmonics, their width is very roughly the circumference of the globe divided by the degree; the sixth harmonic

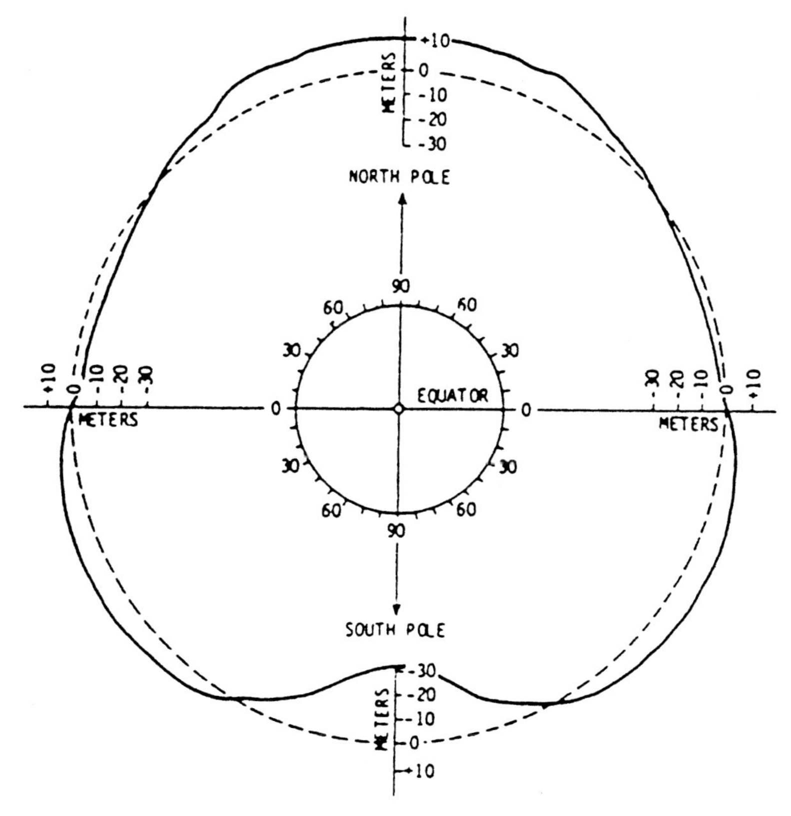

Fig. 2.10 Pear-shaped component of the Earth's shape, shown in section through poles; deviation in meters.

thus corresponds to a wavelength of between six and seven thousand kilometers. Lemoine *et al.* (1998a, b) define resolution as one-half the wavelength, or 20,000 km divided by the degree of the harmonic. The increasing lateral resolution of these findings permitted more specific correlations with crustal features, warranting separate discussion in the following section.

2.4 Gravity anomalies and global tectonics

Before discussing the application of satellite geodesy to tectonics, it should be stressed again that interpretation of gravity data is inherently ambiguous. As put concisely by Rubincam (1982), ". . . there is an infinite number of density distributions which can generate the observed gravity field." Gravity data by themselves are of little value for geophysics and geology, but must be interpreted in combination with information from as many other sources as possible. Gravity data put constraints on interpretations, as well as suggesting interpretations impossible from exposed geology. Gravity data are difficult to use, but not using them will lead to even more difficulties.

The very first gravity anomalies studied from satellites, those of the third harmonic, led to the discovery, as we have seen, of correspondingly large features: the "pear-shaped" bulge circling the Earth in the southern hemisphere. By 1971, Gaposchkin and Lambeck (1971) were able to construct a global gravity map from satellite data through the sixteenth harmonic, corresponding to a spatial resolution of about 1200 km. This map was used by Kaula (1972) to study the relationship between gravity and global tectonics. Kaula's interpretations have been largely supported by later studies (Lambeck, 1988). The major relationships noted by these authors can be best summarized with use of newer gravity and tectonic maps (Lowman and Frey, 1979).

The gravity map presented as Fig. 2.5 shows low-degree (i.e., very broad) free-air anomalies, i.e., anomalies calculated with corrections only for altitude, as explained in Section 2.2. (Good elementary accounts of gravity anomalies can be found in Garland, 1965, and Wyllie, 1971.) Another map (see Fig. 2.6) shows the geoid corresponding to the gravity anomaly map. The most obvious positive correlation is between gravity values and topography, as over the Andes, North American Cordillera, and the Tibetan Plateau (see Fig. 2.7). We encounter at once one of the main complications in gravity interpretations, the degree to which the topography is isostatically compensated. It has been known for more than two centuries that mountains are not simply additional mass on the crust, but features in which the apparent excess mass is compensated by a deficiency of mass below them. This compensation may result from variations in crustal thickness (Airy compensation) or lateral variations in crustal density (Pratt compensation). Other types of compensation have been proposed, and it is generally agreed that for any large area, several mechanisms may interact (Lambeck, 1988).

Passing over these complications, we see that the main positive anomalies appear to correlate with areas of what are, in plate tectonic theory, zones of crustal convergence (see Fig. 2.7). The Andes, for example, owe their elevation (and their correspondingly deep roots) to the convergence of the Nazca with the South American Plate. The Tibetan Plateau is thought by many to result from the convergence of Peninsular India with Asia. A similar positive correlation (Kaula, 1972) is between positive gravity anomalies and Quaternary volcanism, as in the Andes, Aleutian Islands, and Indonesia. Plate tectonic theory provides a consistent interpretation of this correlation as well, in that these volcanic areas can also be explained as resulting from plate convergence.

There are obvious exceptions to this correlation that readers can verify for themselves by reference to the maps. For example, the gravity map shows no correlation in sign or direction with the volcanic fields of the East African Rift Valleys (see Fig. 2.7). Kaula noted the absence of correlation between gravity and "temperature indicators" such as high heat flow, inferring that horizontal variations in rock type were important. The Rift Valleys are evidently zones of incipient plate divergence, as are the mid-ocean ridges, which over large areas also show little relation to the gravity.

Analyses such as that just discussed have in effect been summarized by Kaula (1989) in a classification of the sources of the Earth's gravity field, with six categories: **deep heterogeneities (mantle)**, 50%; **plate tectonics,** 20%; **thermal isostasy**, 10%; **crustal isostasy**, 5%; **lithospheric strength**, 5%; and **surface loads**, 5%.

Such relationships can be interpreted, very broadly, in terms of mantle convection (Peltier, 1985; Lambeck, 1988). The reality of mantle convection is essentially unquestioned, but its nature – whole mantle vs. two layer, boundary layer vs. penetrative – is still not known. The study by Silver *et al.* (1988) shows how geoid models can serve as a constraint on seismic and geochemical data. The complexities of such interpretations are formidable and can not be pursued further here. The difficulty of interpreting the satellite gravity data will almost certainly lead to a better understanding of mantle dynamics and crustal movements. We will turn now to this subject as approached by a higher-resolution satellite gravity technique, radar altimetry.

2.5 Marine gravity and ocean-floor topography

One of the most unexpected results of space flight has been the ability to map the floors of the oceans from space, by satellite radar altimetry. This technique was first demonstrated from *Skylab* in 1973. This date is somewhat ironic, for in the same year the plans for the US Geodynamics Program were announced, including all known methods for study of the solid earth but omitting satellite altimetry.

As previously discussed (see Fig. 2.4), the mean sea surface is a subdued replica of the ocean-floor topography. The geoid as mapped from satellite tracking has already been illustrated; it shows only the very broad features, i.e., the low harmonics. Satellite sea-surface altimetry gives us a much higher resolution look at the geoid, showing far more detail than does satellite tracking.

A global satellite altimetry survey was carried out by *Seasat*, and the data used by Marsh *et al.* (1985) to map the physical geoid and hence the main bathymetric features of the world's oceans. Since then, radar altimetry has been carried out by other satellites. The US Navy *Geosat* mission, launched in 1985, generated nearly 5 years of coverage, launching what Douglas and Cheney (1990) termed "a new era in satellite oceanography." The data from the first 18 months, with ground track spacing averaging 4 km, were necessarily classified. However, after this period, the satellite was put in a 17-day exact repeat orbit for oceanographic research, and the data from this part of the mission have been made freely available (Sandwell, 1991). Satellite altimetry from this and other missions has proven remarkably valuable for mapping the ocean floor and in fact oceanic crustal structure, as already demonstrated. The following summary will include only a few additional examples of these applications, both scientific and economic (Bostrom, 1989).

A survey of marine geology using *Seasat* altimetry was carried out by Craig and Sandwell (1988), using along-track profiles. They found that seamounts – inactive underwater volcanos – could be detected from the slight elevation of the overlying sea surface, but in addition their size could be estimated. A total of 8556 seamounts were mapped, about a quarter of them previously unknown. The map (Figs. 2.11, 2.12) produced in this way is thus an essentially new look at the major expression of intraplate volcanism, which must be understood for studies of plate tectonics, mantle chemistry, and the terrestrial geothermal flux. Craig and Sandwell point out several of the most interesting features of this map: the scarcity of seamounts in the Atlantic, the generally small size of those in the Indian Ocean, and the prominent linear trends in Pacific seamounts. The map stimulates obvious speculative questions. For example, the line of seamounts northeast of New Zealand, the Louisville Ridge, appear clearly related to the Eltanin Fracture zone on this map (and on that of Haxby, 1987), yet detailed surveys and dating of the volcanos along the Ridge indicate no direct connection for at least the newer part of the chain. The complexities of this problem are reviewed by Gordon (1991).

Marine volcanos are often interpreted as hot-spot trails, but satellite altimetry has found at least one area where this does not seem likely. Filmer and McNutt (1989), using conventional bathymetry in combination with geoid heights from *Seasat* and *GEOS-3*, have studied the Canary Islands. The smooth progression of ages of volcanic rocks in this group strongly suggests a hot-spot trail, presumably with a mantle plume under one end. However, after making

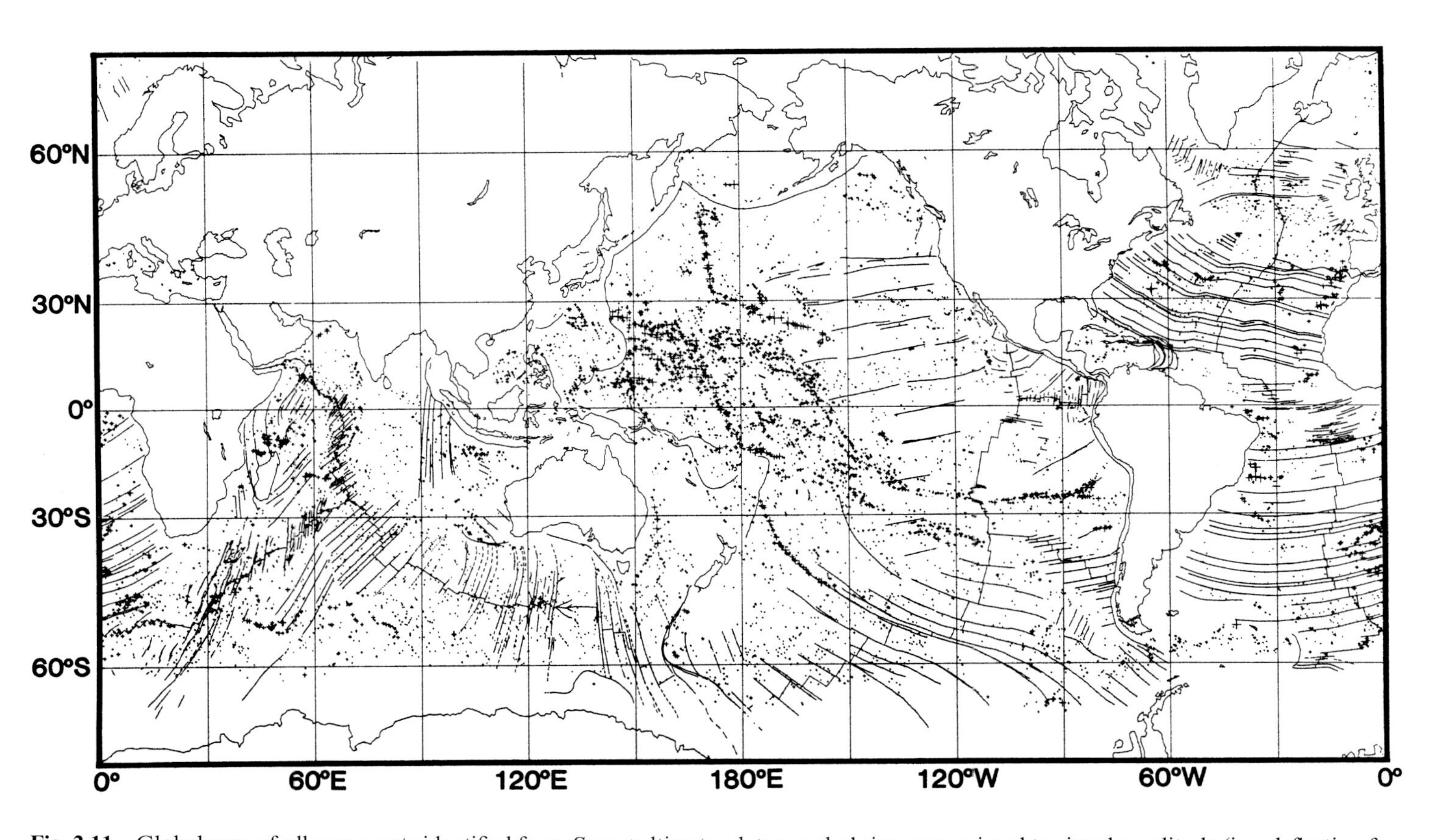

Fig. 2.11 Global map of all seamounts identified from *Seasat* altimetry data; symbol size proportional to signal amplitude (i.e., deflection from vertical). From Craig and Sandwell (1988).

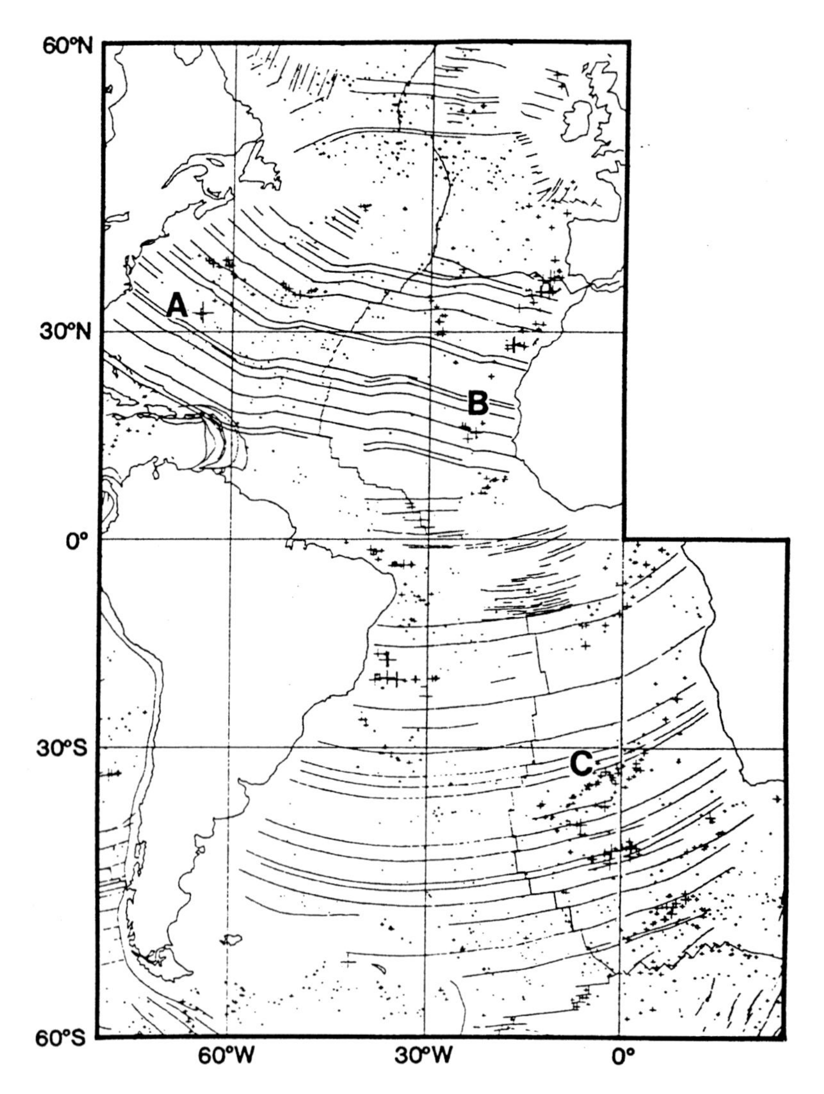

Fig. 2.12 Seamounts identified from *Seasat* altimetry in Atlantic Ocean; symbol size proportional to signal amplitude. A: Bermuda Rise; B: Cape Verde Islands; C: Walvis Ridge. From Craig and Sandwell (1988).

various corrections, Filmer and McNutt found no evidence of a swell expressing a plume or any thermal disturbance of the lithosphere, concluding that if the Canary Islands are a hot-spot trail, they represent a very different expression of a mantle plume from any other well-known hot spot. This anomaly is of some interest in view of the arguments by Lowman (1985a, b) that there are no valid hot-spot trails on continents; the Canary Islands example may represent control by crustal structure rather than crustal motion.

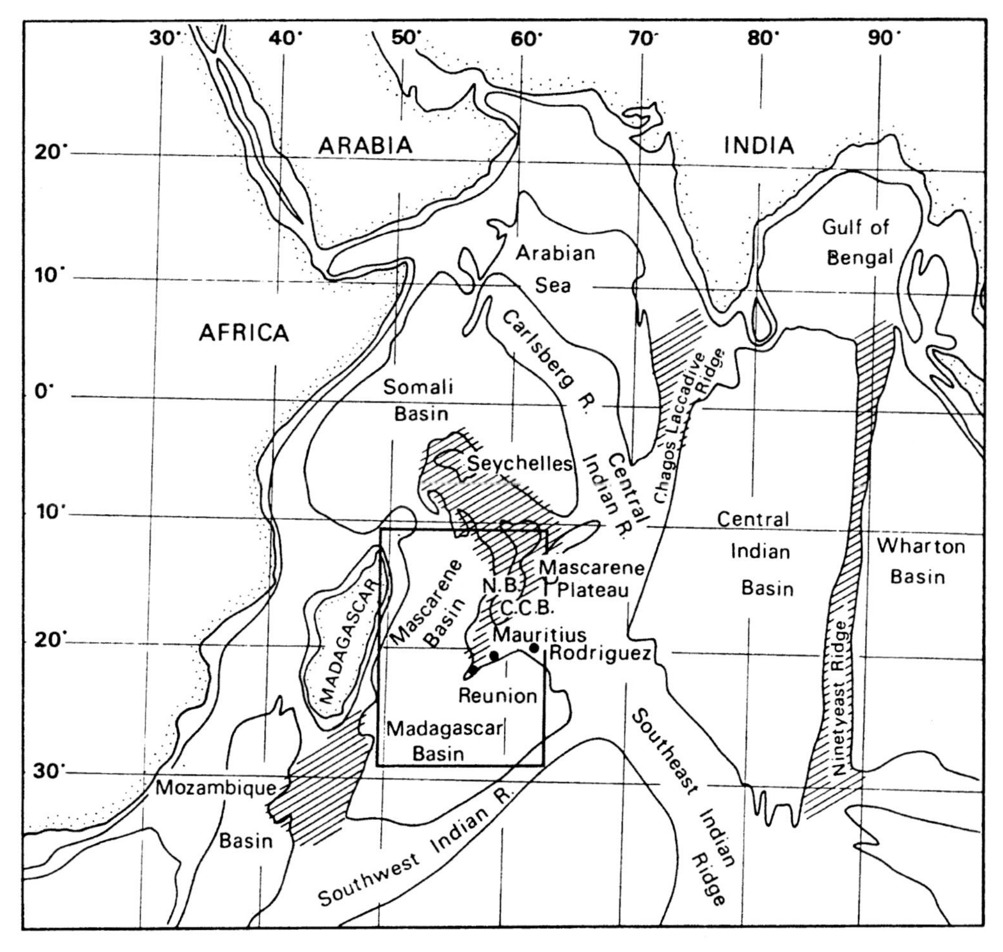

Fig. 2.13 Structural sketch map of Indian Ocean; study area in rectangle. N.B.: Nazareth Bank; C.C.B.: Cargados-Carajos Banks. From Bonneville *et al.* (1988).

A somewhat similar study using *Seasat* and *GEOS-3* altimetry was done by Bonneville *et al.* (1988) for the Indian Ocean. The origin of the Mascarene Plateau (Fig. 2.13) is not understood, although it appears continuous with the continental composition Seychelles Bank (see the Digital Tectonic Activity Map, Fig. 1.2). Bonneville *et al.* used the radar altimetry to compile a geoid map (Fig. 2.14) of the area, removing long-wavelength anomalies with the aid of other satellite data. They then calculated crustal rigidity for the area, finding the increasing rigidity to the south consistent with a hot spot or mantle plume origin for the southern Mascarene Plateau and the Mascarene Islands.

In addition to studies of ocean-floor topography, there are even

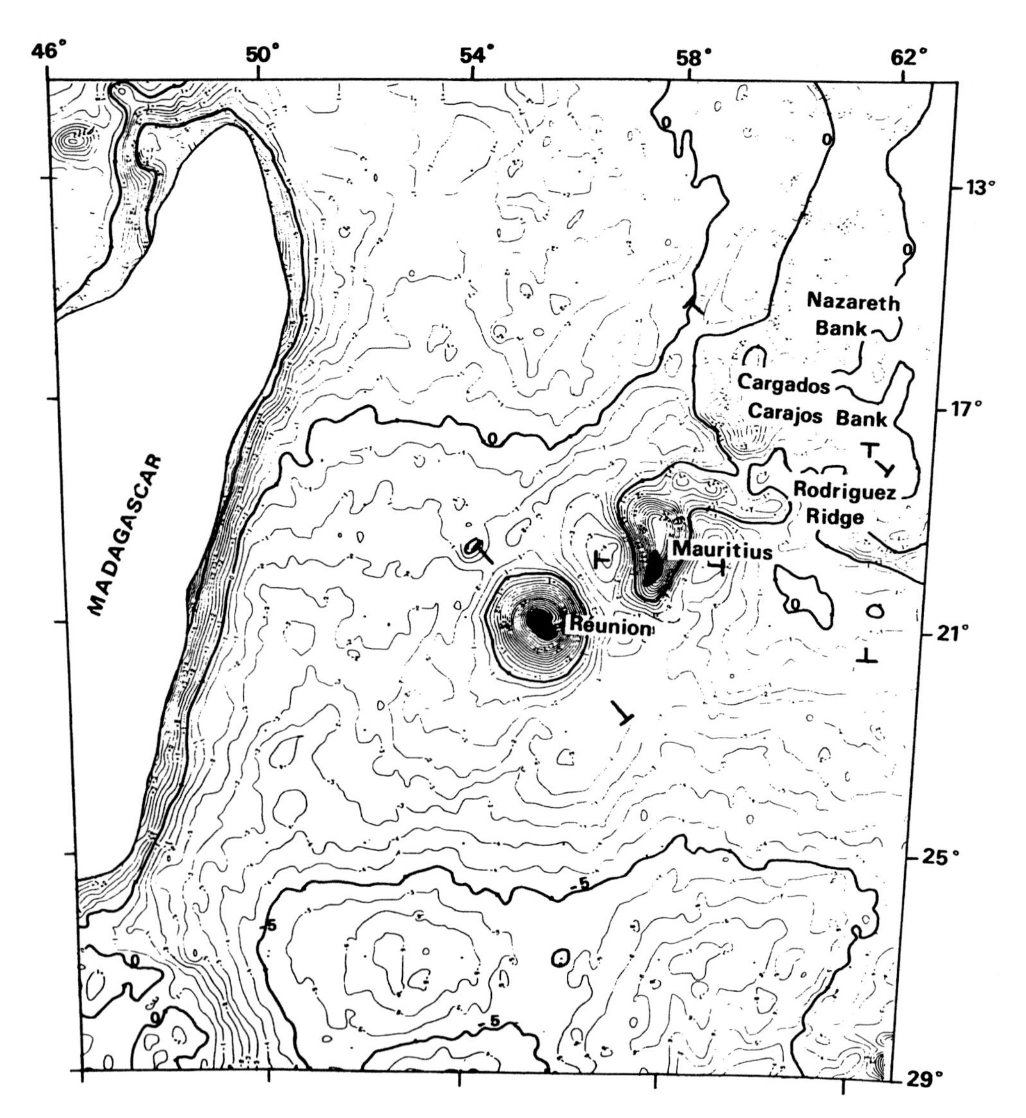

Fig. 2.14 Geoid level, 0.5 m contour interval. Long-wavelength anomalies (>3000 m) removed using GRIM3B model. From Bonneville *et al.* (1988).

now dozens of papers using satellite altimetry to study the upper mantle as expressed in the marine geoid (Sandwell, 1991), and in particular short-wavelength mantle convection. Only a few of these will be mentioned as examples. Haxby and Weissel (1986) reported evidence for small-scale mantle convection from *Seasat* data over the Pacific. Sandwell and Renkin (1988) in contrast found "no direct evidence" in the altimetry for mantle convection. McNutt and Judge (1990) noted the anomalous situation over the "superswell" around

French Polynesia, which is strongly negative gravitationally; high heat flow and a thinned lithosphere may contribute to this situation but the relationships are unclear. It is obvious that by itself satellite altimetry will not settle any of these problems, but it will obviously contribute to their solution in combination with data from seismology, marine magnetic surveys, and other investigations.

2.6 Plate motion and deformation

One of the most dramatic accomplishments of space geodesy has been the direct confirmation of oceanic crustal motions predicted by plate tectonic theory as the sea-floor spreading concept. As shown on the tectonic activity map (see Fig. 2.7), these motions are thought to be a few centimeters per year, roughly the rate at which fingernails grow. Direct measurement of such motions, over continental and especially oceanic distances, was utterly impossible before the development of space geodesy techniques. As pointed out by Flinn (1981), the cumulative errors of trilateration in land surveys introduce prohibitive errors into land surveys for distances over 100 km, and trilateration over large oceans is impossible. Wegener (1966) claimed direct measurement of continental drift by trans-Atlantic longitude measurements using radio time signals, but as his reported Greenland–Europe rate – 36 *meters* per year – suggests, this method was far too inaccurate to succeed. Even the monumental study by Proverbio and Quesada (1974), using several decades worth of data from the International Latitude Service, was not decisive (Lowman, 1985a, b). However, Wegener's objective has been partially achieved in that the rate and direction of plate movements in and bordering the Pacific Ocean have now been measured by satellite laser ranging (SLR) and very long baseline interferometry (VLBI).

As previously discussed in Chapter 1, plate tectonic theory can be reduced to three essential elements: **ridges**, or spreading centers, where new crust is created; **trenches**, or subduction zones, where crust is destroyed or recycled by return to the mantle; and **transform faults**, fractures with dominantly horizontal movement connecting ridges or trenches. These elements bound **plates**, relatively rigid segments of the Earth's lithosphere (which includes the crust and upper mantle above the asthenosphere). Plates may include oceanic and continental crust; the Eurasian Plate, for example, includes all crust between the Verkhoyansk Range of Siberia and the Mid-Atlantic Ridge in the North Atlantic Ocean; Fig. 2.15 from Stein (1993) shows the 12 main plates conventionally recognized and used for plate motion models. Plate movement of several centimeters per

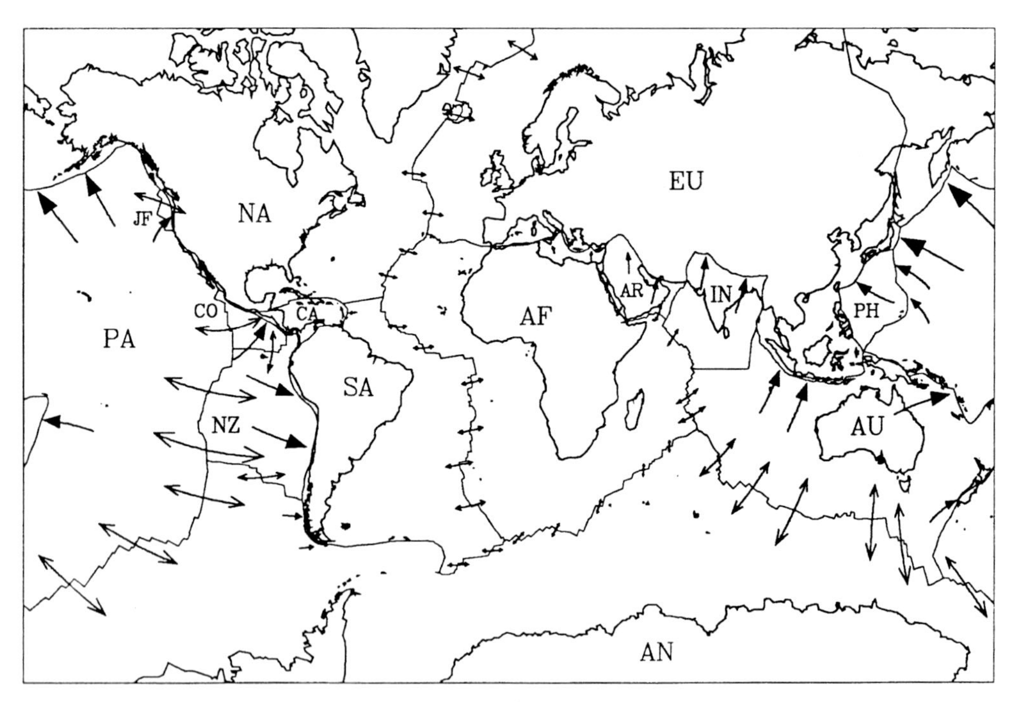

Fig. 2.15 Standard plates used for NUVEL-1 model. Relative plate velocities shown by arrows, length proportional to displacement if plates were to maintain their present angular velocities for 25 million years. Plate convergence with single solid arrow-head shows zones where convergence is asymmetric and polarity known. From Stein (1993).

year and continental drift are central to plate tectonic theory. All plate movements, taking place on a sphere, can be described as rotations (Dewey, 1975).

To demonstrate plate movement geodetically, three requirements must be met. First, the plates in question must be proven *rigid enough* to ensure that apparent baseline changes are not the result of local movements, crustal deformation in diffuse plate boundaries, or intraplate deformation in general. Sato (1993) has published a penetrating discussion of this problem, with reference to Japan and the western U.S., where non-rigid behaviour is well demonstrated. Second, the baseline changes, obviously vector rather than scalar quantities, must agree in *magnitude* with those predicted by plate tectonic theory, chiefly estimated from spacing of dated marine magnetic anomalies. Third, the apparent plate motions must agree in *direction* with those predicted by plate theory, generally normal to the magnetic anomalies and parallel to transform fault azimuths. The baselines and sites initially proposed for the NASA Crustal

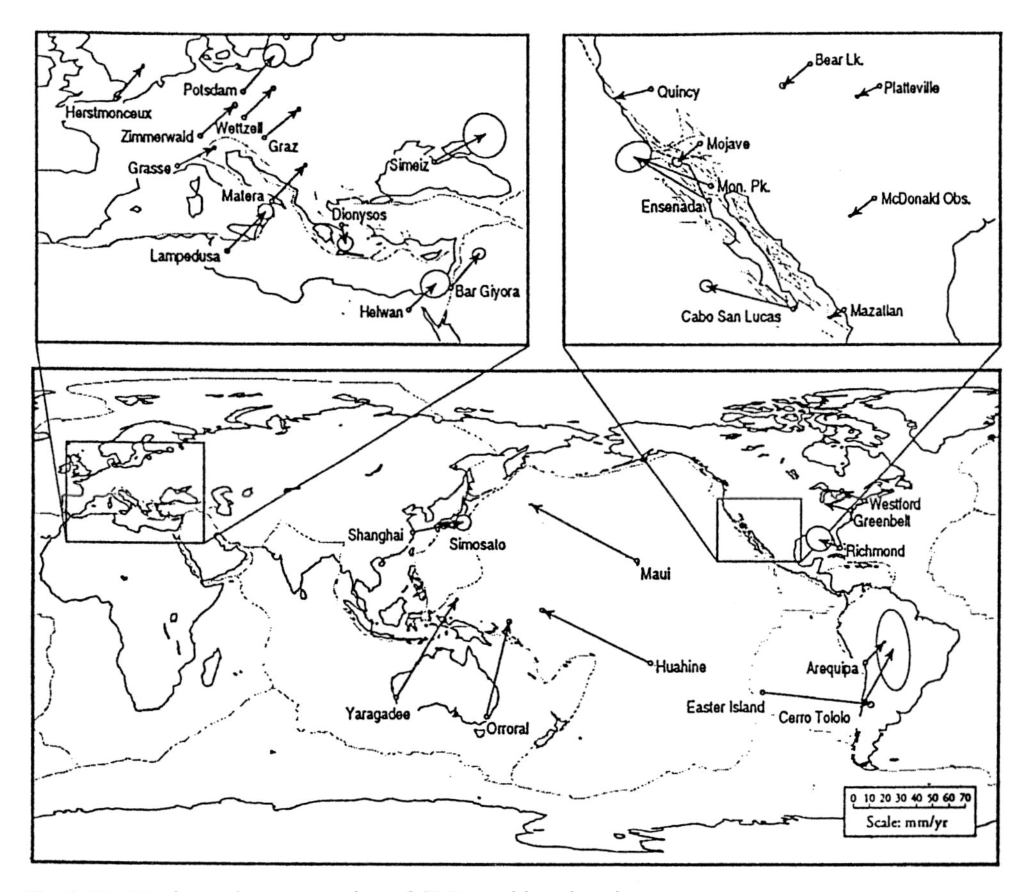

Fig. 2.16 Horizontal vector motion of SLR tracking sites; inset maps to same scale as main map. From Smith *et al.* (1994).

Dynamics Project (Lowman *et al.*, 1979; Allenby, 1983) were planned to meet these requirements. Measurements have been carried out now across many plate boundaries, and it appears that plate movements have now been successfully measured, independently by SLR and VLBI, in and around the Pacific Basin (Fig. 2.16). Detailed tabulations of these measurements have been presented by Robbins *et al.* (1993) (*LAGEOS*), Ma *et al.* (1992) (VLBI), and Ryan *et al.* (1993) (VLBI). Useful summaries and discussions of the results have been published by Sato (1993) and Smith *et al.* (1990).

The rigidity of the Pacific Plate has been demonstrated by the Maui–Huahine baseline (Fig. 2.17), showing no significant changes. A similar result is shown for the baselines from these islands to Monument Peak, California, showing very small changes. Since Monument Peak is just west of the active San Andreas fault system,

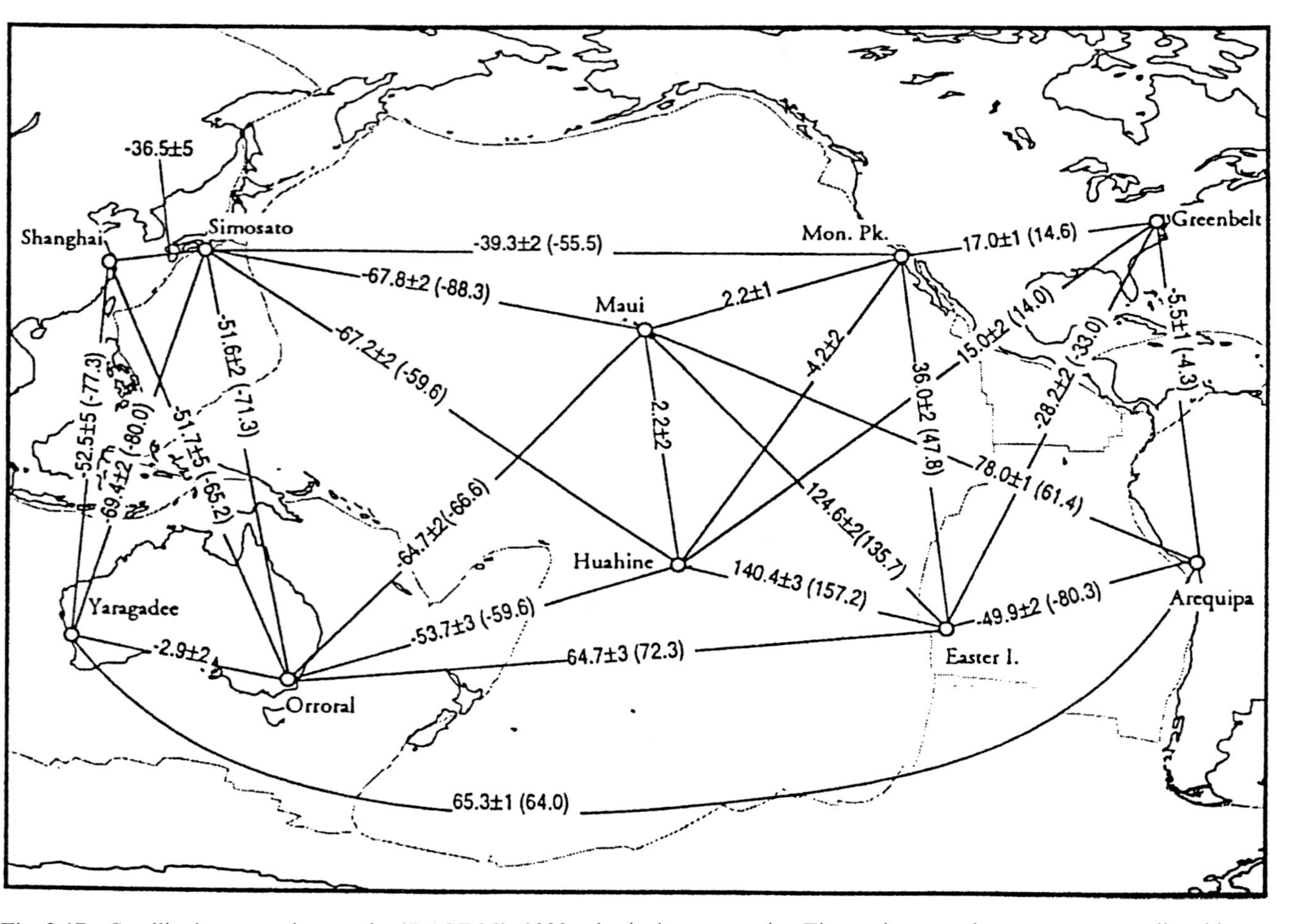

Fig. 2.17 Satellite laser ranging results (*LAGEOS*). 1993 spherical rates, mm/yr. Figures in parentheses are rates predicted by NUVEL-1. From Smith *et al.* (1994).

this result might surprise us, but in fact these three stations are all on the same (Pacific) plate. Distribution of shear between the Pacific and North American Plates, a long-standing problem, has been treated by Feigl *et al.* (1993).

Rigidity of intracontinental baselines has been demonstrated to date for North America (Jordan and Minster, 1988; Argus and Gordon, 1996), even though they cross areas of known activity. However, much remains to be done in this category, as will be discussed later in this chapter, especially for the Eurasian Plate.

Movement of the Pacific and Nazca Plates has been demonstrated: Kauai, for example, has been shown to be moving toward Japan at 8.7 cm/yr, compared with the predicted rate of 9.9 cm/yr; Maui is moving away from Arequipa. The apparent movement here is significantly greater than that calculated on the basis of rigid plates, the difference probably resulting from deformation of the South American Plate over the Peru–Chile Trench subduction zone (Robbins *et al.*, 1993). Similar effects have been seen in Alaska, a tectonically analogous area, over the subduction zone. Sato (1993) has shown that the Japanese station, Kashima, is moving to the northwest at about 2 cm/yr, a significant difference from rigid-plate behaviour. Collectively, however, the space geodesy measurements obtained by two independent methods – SLR and VLBI – appear to confirm the plate rigidity, movement direction, and rate required by plate tectonic theory.

The movement of a small continent, Australia, also appears to have been demonstrated although, as the geodetic results cited by Lambeck (1988) indicate, several more intracontinental baselines and several years of measurement are needed to demonstrate that the continent is in fact moving as a unit. Baja California is also demonstrably moving, as has been known from conventional geodesy for some time, although one would hardly characterize this small slice of crust as a "continent."

Right lateral regional shear along the east margin of the Pacific Plate, in North America, has been measured by SLR and VLBI, in additional to conventional surveys, since the early-1970s (e.g., Sauber *et al.*, 1986). There is general agreement that the regional shear movement is several centimeters per year, but the way it is distributed is not yet clear. Contrary to popular belief, the San Andreas fault is not "the" plate boundary, but only one of many active faults along which the shear is distributed (Sauber *et al.*, 1994). There are also aspects not understood, such as the degree of vertical movement as demonstrated by the Loma Prieta earthquake of 1989. It has been argued by Martin (1992) that vertical movements may dominate over geologic time, despite the horizontal movement demonstrably occurring now. However, the general pattern of contemporary plate motion seems well demonstrated by space geodetic techniques.

The precision of these results is, even to those familiar with the techniques, nothing less than astonishing: orders of magnitude greater than expected from satellite methods as late as the early-1960s. But beyond their precision, the results are also astonishing in that they agree with plate motions inferred from totally independent lines of evidence. Several numerical models of global plate motion, such as those of Minster and Jordan (1978), Chase (1978), and DeMets *et al.* (1990) have been based solidly on plate tectonic theory. The NUVEL-1 model of DeMets *et al.* illustrates the general procedures used. Plate directions for twelve assumed-rigid plates have been inferred from transform fault azimuths and earthquake slip vectors, and plate rates have been determined from the ridge spreading rates as measured from dated magnetic lineations. The spreading rates are particularly important in this context, for the lineations generally used cover about three million years and hence should give only average rates for this time. But as we have seen, the rates directly measured over only a decade or so are remarkably close to those estimated from the ridge spreading rates (Carter and Robertson, 1986; Gordon and Stein, 1992). Given the many problems in dating marine magnetic anomalies (Agocs *et al.*, 1992), this agreement must be considered strong support if not confirmation of sea-floor spreading. There are of course some glaring exceptions to simple sea-floor spreading from contemporary spreading centers, as in the north Pacific. The dated magnetic anomalies become younger as they approach the coast of Alaska and British Columbia, generally agreed to be a subduction zone where the oldest oceanic crust should be descending into the mantle. However, even this anomaly was ingeniously explained by Atwater (1970) as resulting from subduction of the spreading centers themselves.

In summary, the basic mechanisms of plate tectonics – sea-floor spreading and transform faulting, operating on rigid oceanic crust – appear to have been confirmed beyond reasonable doubt by space geodesy. However, we must consider the universally used term "plate tectonics and continental drift," and ask if the new geodetic methods have confirmed continental drift as well. The classic evidence for drift, such as similar fossils on now-separated continents, has been repeatedly challenged by many geologists, such as Simpson (1947), Cloud (1968), Meyerhoff and Meyerhoff (1972), and Lowman (1995), and will not be further covered here.

2.7 Plate tectonics and continental drift

Continental drift is in plate tectonic theory considered simply as a corollary of plate movement (Hallam, 1983), and it is widely believed that continental drift has finally been confirmed by space geodesy. A *Science* headline "Continental drift nearing certain detection" (Kerr, 1985) gives the flavor of this belief. "Opening of the Atlantic" is referred to in effect as an observed event.

This judgement appears premature, at least for the classic continental drift cited by Wegener (1929), the separation of the Americas from Africa and Eurasia. One problem stems from the fundamental nature of continental crust as contrasted to oceanic crust. As discussed by McKenzie (1969), Molnar (1988), England and Jackson (1989), and Thatcher (1995), continental crust is inherently much more deformable than that of the ocean basins, as demonstrated by the broad areas of tectonic activity found in intracontinental plate boundaries shown on the tectonic activity map (see Fig. 2.7). The difficulty for space geodesy is that intraplate rigidity on these supposedly moving continents has not been demonstrated in several areas. For North America, the baselines across areas of known activity, such as the Basin and Range Province and the New Madrid seismic zone, appear stable within the measurement period (Jordan and Minster, 1988; Argus and Gordon, 1996). However, the problem is still unsolved in Eurasia, treated as a single plate in all plate-motion models cited (see Fig. 2.15). Taken at face value, continental drift as a result of plate movement implies that the crust between the Mid-Atlantic Ridge and Siberia is rotating as a unit in a counter-clockwise sense away from the Ridge. The apparent increase in space geodesy baselines between North America and western Europe (Smith *et al.*, 1994), for example, suggests such movement, to be discussed below. However, the original intracontinental baselines proposed for the Crustal Dynamics Project to demonstrate plate rigidity (Lowman *et al.*, 1979) have not yet been established although *GPS* nets are beginning to fill this need.

As shown on the tectonic activity maps, and the seismicity map in Chapter 1, there are sizable zones of major activity between the European geodetic sites and the interior of Eurasia. The catastrophic Rumanian earthquake of 1977, for example, was in plate theory the result of plate convergence. The occasional earthquakes of the Rhine Graben, some strong enough to be damaging, similarly must express crustal divergence. In plate tectonic theory, such orogeny, seismicity, and volcanic activity are the result of horizontal crustal movement. There is thus a priori evidence that apparent

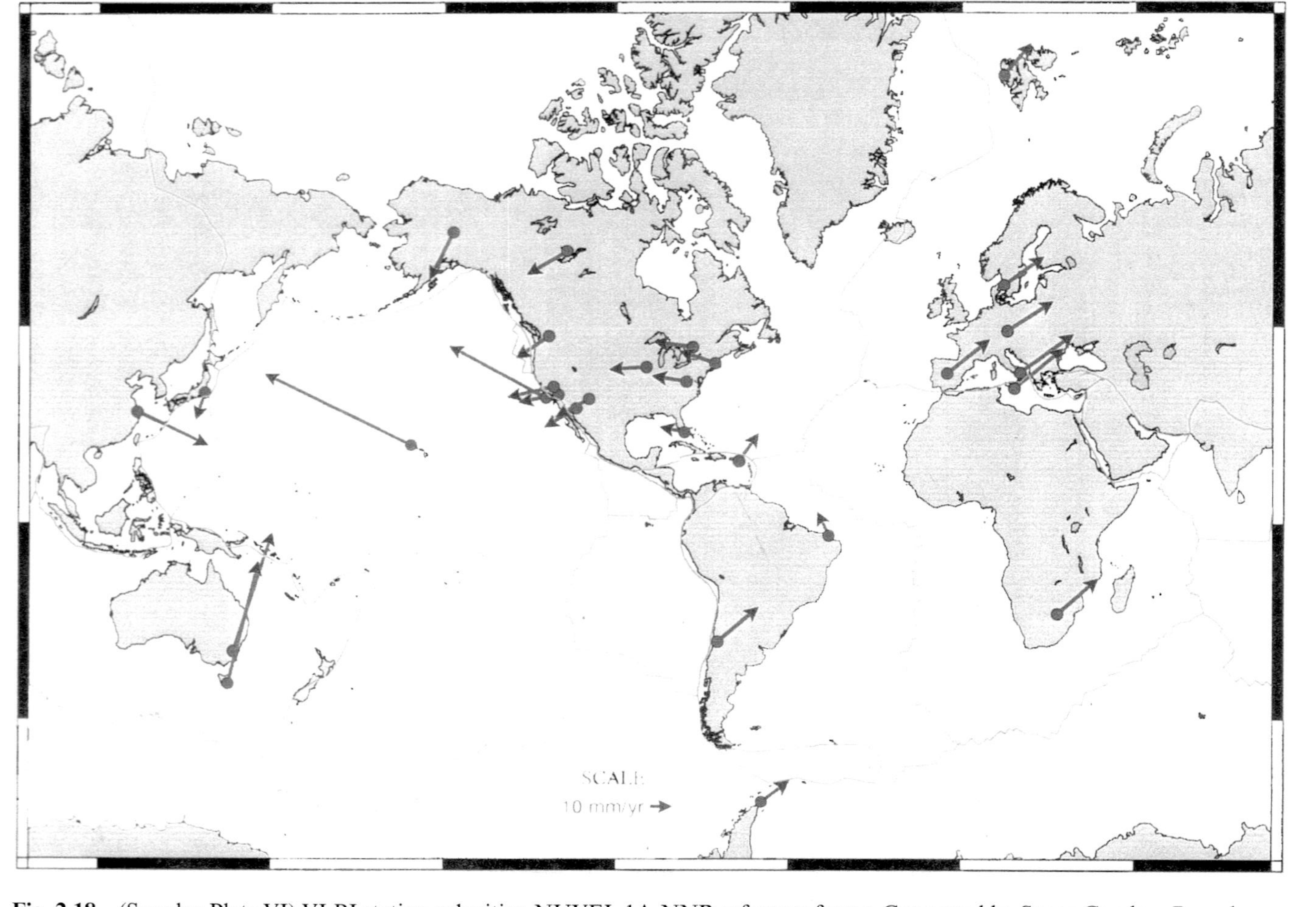

Fig. 2.18 (See also Plate VI) VLBI station velocities, NUVEL 1A-NNR reference frame. Computed by Space Geodesy Branch, Goddard Space Flight Center, 1998. Note similar azimuths of all European stations.

horizontal movements in western Europe do not necessarily reflect movement of the Eurasian Plate as a whole.

It is generally assumed that space geodesy measurements show plate movement over the mantle. This is clearly true for the Pacific Plate (see Fig. 2.16). However, as discussed by Lowman (2000), there is an apparent contradiction in western Europe between space geodesy results (Fig. 2.18) and the *World Stress Map* (Zoback, 1992), a simplified version of which is presented in Fig. 2.19. (A similar pattern for western Europe was found by Bird and Li, 1996.) Tectonism and seismicity in this area, as along the Rhine Graben, are considered to be caused by ridge push from the Mid-Atlantic Ridge, since the maximum horizontal stresses are generally parallel to the plate velocity trajectories implied by the AM-2 model of Minster and Jordan (1978). The problem is that space geodesy stations appear to be moving at almost right angles, to the northeast (Fig. 2.18), to the movement implied by the *World Stress Map*, to the southeast. A later investigation of global plate velocities using *GPS* data and the NUVEL-1A model by Larson *et al.* (1997) found similar velocities and directions.

There are several possible explanations for this contradiction. One is that the AM-2 model, based on movement over assumed-fixed mantle hot spots, is not applicable to Europe, in contrast to the Pacific Plate where it is successful. A more fundamental one may be that the assumption of fixed hot spots on which the AM-2 model is based is incorrect, as suggested by Molnar and Atwater (1973). The most obvious explanation is that the motions calculated for Fig. 2.18 – which are similar to those shown on other maps, including those based on GPS measurements – are dependent on choice of terrestrial reference frames. When motions of European sites are plotted (Ryan *et al.*, 1993) using NUVEL-1 but with the North American Plate held stationary, the site motions are to the southeast, in the direction implied by the stress directions. The geodetic sites are, in other words, being pushed to the southeast by the Mid-Atlantic Ridge, as one would expect both from the *World Stress Map* and the tectonic activity map. The anomalous results shown in Fig. 2.18 apparently result from use of NUVEL-1 but with the Pacific Plate held stationary. The implication of this anomaly is that numerical plate-motion models must be interpreted with caution, and that they are highly dependent on choice of terrestrial reference frames as discussed by Ma *et al.* (1992).

Another problem is new evidence that Europe is not the "passive margin" it appears to be. It has been suggested (Lowman, 1991) that western Europe is undergoing slow subduction, along the eastward

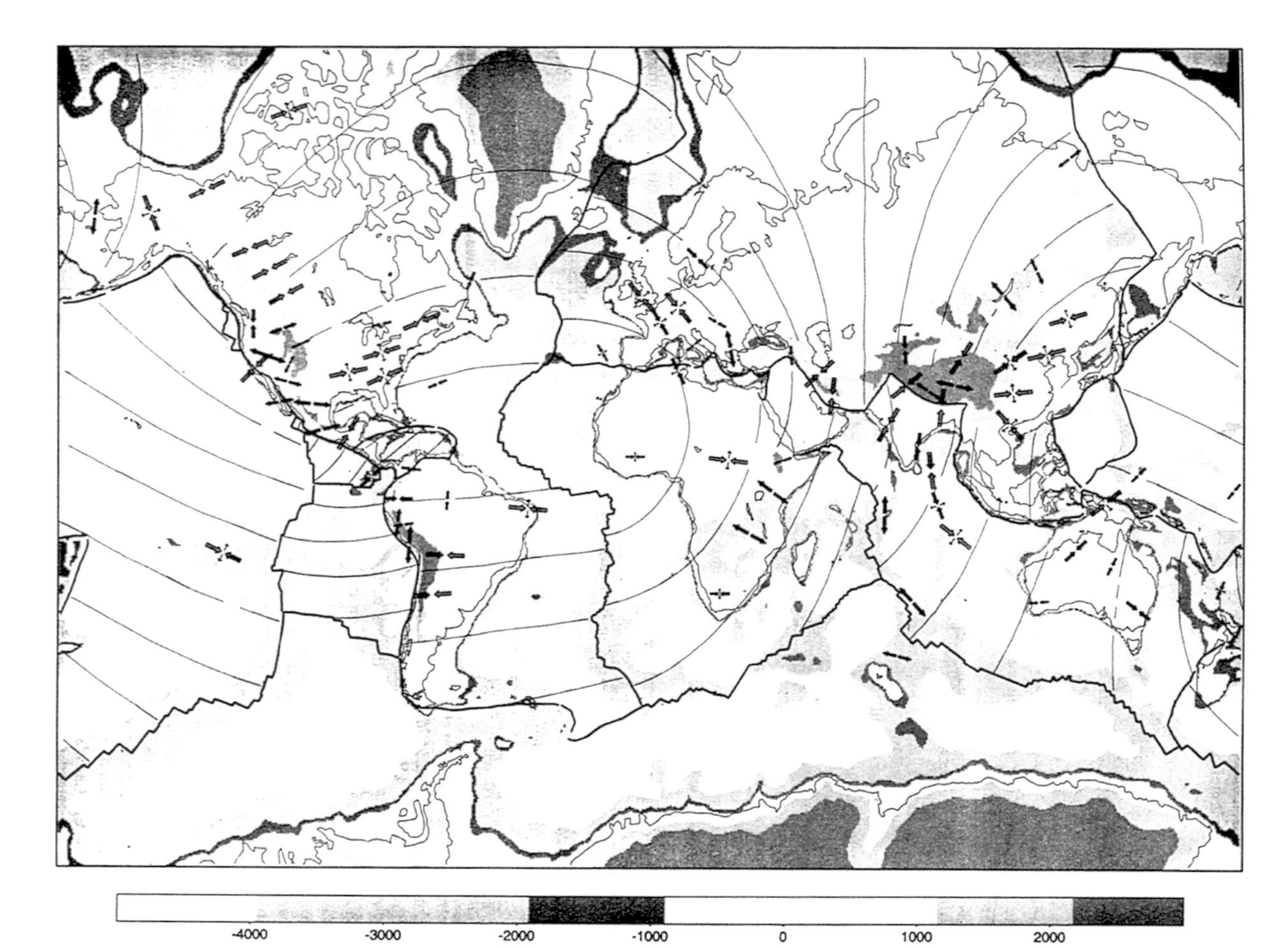

Fig. 2.19 Generalized version of *World Stress Map*, from Zoback (1992). Values shown in shading are topographic elevations or depressions in meters above or below sea level, respectively. Refer to original paper for details.

dipping Flannan Reflector under Scotland (Flack and Warner, 1990). The apparent drift of Europe may express a diffuse plate boundary analogous to that of southern Alaska where VLBI measurements have detected movement well north of the subduction zone (Ma *et al.*, 1990). It has been shown by Smith *et al.* (1990), from *LAGEOS* data, that underthrusting in subduction zones can produce regional deformation several hundred kilometers from the associated trenches. However, this speculation is also contradicted by the horizontal stress directions shown by the *World Stress Map*; as just discussed, motion caused by passive margin subduction should produce site motions to the southeast, not the northeast.

Another space geodesy measurement is at least suggestive of intracontinental deformation across the Eurasian Plate. Three baselines (Smith *et al.*, 1994) between western Europe and Shanghai for several years indicate consistent decreases of about one centimeter per year, despite the eastward movement of China implied by the "escape tectonics" theory of Molnar and Tapponier (1975). There is no obvious way, in the absence of a much denser geodetic net, to tell which station is actually moving, but a one centimeter annual rate for the European stations would account for most of the apparent trans-Atlantic increase. It must be pointed out that even if the western European stations are eventually shown to be moving eastward with the rest of the Eurasian Plate, this will produce a major problem for plate tectonic theory in that the European tectonic activity just cited could hardly be the direct result of horizontal plate movement. It would then be necessary to reassess the possible roles of vertical movement and magmatism in orogeny, as in the "surge tectonics" hypothesis of Meyerhoff *et al.* (1992).

To these problems must be added others inherent in plate tectonic theory. For trans-Atlantic drift, the chief difficulty is a driving force for the North American and Eurasian Plates (Lowman, 1985b). Purely oceanic plates can be driven by the well-understood slab pull: subduction of oceanic crust under the influence of increasing lithospheric density and the basalt–eclogite transition in subduction zones. Neither phenomenon can apply to plates whose leading edges are continental crust, and the edges of North America and Eurasia are obviously not being subducted. Added to the other evidence cited elsewhere (Lowman, 1985a) indicating that the trans-Atlantic continents are fixed above the mantle, and that there is no low-velocity zone under cratons (Knopoff, 1983), these problems must be considered major obstacles to confirmation of continental drift in the classic region where it must occur if it occurs at all.

It is suggested, in summary, that space geodesy can in principle provide a decisive test of continental drift in a plate tectonics context, but that it has not yet done so. The establishment of much more extensive nets will be necessary to demonstrate that the continents in question are as rigid as plate theory requires. It should be remembered that although Einstein's theory of general relativity, published in 1916, was verified within three years by Eddington's 1919 eclipse observations, relativity is still tested at every opportunity many decades later. Plate tectonics and continental drift deserve similar testing.

2.8 *GPS* measurements of crustal deformation

The *Global Positioning System* is rapidly beginning to dominate space geodesy for distances of several hundred kilometers and even more. Applications of *GPS* have in fact become a sizable industry by themselves, with many compact receivers on the market. Hundreds of papers have been presented at scientific meetings, far too many to summarize here. Useful reviews have been presented by Lisowski (1991), Colombo and Watkins (1991), and for California, Feigl *et al.* (1993). A few examples will give some idea of the value of this system.

Perhaps the most general problem that has been approached with *GPS* is one discussed briefly in Chapter 1, that of how deformation of continental crust is best described: as collections of microplates, or by a continuum model. The discussion of Thatcher (1995) will again be cited, starting with an instructive diagram (Fig. 2.20) of the differences between these two kinematic models. Two further diagrams (Figs. 2.21, 2.22) show deformation patterns in two seismically active areas, the southwest US and the Middle East. As shown in a global context on the tectonic activity map (see Fig. 2.7), these areas are broad and irregular; they could, in principle, be explained by either continuum or microplate models. Thatcher's excellent discussion, which can not be reproduced here, brings out difficulties in each one, specifically calling for future *GPS* surveys to help settle the problem.

The study by Le Pichon *et al.* (1995) is focussed on the problem discussed by Thatcher, in this case the crustal deformation of Greece, Turkey, and the Aegean Sea. This area is intensely active seismically and volcanically, and has been interpreted as an example of extrusion tectonics by McKenzie (1972), the concept being that the Anatolian block is being squeezed westward by the northward movement of Africa and the Arabian Peninsula. However, given the density of faulting and seismicity, it is valid to interpret this as continuum deforma-

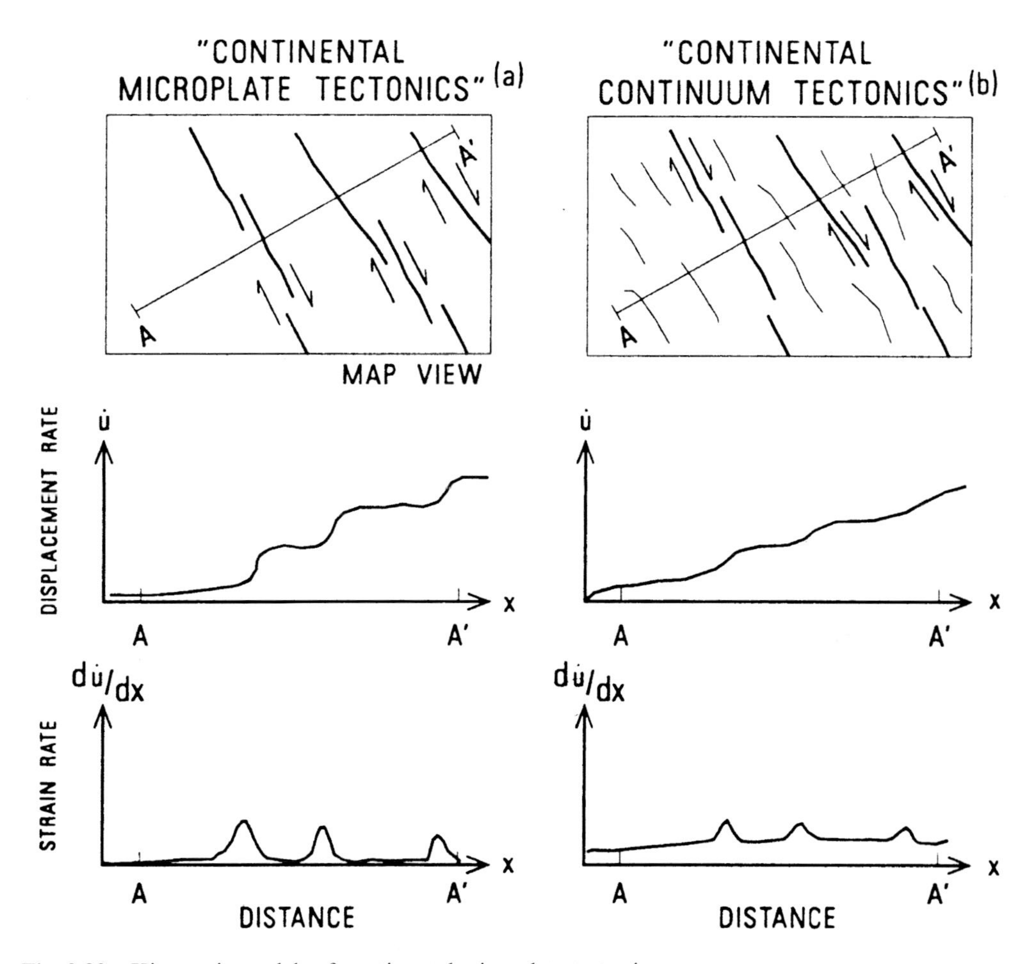

Fig. 2.20 Kinematic models of continental microplate tectonics vs. continuum tectonics. Fault motion assumed similar in both. From Thatcher (1995).

tion (Dewey and Sengor, 1979). Le Pichon *et al.* used both SLR (ranging to *LAGEOS 1*) and *GPS* methods to compile a map of the velocity field of Anatolia–Aegea relative to Europe (Fig. 2.23). As shown, the motion can be "approximated" by rigid rotation around a pole near the Nile delta, although the authors point out that the movement might have been started by phenomena such as gravitational collapse and trench retreat. The important point in the context of this chapter is simply that this study provides an excellent example of how space geodesy can be applied to the details of tectonic movements.

The *Global Positioning System* can also measure vertical movements precisely, as the following example will show.

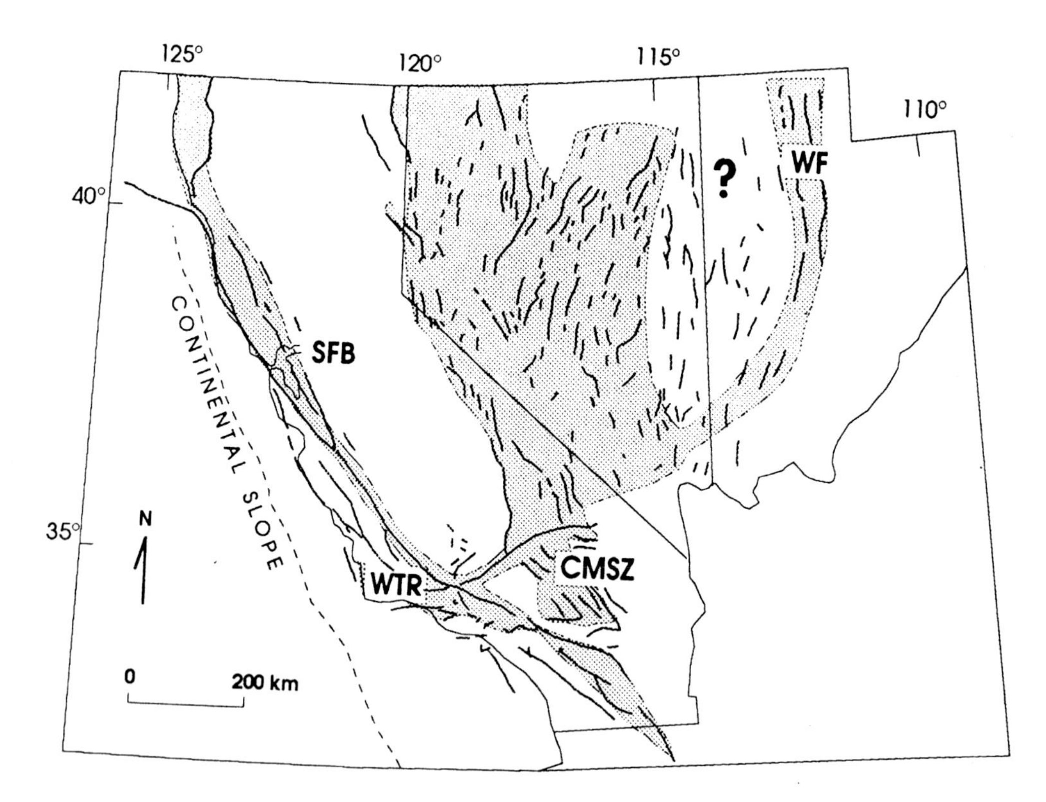

Fig. 2.21 Active faults and inferred deformation in western US. WF: Wasatch fault; CMSZ: Central Mojave Shear Zone; WTR: western Transverse Ranges; SFB: San Francisco Bay. From Thatcher (1995).

The general mechanism of earthquakes of the Pacific Ocean subduction zones, such as southern Alaska and the Aleutian Islands, is now well established as sudden slippage along dipping faults. However, the deformation during and after such earthquakes is too complex to be explained by simple rigid-plate models. A *GPS* study of the Kenai Peninsula, Alaska, by Cohen *et al.* (1995) shows how this technique can throw light on the nature of this deformation (Fig. 2.24). The area was hit by the 1964 Prince William Sound earthquake, which with a moment magnitude, **Mw**, of 9.3, was the largest event in North America in historic times. The Kenai Peninsula subsided as much as 2 meters during the earthquake, and has been rising since. Cohen *et al.* used *GPS* receivers to measure the post-seismic uplift for the 1964–1993 period, comparing the *GPS* measurement with those of previous surveys. They found that the post-seismic uplift was decelerating, and deduced the probable post-seismic slip along the subduction zone. This turned out to be

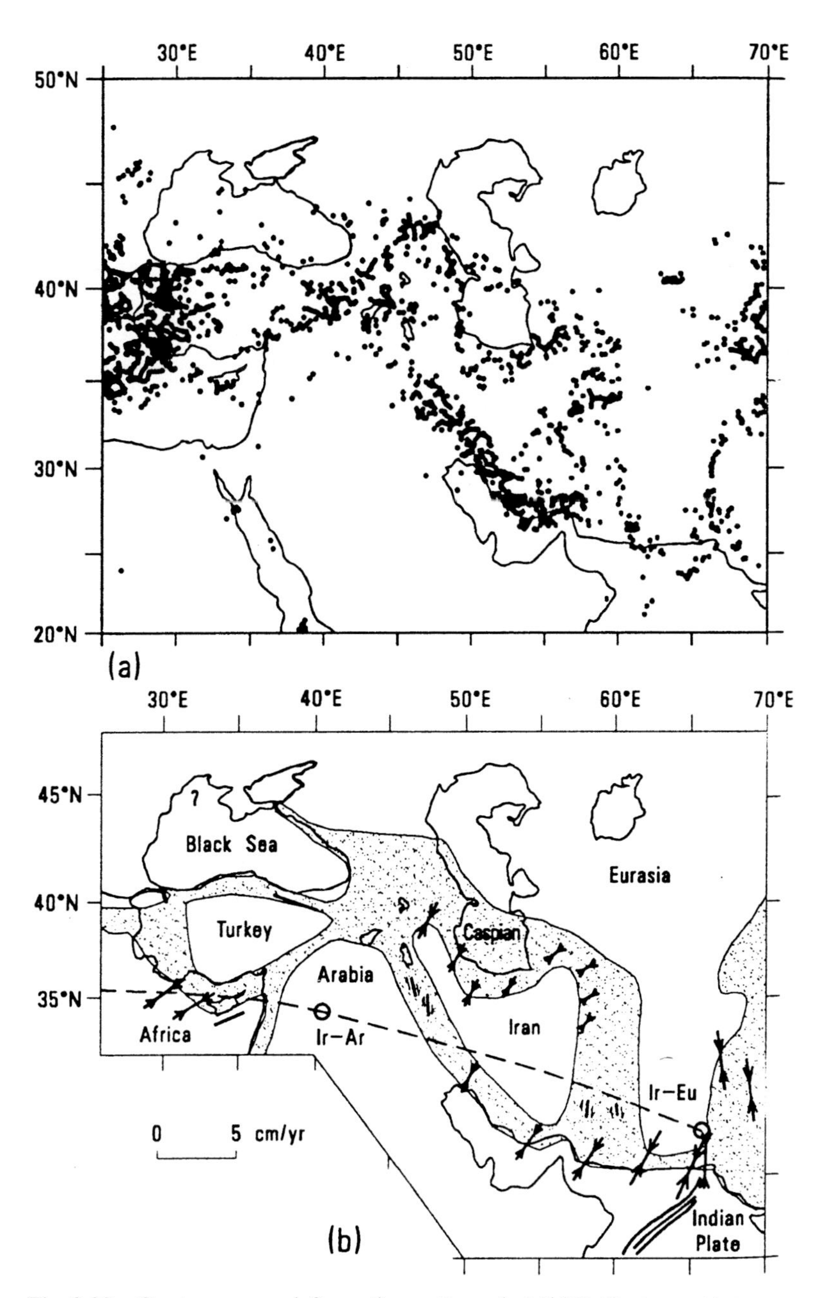

Fig. 2.22 Contemporary deformation patterns in Middle East; most intense deformation stippled. Dots: shallow earthquake epicenters, 1961–1980. Slip vectors show relative plate motion; circles are rotation poles. From Thatcher (1995).

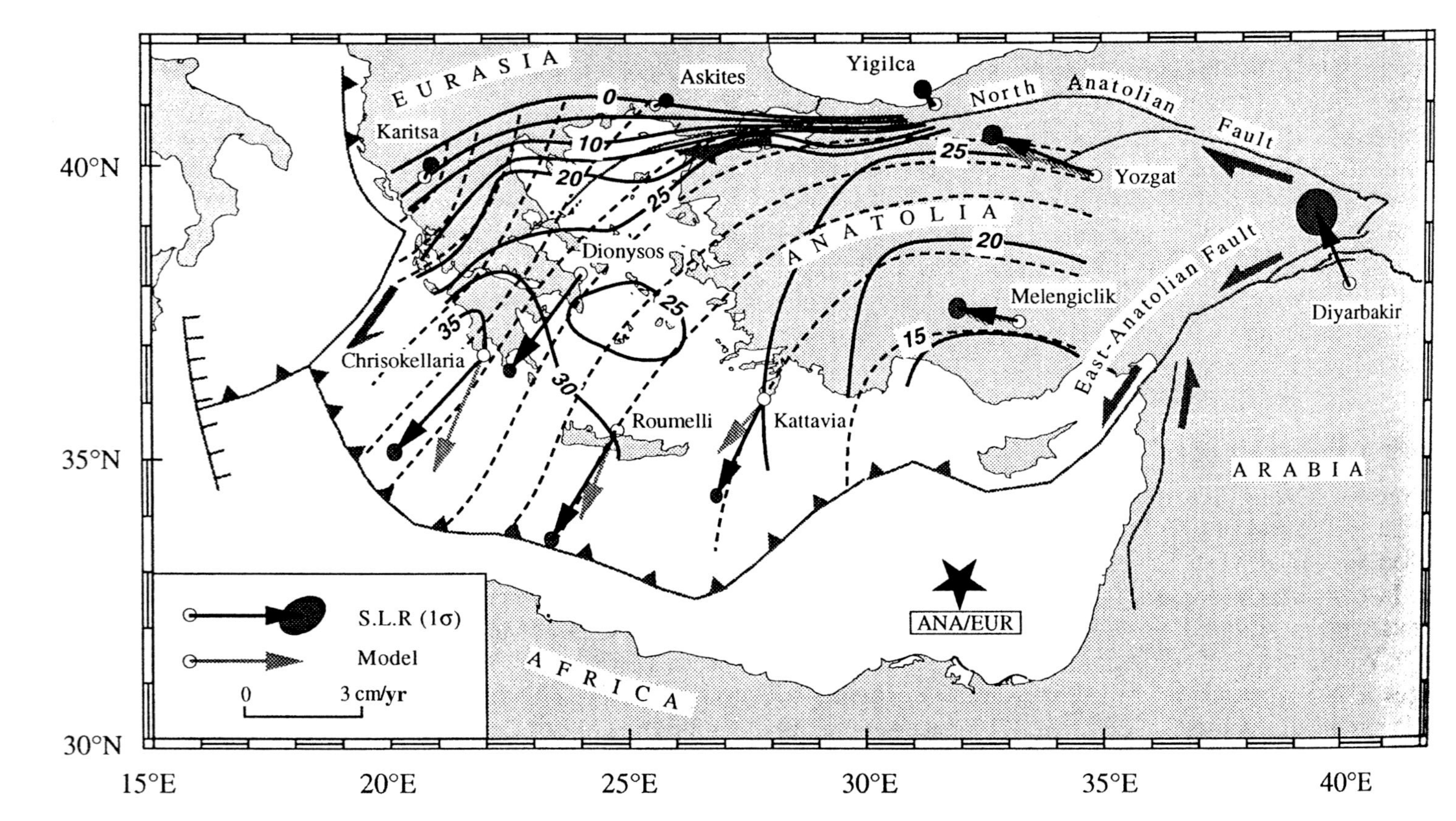

Fig. 2.23 Velocity field of Anatolia–Aegea with respect to Europe. Solid lines are equal velocity contours (5 mm/yr interval); dashed lines are flow lines. From Le Pichon *et al.* (1995).

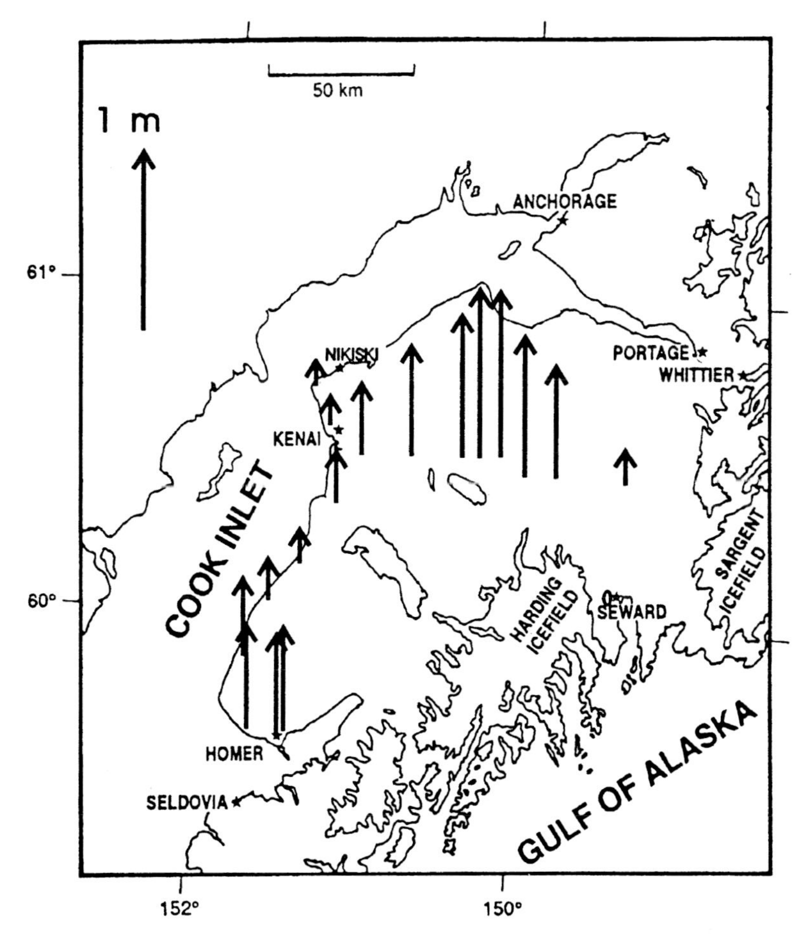

Fig. 2.24 Cumulative post-seismic uplift between 1964 and 1995, Kenai Peninsula. From Cohen and Freymueller (1997).

(Cohen, 1996) considerably greater than the 1.4 meters expected from rigid-plate motion of the Pacific Plate. The study provides a good example of how *GPS* measurements of vertical motion can in principle clarify the mechanism of plate motions.

Southern Alaska was also studied with *GPS*, this time concentrating on horizontal motions, by Sauber *et al.* (1997). This study carried out *GPS* measurements in 1993 and 1995 between the coast and the Denali fault. The area is extremely complex because it represents the transition between subduction under the Aleutian Islands and strike slip on major faults parallel to the coast, as shown in a generalized way on the tectonic activity map (see Fig. 1.1). The results are too complex to be reported here; the study is cited as an example of how space geodesy can be applied to the problem of dangerous earthquakes, by clarifying the location and amount of strain accumulation.

Of immediate interest for the active and dangerous tectonism of the Pacific coast of North America were the many *GPS* studies of faulting and crustal deformation in California. There is ample geodetic evidence, going back several decades, that California is undergoing regional right lateral shear of several centimeters per year, as mentioned previously. However, the detailed pattern of strain accumulation and strain release, and its variation with time, is extremely complicated (Hager *et al.*, 1991). Some typical studies of this problem follow; the most general is that of Feigl *et al.* (1993).

A program of *GPS* measurements by Donellan *et al.* (1993) across the Ventura Basin found, in a 2.7-year period, shortening across the Basin of about 7 mm/yr. The authors suggested that this strain is occasionally released by earthquakes along the faults bounding the basin. Burgmann *et al.* (1992) carried out a *GPS* program across the San Andreas fault near the center of the 1989 Loma Prieta earthquake, finding very rapid aseismic slip on a nearby fault that might be precursory to a rupture. Similar *GPS* measurements were done by King *et al.* (1992) across the Hayward Fault, detecting within one year variations in horizontal slip rate and an anomalously high rate of vertical movement. Vertical movements are much more important than they might seem, for it has been realized since the Whittier Narrows earthquake of 1987 that much strain release may be along unexposed thrusts.

These studies illustrate the great advantage of *GPS* for dense nets of precise geodesy that can be established quickly, as they were after the 1989 Loma Prieta **Ms** 7.1 earthquake on the San Andreas fault. Such a program was carried out by Arnadottir and Segall (1994), as shown in Figs. 2.25 and 2.26. Apart from the substantial damage on the San Francisco peninsula, the Loma Prieta earthquake was scientifically interesting because there was little surface rupture , unlike the classic 1906 San Andreas event. Arnadottir and Segall combined geodetic data from several techniques including *GPS* and VLBI, inverting them to determine the actual nature of movement along the fault plane. They found the slip direction to be surprisingly complex, varying from right-lateral to oblique right-reverse at different locations. Such studies should eventually bring about a big improvement in our understanding of earthquake mechanisms, with obvious importance for geologic hazard mitigation.

The previous examples of *GPS* use in California have been essentially scientific. However, another example demonstrates the close connection between tectonics and everyday life. Southern

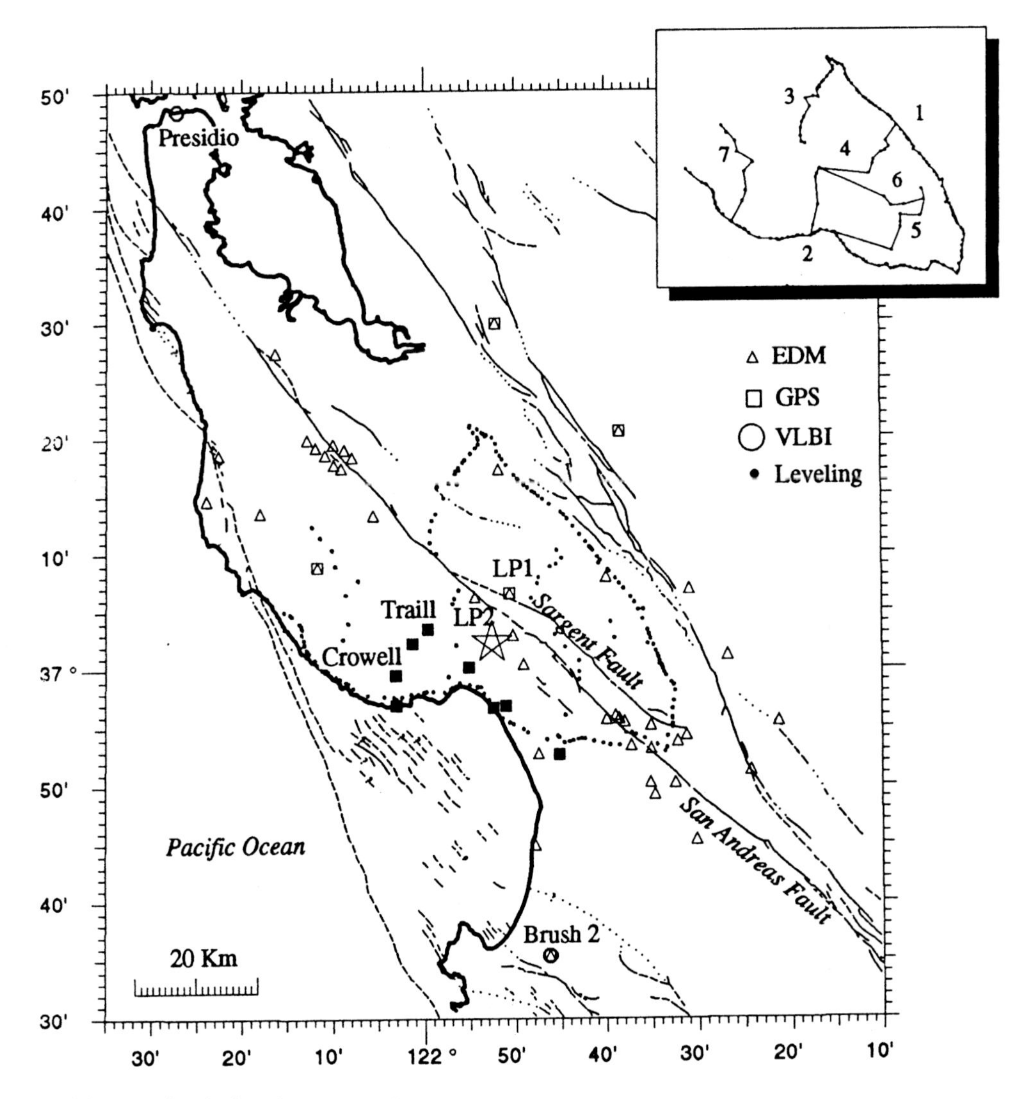

Fig. 2.25 Major faults of San Francisco Bay area, with locations of sites occupied for EDM, *GPS*, VLBI, and leveling. Insert shows leveling line. Star shows epicenter of main shock of Loma Prieta earthquake of 1989. From Arnadottir and Segall (1994).

California is a thicket of faults, most of them active or potentially so (Figs. 2.27, 2.28). This creates very real problems for surveyors and civil engineers, for the continual regional interplate shear, about 4 cm/yr, is literally bending geodetic nets, roads, pipelines, aqueducts, and railroads out of shape. Some of this motion, occurring during earthquakes, is sporadic, the rest secular. Satalich (1993) has described a large *GPS* program of precision surveying being carried

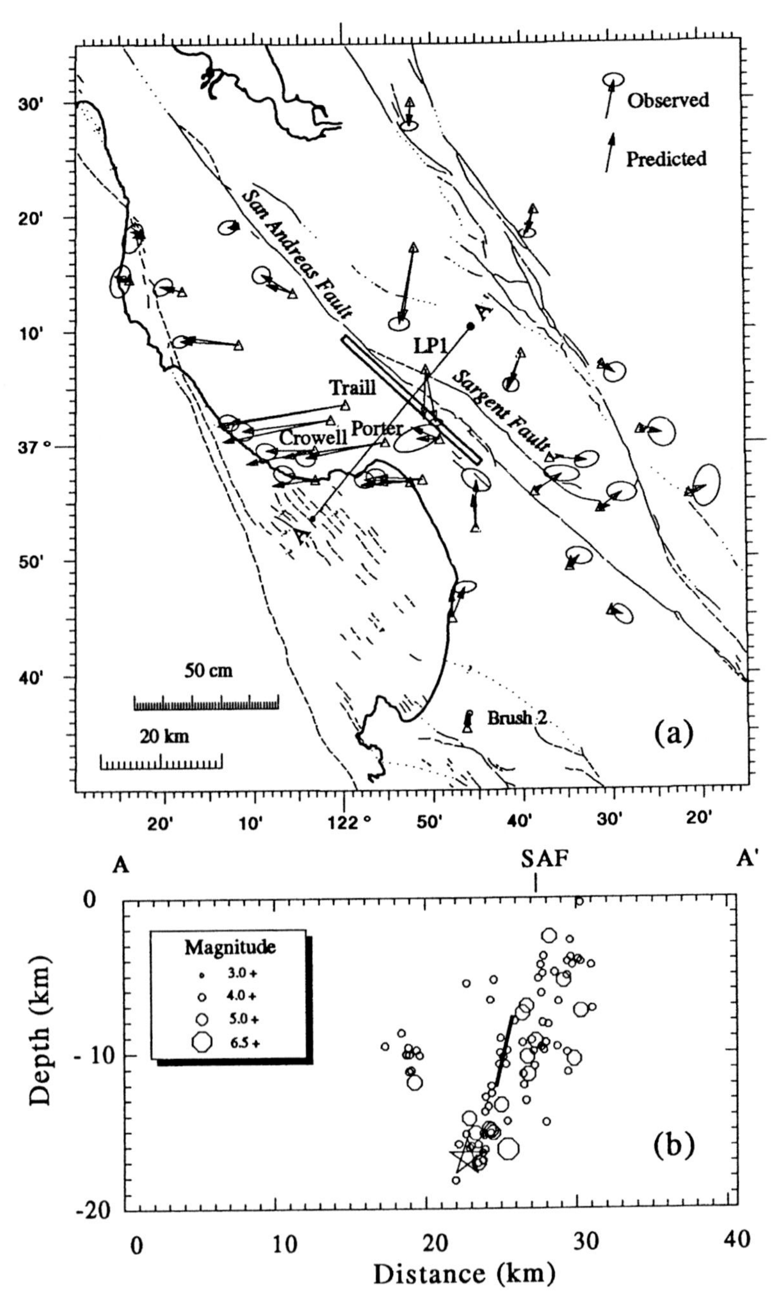

Fig. 2.26 (a) Observed and predicted horizontal displacements from the best uniform slip dislocation model. Rectangle shows surface projection of fault model. (b) Location of dislocation in cross section AA′. Main shock epicenter shown with star; hexagons are aftershocks from 18–31 October 1989. From Arnadottir and Segall (1994).

Fig. 2.27 *Landsat 1* Band 6 image of Los Angeles; 1972. For scale, see Fig. 2.28.

out by the California Department of Transportation (Caltrans). Impelled by a series of bond issues intended to modernize California's transportation systems, Caltrans established a 90-station *GPS* control net. The objectives were to unify the regional control nets and to update them continually to allow for tectonic movement. Particular attention was paid to known active faults, and the effort coordinated with various agencies conducting *GPS* tectonic research, as described previously. Vertical as well as horizontal measurements were made, since in Los Angeles and Ventura Counties there is not only tectonic movement but elevation changes caused by oil or water withdrawal, water re-injection, and similar

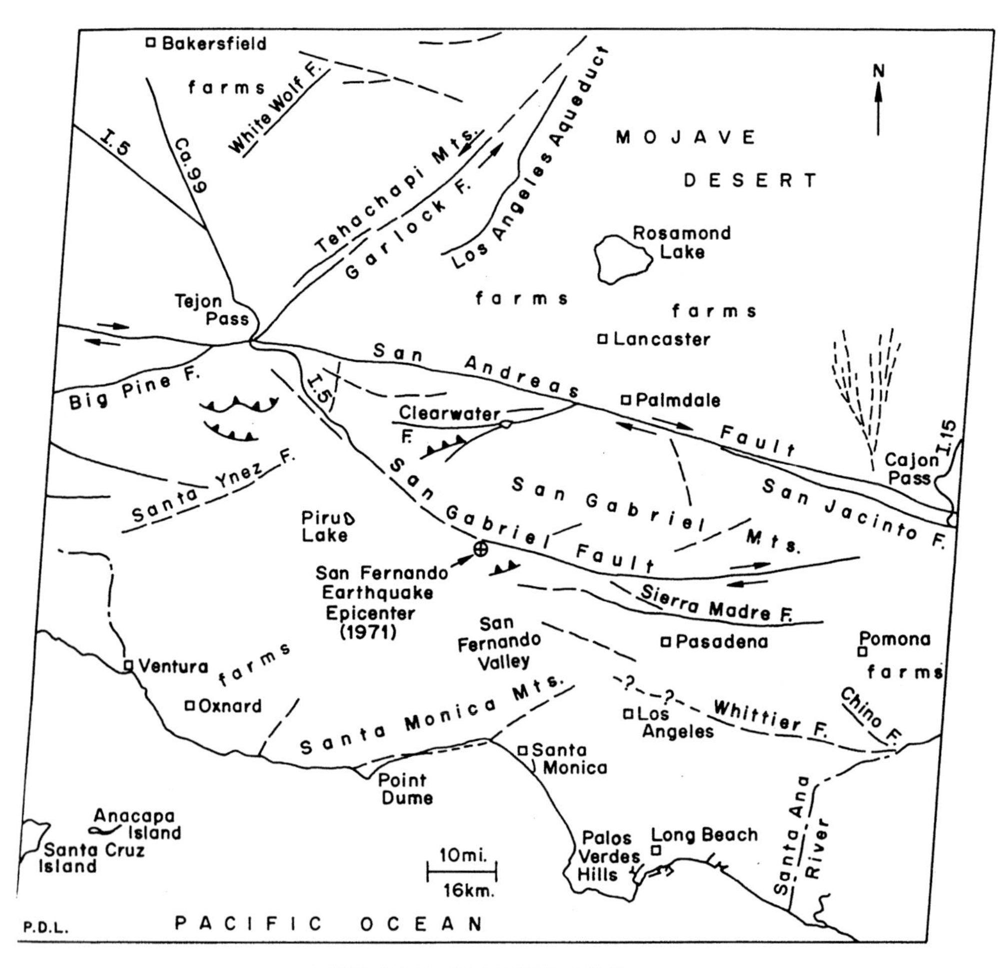

GEOLOGIC SKETCH MAP
TRANSVERSE RANGES, CALIFORNIA
FROM LANDSAT-1 IMAGE 1090-18012

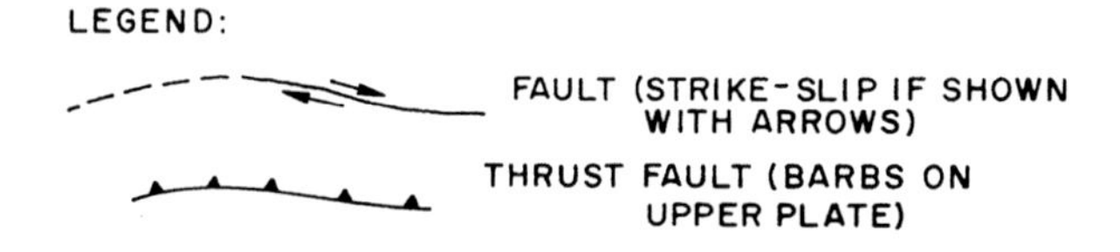

Fig. 2.28 Geologic sketch map of Fig. 2.27.

phenomena. Vertical movements are especially important for new rail lines. Collectively, the Caltrans project provides a dramatic demonstration of the dollar-value of space geodesy.

The *Global Positioning System* is a military system, whose full power can not always be utilized because of necessary security restrictions, such as selective availability. However, it is clear that even with these restrictions *GPS* is now playing a major role in space geodesy. It has proven particularly useful for attacking the problem of diffuse plate boundaries and intracontinental deformation with which existing VLBI and SLR nets are beset. The tests of plate rigidity discussed before, necessary before continental drift can be directly demonstrated, are an obvious application of *GPS*.

2.9 Earth rotation and expansion tectonics

Space geodesy has implicitly provided a rigorous test of Earth expansion as a mechanism for tectonic activity, "expansion tectonics" henceforth. This theory, in brief, holds that the Earth has expanded greatly – by a large fraction of its initial radius – over geologic time, and is doing so today, accounting for sea-floor spreading, the distribution of continents, and other major features of terrestrial geology. Expansion tectonics is very much a minority view, but one held by many highly-qualified geologists and geophysicists. A leader in the field is S. W. Carey (1976, 1981a), who has compiled several well-documented lines of evidence supporting the "necessity" for Earth expansion. H. G. Owen (1981) has presented detailed plate reconstructions pointing in his view to expansion as a mechanism for "ocean-floor spreading." Schmidt and Embleton (1979) found common apparent polar wander paths for several continents during the Precambrian, concluding that an increase of 45% for the Earth's radius during the time involved was implied.

An important point in favor of expansion tectonics, whatever else can be said, is that it leads to testable predictions (Runcorn, 1981). One prediction implicit in a theory of major and continuing expansion is an increase in the length of day (LOD) because of the absolute necessity to conserve angular momentum. The principle is illustrated in elementary physics books by the familiar pirouetting skater, whose spin rate increases as she folds her arms, thus conserving angular momentum. As applied to the Earth, it means that the rotation rate must decrease if the solid part of the planet expands. (An elementary mathematical summary has been presented by Stewart, 1981.) The consequence of this is that LOD must increase if the Earth expands.

The constraints of angular momentum conservation are fundamental (Carey, 1981b), and they lead to several tests of expansion tectonics to which space geodesy can be applied. An obvious one, already discussed, is increasing baseline lengths among various space geodesy stations – among all of them, in principle. Expansion has failed this test so far in that the changing baseline lengths generally fit the NUVEL-1 model, with serious but explainable exceptions as discussed previously, in which the radius of the Earth is implicitly constant.

The most definitive test would be a continuing increase in LOD. It must be pointed out here that scientists who have concentrated on the problem of Earth rotation, of which LOD is a major component, do not even mention the Earth-expansion hypothesis (e.g., Munk and MacDonald, 1960; Lambeck, 1980). This is probably because they consider the evidence for both paleorotation and present rotation rates to conclusively rule out major expansion.

As mentioned previously, space geodesy provides several ways to measure LOD with precision on the order of a few milliseconds (Dickey *et al.*, 1993). The literature on this is large and can not be even summarized here, but a tabulation by Chao (1994) (Table 2.3) shows that the precision now attained is good enough to detect mass redistributions as small as those caused by reservoir filling and emptying. Measurements of LOD have been going on for several decades by precise astronomical techniques, and in recent years by gravity-independent methods involving atomic clocks. Viewed together with the space geodesy results, they have essentially disproven major earth expansion by default: there is simply no sign of a steady trend in LOD pointing to such expansion.

A possible cause for Earth expansion suggested by several scientists, including Dirac, Jordan, Dicke, and Holmes, is secular decrease in the universal gravitational constant G. In principle, such a fundamental assumption would permit expansion, possibly to a very slight degree. However, space geodesy again appears to rule out this proposal. The value for G is fundamental to interpretation of, for example, *LAGEOS* orbits (Tapley, 1993). But the entire mathematical infrastructure of space geodesy is a rigid one. If fundamental quantities such as G were changing significantly now – as implied by expansion tectonics – such changes would show up within a few years at most. They have not.

Expansion tectonics is advocated by many scientists, who repeatedly demonstrate serious anomalies in conventional tectonic theory, many of which deserve intensive study. It was commented by one of Einstein's collaborators, Leopold Infeld, concerning the later-retracted "cosmological constant," that an incorrect solution of an important problem can be more valuable than a correct solution of a trivial problem. Expanding-Earth advocates have made a real contribution to tectonics, but the hypothesis appears to have been conclusively disproven by space geodesy.

Table 2.3 *Geophysical causes of variation in length of day (LOD). From Chao (1994).*

		Amplitude (peak-to-peak)			
Geophysical source	Temporal signal	ΔJ_2 ($\times 10^{-10}$)	ΔJ_3 ($\times 10^{-10}$)	$\|\Psi\|$ (mas)[a]	ΔLOD (ms)
Tidal deformation					
	Long-period	Up to 20	?	0	Up to 0.8
Solid earth	Diurnal	0	0	Up to 4 (in $\|m\|$)	0
	Semi-diurnal	0	0	0	0
Oceans	All tidal period	Up to 4		Up to 1 (in $\|m\|$)	0.08
Atmosphere					
IB[b]	Days–Seasonal–Interannual	8 (peak)	10 (peak)	100 (peak)	0.15 (peak)
		3 (annual)	5 (annual)	55 (annual)	0.05 (annual)
		1 (interannual)	1 (interannual)	<5 (interannual)	0.02 (interannual)
Non-IB	Days–Seasonal–Interannual	15 (peak)	20 (peak)	200 (peak)	0.3 (peak)
		5 (annual)	6 (annual)	82 (annual)	0.1 (annual)
		2 (interannual)	2 (interannual)	<10 (interannual)	0.03 (interannual)
Continental water					
Snow	Seasonal–Interannual	2 (annual)	1 (annual)	20 (annual)	0.04 (annual)
Rain	Seasonal–Interannual	1 (annual)	1.7 (annual)	16 (annual)	0.02 (annual)
Glaciers	Secular	0.02 per year	0.01 per year	0.04 per year	4×10^{-4} per year
Reservoirs	Cumulative since 1950	−0.4	0.3	10	−0.006
Ice sheet	Secular	?	?	?	?
Groundwater	Seasonal–Secular	?	?	?	?

Table 2.3 (*cont.*)

Geophysical source	Temporal signal	Amplitude (peak-to-peak)			
		ΔJ_2 ($\times 10^{-10}$)	ΔJ_3 ($\times 10^{-10}$)	$\lvert\Psi\rvert$ (mas)[a]	ΔLOD (ms)
Ocean					
Sea level	Secular	0.03 per year	−0.02 per year	0.05 per year	5×10^{-4} per year
Circulation	Seasonal–Interannual	?	?	?	?
Earthquake	Episodic: (1) 1960 Chile (2) 1964 Alaska	0.5 (2)	0.3 (1)	23 (1)	−0.008 (1)
	Cumulative secular (1977–90)	−0.002 per year	0.008 (peak)	0.03 per year	-10^{-4} per year
Post-glacial reb.[c]	Secular	−0.3 per year	?	3 per year	?
Tidal braking	Secular	−0.005 per year	0	0	-10^{-4} per year
Mantle convection/ Tectonic movement	Secular	?	?	?	?
Core activity	Secular	?	?	?	?

Note:
[a] mas = milliarcseconds; [b] IB = inverted barometer correction; [c] reb. = isostatic rebound.

2.10 Extraterrestrial gravity fields

The ability to cross space has given us the first opportunity to study in detail the gravity fields of other planets. At this writing, every planet but Pluto has been visited by spacecraft, in addition to many satellites. It will be instructive to compare the gravity fields of those bodies most similar to the Earth – the Moon, Mars, and Venus – with that of the Earth. Several useful reviews have been published on this and related subjects. The most comprehensive modern paper is by Phillips and Lambeck (1980), covering the gravity fields and tectonics of the terrestrial planets. However, the now-outdated paper by MacDonald (1963), essentially the last astronomical treatment of planetary geophysics, is still a valuable introduction to the principles and the first-order problems.

2.10.1 Gravity field of the Moon

As one would expect, we know far more about the lunar gravity field than any other, except that of the Earth itself. Reviewed by Kaula (1975) and Ferrari and Bills (1979), measurements related to lunar gravity come chiefly from Doppler tracking of spacecraft in lunar orbit, but also from laser ranging to the retroflectors on the surface, laser altimetry from *Apollo* spacecraft, and instruments landed on the surface during the *Apollo* expeditions. There is consequently a very large literature on this subject, and one which is growing as new data are acquired from new lunar missions, *Clementine* (1994) and *Lunar Prospector* (1998) in particular. Accordingly, only the main findings from these sources will be summarized here, more or less in historic order.

As we have seen, the gravity field of a planet is closely related to the planet's shape. We should therefore begin by mentioning the pre-spacecraft discovery by ground-based astronomers (Baldwin, 1963; Kopal, 1971) that the Moon is a triaxial ellipsoid, roughly pear-shaped with the small end pointed away from the Earth. This shape is far too irregular to express hydrostatic equilibrium, and Urey (1952) cited it as evidence that the Moon must be cool and rigid throughout at present. The opposite view, that the Moon's shape represented mantle convection, was presented by Runcorn (1967), but this was before new information became available from Moon-orbiting satellites. The opposing views were somewhat analogous to those about the Earth's shape and gravity field, which as we have seen were not resolved until satellite tracking was possible.

The series of five *Lunar Orbiter* missions at different altitudes and inclinations revolutionized our knowledge of the Moon's gravity (as well as its geology). The most immediate discovery from Doppler tracking of the *Lunar Orbiters* was a new value for the mass of the Moon and a map of its gravity field to degree and order 13 (Michael and Blackshear, 1972). However, the most surprising subsequent discovery from the tracking data, by Muller and Sjogren (1968), was of large positive gravity anomalies over the circular maria. They termed these "mascons," since they were clearly mass concentrations. Muller and Sjogren suggested that these features were the buried iron bodies that had formed the mare basins, but this suggestion was immediately challenged by several authors all of whom acknowledged the great achievement of discovering the mascons. An issue of *Science* published shortly after included several of these discussions; the most useful is that of O'Keefe (1968). O'Keefe pointed out that Muller and Sjogren's "astonishing" feat actually demonstrated that the Moon, in particular the lunar highlands, was roughly in isostatic equilibrium and, by implication, was probably differentiated, an interpretation confirmed after the *Apollo* landings began. A more general treatment of the mascon interpretation was published by Kaula (1969), discussing several issues such as the nature of the basin-forming impacts. The proposal that the mascons were caused by buried impacting bodies was generally rejected, one reason being the knowledge that such bodies are almost always destroyed and largely ejected from the crater.

Wise and Yates (1970) proposed the explanation that is still widely accepted, although modifications seem called for since the new data from *Clementine* and *Lunar Prospector*. The mascons are actually Bouguer anomalies, caused by unusually dense material. Wise and Yates suggested that after the basin-forming impacts, a plug of lunar mantle material was pushed upward to occupy the initial crater, approaching isostasy. At some later time, the basins were flooded by the mare basalts, which as added mass to the lunar crust produced the positive gravity anomalies. Later studies by various workers, taking into account mare structure and other geologic factors, has led to much more detailed reconstructions of the history of the lunar maria (Melosh, 1978; Solomon and Head, 1980; Bratt *et al.*, 1985).

The resumption of lunar exploration, unfortunately unmanned, in the 1990s has generated new models of the lunar gravity field. The *Clementine* mission of 1994 (Nozette *et al.*, 1994) provided more than 2 months of tracking and laser altimetry from a nearly circular low orbit, excellent for mapping. As reported by Zuber *et al.* (1994), *Clementine* confirmed O'Keefe's (1968) conclusion, from the

Muller–Sjogren analyses, that the highlands were isostatically compensated. However, the mare basins show a much more complicated behavior in this respect. The classic mascon basins, the near-side circular maria, are uncompensated, consistent with the interpretation of Solomon and Head (1979) that topography is supported by the strength of the cold and rigid lithosphere. An interesting early study of the lunar Apennines (Fig. 2.29), the rim of the Imbrium Basin, by Ferrari *et al.* (1978) concluded that these mountains were similarly supported, and were not in isostatic equilibrium. This points up one of the many differences between the Moon and the Earth, where major mountain ranges are in isostasy.

Using the *Clementine* tracking and laser altimetry data, Zuber *et al.* (1994) constructed a series of geophysical maps of the entire Moon. Another finding from the *Clementine* laser altimetry (Spudis *et al.*, 1994) was confirmation of several extremely old multi-ring basins and discovery of possible new ones. Of particular interest is the South Pole–Aitken Basin, 2500 km wide and hence the largest known impact basin in the solar system.

The *Clementine* data have been used by Wieczorek and Phillips (1997) to study the structure and composition of the lunar highland crust. Their work provides an excellent example of how geophysical and geochemical data can be combined. A detailed summary of their results is not possible here, except to note that they found Airy compensation to be dominant, and interpretable in terms of a stratified highland crust.

The *Lunar Prospector* mission (Binder, 1998) has been even more successful than *Clementine* in a geophysical sense because of its low-altitude (ca. 100 km high) circular near-polar orbit. Doppler tracking data have permitted construction of a greatly improved 75th degree lunar gravity model (Fig. 2.30) (Konopliv *et al.*, 1998), when combined with data from *Lunar Orbiter*, *Apollo*, and *Clementine*. The structure of the mascons, now seen with gravity data having a spatial resolution of 75 km, seems now to be more complex and less easily interpreted than previously thought, a conclusion reached earlier by Zuber *et al.* (1994). Seven new mascons have been discovered, some of which have little or no mare basalt fill. This implies that the mantle plug hypothesized by Wise and Yates (1970) may be more important for such features. The new data have generated new controversy on the degree and mechanisms of isostatic compensation in the Moon, a refreshing development. Further discussion of these would be out of place in a book about the Earth, except for mention of the mascon basin study by Golombek (1979).

After a perceptive discussion of faulting, Golombek analyzed

Fig. 2.29 Mount Wilson 100 inch (2.5 m) telescope view of the 1300-km-wide Mare Imbrium; north at top. Apennines at lower right.

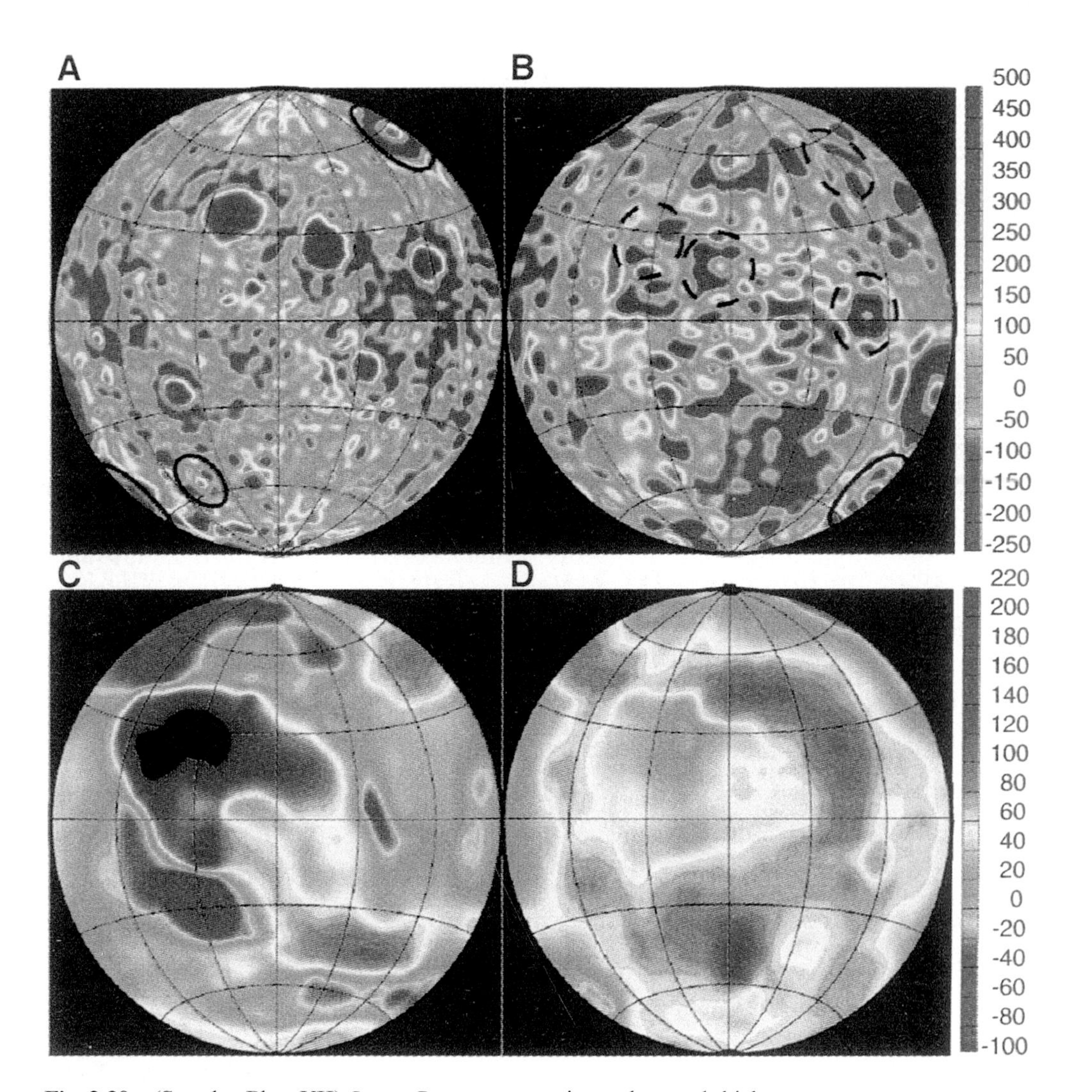

Fig. 2.30 (See also Plate VII) *Lunar Prospector* gravity and crustal thickness maps (Konopliv, *et al.*, 1998), Lambert equal-area projection. A, C: near side; B, D: far side. *Top*: Vertical gravity anomalies, in milligals. Newly-discovered near-side mascons shown with solid circles, far-side ones with dashed circles. *Bottom*: Crustal thickness in kilometers, calculated with an Airy compensation model (constant density) without principal mascons.

the structure of mascon basins. Among his conclusions relevant to terrestrial geology is the inference that fractures propagate upward, and that stresses can not be correctly calculated from the surface expression of fractures. He further explained the scarcity of strike-slip faults on extraterrestrial surfaces as resulting from fracture formation at depth, where maximum compressive stresses are vertical.

As discussed in Chapter 4, the nature and origin of crustal fracture patterns – lineaments – is still poorly understood. Golombek's analysis, based originally on lunar gravity data, is a good example of how extraterrestrial geology can have implications for the geology of the Earth.

Even more fundamental inferences about the Moon have been made by Konopliv *et al.* (1998) from the new tracking data. These data have permitted calculation of a new value for the polar moment of inertia, close to a previous estimate by Hood and Jones. These values imply that the Moon may have a core, presumably of iron, as large as 900 km in diameter. If confirmed, this would put constraints on theories of the Moon's origin, in particular the giant-impact model in which the Moon formed from debris ejected from the Earth by collision with a Mars-sized object. However, the latest version of the giant-impact theory (Cameron, 1997) has the impact happening in the late stages of accretion of the Earth.

In summary, the gravity field of the Moon has proved to be inherently interesting, by itself and by comparison with that of the Earth. The Moon's geologic evolution has been shown to be much simpler than terrestrial geology, the basic reason being that the Moon ran out of internal energy at an early stage. Its crust accordingly has not undergone the repeated re-working experienced by the terrestrial crust. The Moon's role as a fossil planet is discussed in Chapter 6. At this point it can simply be noted that its "fossil" nature has helped interpret the gravity data, in that structures produced billions of years ago have been essentially frozen in. Finally, the application of gravity methods to lunar exploration typifies their strengths and weaknesses: indecisive by themselves, but highly useful when interpreted in the context of other data.

2.10.2 Gravity field of Mars

Most of our knowledge of the gravity field of Mars comes from satellite tracking, which began in 1895 when Struve (1895) used the 30 inch (0.8 m) refractor at Pulkowa Observatory to monitor the orbits of Phobos and Deimos. He applied Clairaut's Theorem, still the standard approach, to obtain a value of 5.210×10^3 for the flattening of Mars. This is essentially the modern value, from Doppler tracking of spacecraft (e.g., 5.216×10^3, Reasenberg *et al.*, 1975).

Since the 1960s, several spacecraft have been put in orbit around Mars; the two *Viking Orbiters*, operating for about 4 years, were unusually valuable for Doppler tracking. The *Mars Global Surveyor* (*MGS*) was put into orbit around Mars in 1997, operating for more

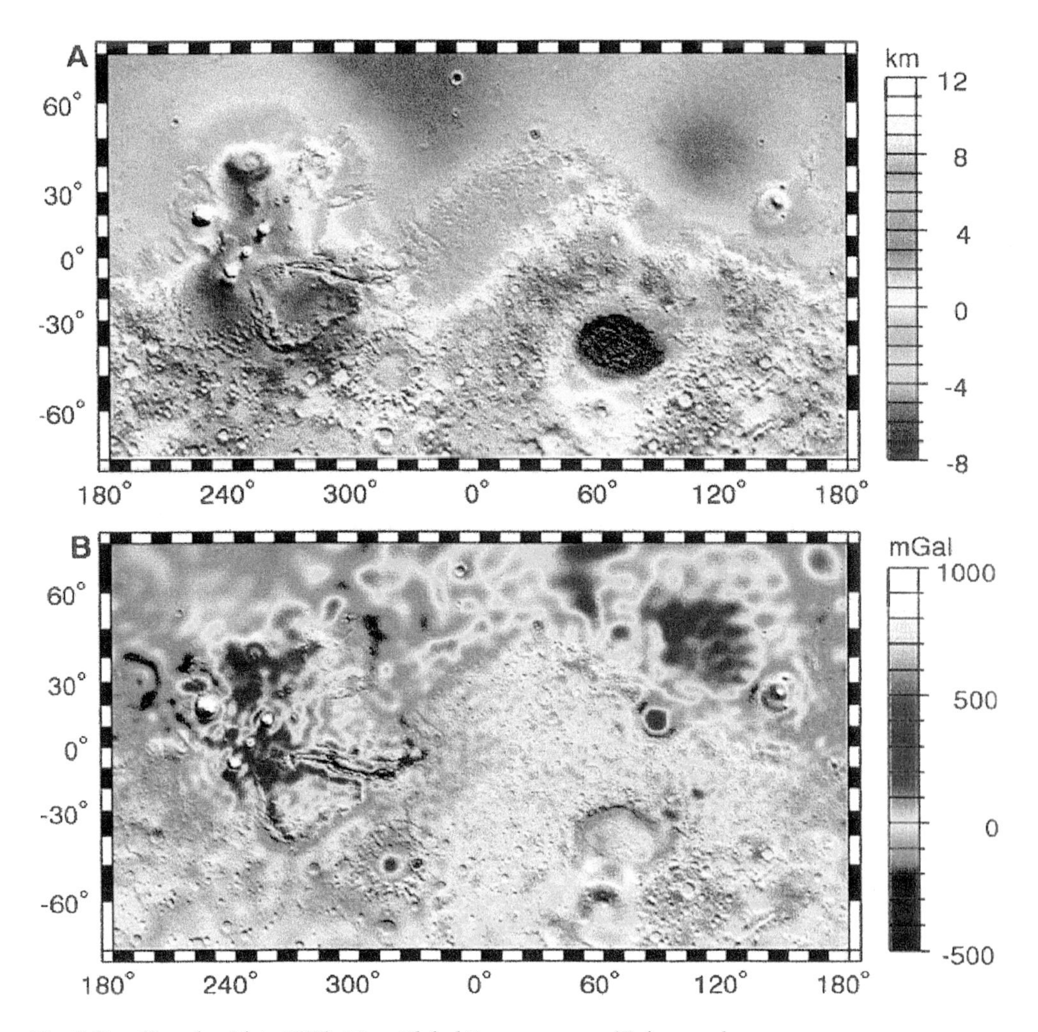

Fig. 2.31 (See also Plate VIII) *Mars Global Surveyor* maps (Zuber *et al.*, 2000) of topography (*top*) and free-air gravity values (*bottom*). Tharsis area near the equator between 220 and 300 deg. E.; Hellas Basin: 45 deg. S, 70 deg. E.; Utopia: 45 deg. N, 110 deg. E.

than 4 years. Tracking data and altimetry from the *MGS* Mars Observer Laser Altimeter (MOLA) have been extremely valuable (Zuber *et al.*, 1998), permitting compilation of detailed topographic, free-air gravity, and crustal thickness maps (Fig. 2.31) (Zuber *et al.*, 2000). Before discussing these, a few major characteristics of martian geology should be noted.

The most important factor is that Mars is much more evolved geologically, and more internally active now, than the Moon, as will be discussed in Chapter 6. Generally speaking, Mars can be considered geologically intermediate between the Moon and the Earth. The present surface environment of Mars is physiologically that of space, with surface pressures of about 6 millibars. However, it is clear that Mars has retained considerable water, since recent fluvial erosion and deposition, stratification, and mass wasting are widespread. Consequently, the physiography of Mars is a complex palimpsest of impact craters, tectonic features, fluvial channels and deposits, volcanic rocks, and wind-deposited material. However, Mars has evidently not reached the stage of plate tectonics, and has occasionally been called a "one-plate planet." Comprehensive treatments of the planet have been published by Carr (1982) and Kieffer *et al.* (1992).

Most discussions of the gravity field of Mars center on the Tharsis Uplift, a volcano-capped plateau several thousand kilometers wide that dominates the topography of the planet. Bills and Ferrari (1978) showed that with reasonable assumed values for crustal density, a negative Bouguer gravity anomaly can be mapped over Tharsis (and other volcanic areas). The latest maps from *MGS* (Fig. 2.31) have confirmed and refined early interpretations, providing a spatial resolution of about 178 km (from maps of degree and order 60). The main conclusions by Zuber *et al.* (2000) can be briefly summarized as follows.

The crust of Mars has been estimated to have an average thickness of 50 km, assuming a uniform density of 2900 kg/m^3 (somewhat denser than the average continental crust of the Earth, 2700–2800 kg/m^3 (Lodders and Fegley, 1998). However, there is a marked thinning from the southern highlands to the northern plains, with the exception of the crust under the Tharsis volcanos, which is estimated to be around 80 km thick. The boundary of the crustal dichotomy between south and north does not in general correspond to the crustal thickness change, and Zuber *et al.* concluded that this implies an essentially tectonic origin (Sleep, 1994) for the northern plains as opposed to an impact origin (Frey and Schultz, 1988). However, there is no agreement on this problem, many subtle impact craters having since been found by Mars Observer Laser Altimeter data (Frey *et al.*, 2000). High-resolution gravity maps (not shown) suggest that the northern plains are underlain by buried outflow channels. This would imply that the unusually flat plains (Smith *et al.*, 1998) are blanketed by a thick sediment layer, supporting views that Mars has had large amounts of water, perhaps even an ocean (Head *et al.*, 2000), at some time in its history.

Mars has several large impact basins, notably Hellas (Fig. 2.31), that display mascons similar to those found under the circular lunar maria. Zuber *et al.* consider these to have formed by a Moon-like mechanism, post-impact isostatic adjustment followed by infilling.

2.10.3 Gravity field of Venus

Because of its thick opaque atmosphere, very little was known about the geology of Venus until radar investigations – from Earth and from orbiting spacecraft – were possible. Since the mid-1980s, two Soviet missions and one American mission have finally given us a good look at the physiography of Venus, as will be discussed in Chapter 6. For the study of its gravity field, the best available data at this writing are Doppler tracking of the *Pioneer Venus Orbiter* (*PVO*) (Bills *et al.*, 1987), which transmitted radar altimetry and other data from orbit for over a decade until it entered the Venus atmosphere in 1993.

Because of its Earth-like size and density, Venus is an unusually valuable comparison planet for geophysical and geological studies. Consequently, much effort has gone into analyses of the *PVO* tracking data and geophysical interpretations. The availability of accurate topographic maps from the *PVO* altimeter, and high-resolution radar imagery from the *Magellan* mission, has made the *PVO* data even more valuable. An important finding from the altimetry is that the gross topography of Venus is unimodal, i.e., most features are at the same elevation relative to a reference spheroid (Masursky *et al.*, 1980). The Earth, in contrast, has bimodal topography, the two peaks corresponding to continents and the ocean floors. Gravity maps similar to those for the Moon and Mars have been compiled from the *PVO* data, complete to degree and order 50 (Fig. 2.32) (Nerem *et al.*, 1993).

The most important characteristic of the Venus gravity anomaly maps is that unlike the Earth, where large (long-wavelength) anomalies show relatively little correlation with topography (Phillips and Malin, 1984), the gravity anomalies of Venus do show such correlation. All the main high regions, such as Aphrodite Terra, are closely mirrored by positive free-air gravity anomalies. This at once raises the question of how these areas are supported, roughly analogous to the same question for the Earth's third harmonic. Given the mass of Venus, its lack of water, and the permanently high surface temperatures, a relatively thin (<20 km) lithosphere is inferred (Phillips and Malin, 1984). Isostatic compensation would have to be very deep, where the mantle has little strength, and isostatic support is thus "implausible" (Bills *et al.*, 1987). The general view (Kiefer *et al.*, 1986; Herrick and Phillips,

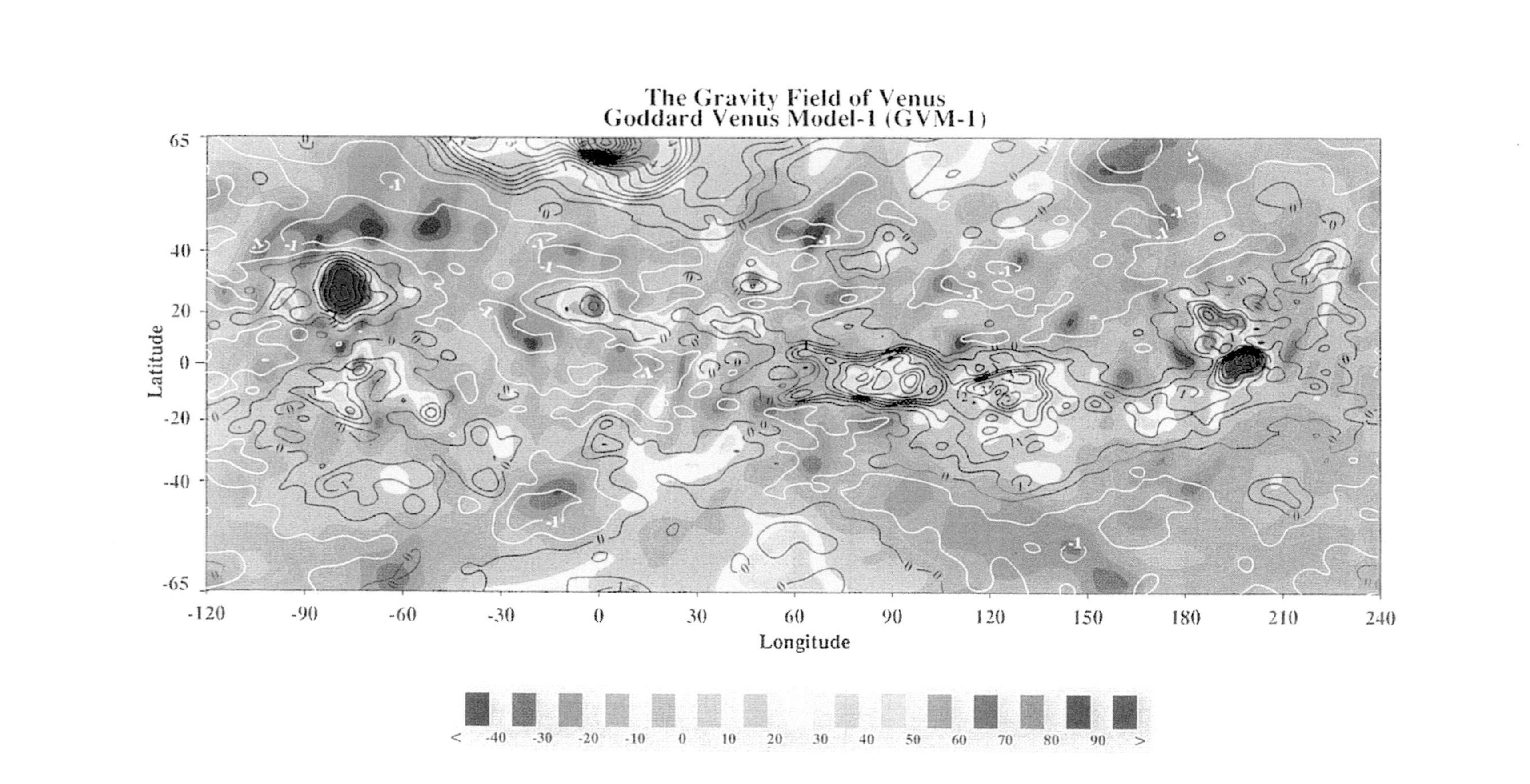
The Gravity Field of Venus
Goddard Venus Model-1 (GVM-1)
65
40
20
0
-20
-40
-65
Latitude
-120
-90
-60
-30
0
30
60
90
120
150
180
210
240
Longitude
< -40 -30 -20 -10 0 10 20 30 40 50 60 70 80 90 >

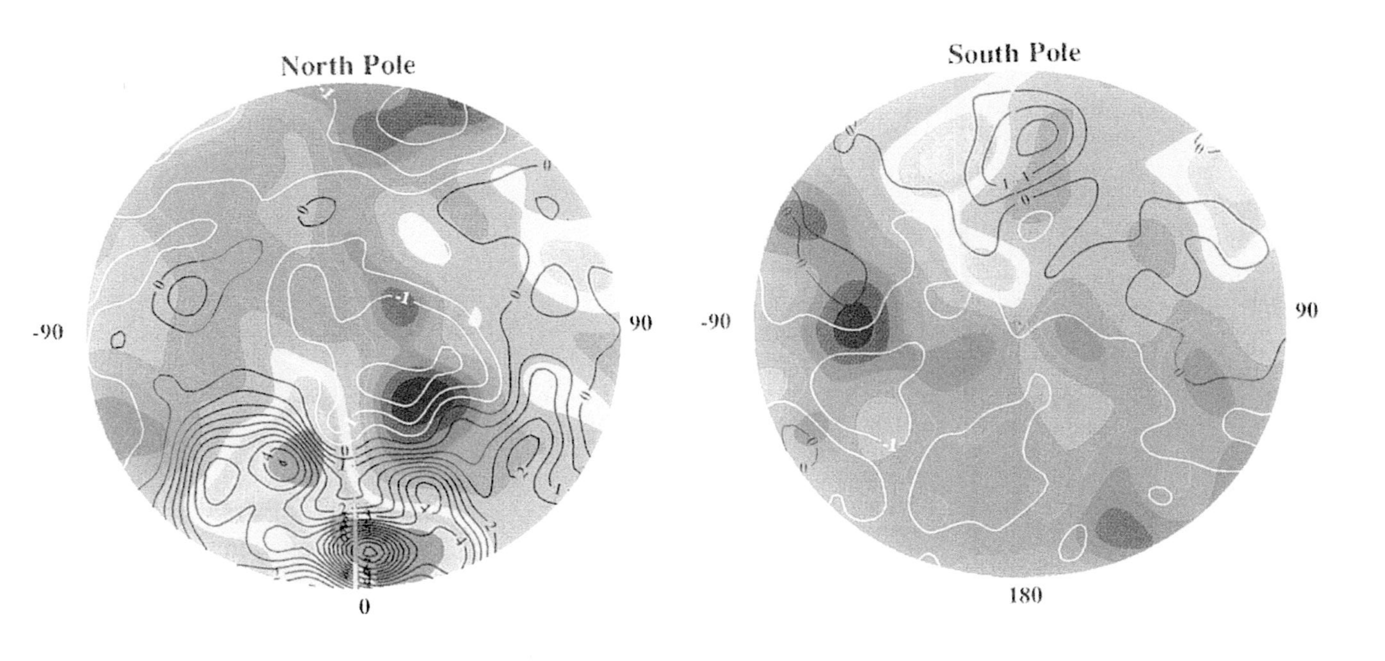

Fig. 2.32 (See also Plate IX) Gravity field of Venus' equatorial and polar segments, in milligals. From Nerem *et al*. (1993).

1992) is that dynamic support is required, i.e., mantle convection. Given the thin lithosphere, mantle convection would be expected to show a much more direct effect on the surface topography and geology than it does in the Earth, where the lithosphere is generally 100 km or more thick under the oceans and much thicker under the continents. Venus thus provides an unusually good planet on which to study the relationship between mantle convection (and mantle plume upwelling) and crustal features. This will be discussed further in Chapter 6, when we have considered the detailed topography of Venus obtained via *Magellan* radar images.

2.11 Summary

Space geodesy uses techniques of a precision and scope that would have seemed almost magical as recently as 1957, when the first artificial satellite was launched. Yet the problems it attacks are largely the same ones of "a world lit only by fire" in Manchester's (1988) memorable phrase: the shape and size of the Earth; the origin of mountains; the cause of earthquakes. Parmenides, Eratosthenes, and John Harrison are looking over our shoulders as we track satellites and place retroreflectors on the Moon.

CHAPTER 3

Satellite studies of geomagnetism

3.1 Introduction

Geomagnetism is a complex but fascinating topic, one with great importance to the proverbial man in the street as well as to the scientist at the computer terminal or on the outcrop.

The Earth's magnetic field is, to begin, an important part of our shield against cosmic and solar particulate radiation, trapping much of it in the well-known Van Allen belts (Heirtzler, 1999). Magnetic storms, occurring when blasts of solar plasma ("coronal mass ejections") hit the ionosphere, can disrupt communications, confuse radar systems, produce spectacular auroral displays even at low latitudes, and knock out electric power grids. In 1989, such an event disabled the Quebec power grid for 9 hours, leaving some 6 million people in two countries shivering in the dark. Laptop computers carried on low-Earth-orbital Space Shuttle missions occasionally crash because of high-energy cosmic ray particles, especially in the South Atlantic Anomaly where the Van Allen belts are closest to the Earth. The magnetic field is known to reverse itself at intervals, about six times every million years (Lanzerotti *et al.*, 1993), during which its strength drops to a fraction of its present value as the magnetic poles meander around the planet (Jeanloz, 1983). Obviously, life itself survives these reversals, but the possible biological effects of a much-weakened magnetic field have begun to cause some concern. In geology, discoveries in geomagnetism – or to be precise, paleomagnetism – have had enormous effect, reviving the once-discarded theory of continental drift in a new incarnation as plate tectonics. Our ability to infer the orientation of the main magnetic field millions of years ago, though fraught with uncertainties, is finding application to a wide range of local geologic problems as well as global ones.

As we have seen, the main outlines of what is now called "space geodesy" had been anticipated before *Sputnik* and its successors. Developments in the study of geomagnetism from space, in contrast, were somewhat unexpected, although they too began with the launch of the early satellites. The very first American satellite to reach orbit, *Explorer 1*, discovered the belts of geomagnetically-trapped radiation, soon named the Van Allen belts. Several successive satellites have carried magnetometers, and our knowledge of the Earth's magnetic fields is now far greater than it was before the Space Age. To fully appreciate the unique value of satellite magnetic field measurements, one must understand something of geomagnetism in general. Let us therefore review not only this subject, but a few key aspects of magnetism itself.

Magnetism is expressed as potential fields, which, like gravity fields, obey the inverse-square law. (Field *strength* decreases as the cube of the distance, being proportional to the derivative of the potential.) However, magnetism is fundamentally different from gravity in several important ways. It is in particular a much more dynamic phenomenon than gravity. A gravity field is produced simply by the presence of mass. A magnetic field, in contrast, is produced by a **changing electric field**, or **moving electric charges**. This phenomenon is expressed mathematically by one of Maxwell's equations, as lucidly explained by Pierce (1956). The most commonly encountered moving electric charges are an electric current, such as the flow of electrons in a wire. Electromagnets are familiar examples of this phenomenon, which may raise the question: What causes permanent magnetization? The answer is that, in a sense, *all* magnets are fundamentally electromagnets, generated by the orbital and spin motion of electrons. In ferromagnetism, the most familiar kind, the individual atoms have magnetic moments that interact strongly. A good intermediate-level discussion of this topic is provided by Butler (1992), who warns that a rigorous understanding of ferromagnetism requires "several years of mind-bending study."

This over-simplified explanation leads at once to the question of why all iron isn't magnetic. The answer is that the magnetism of an iron magnet arises from alignment of magnetic dipoles in "domains," small regions of the magnet with about 10^{18} molecules. Such alignment always exists, but can be altered by the influence of another magnetic field, i.e., by magnetic induction. In ferromagnetism, the induction path is not reversible, following a hysteresis curve. This means that when the magnetizing field is removed, the magnetization of the material does not return to zero, but retains a record of the applied field (Butler, 1992). The common analog tape recorder passes the gamma-iron oxide-bearing tape through a fluctuating magnetic field, producing corresponding fluctuations in the domains on the tape. Computer disks use the same basic phenomenon for digitized information, forming sequences of magnetization with opposite polarities.

Rocks in the Earth's crust may have magnetization impressed on them by the main field. There are two general classes of rock magnetism (Verhoogen, 1969): induced and remanent (or residual) magnetism. Induced magnetism is that produced instantaneously by an external field, and having the same orientation as that field even when the field is turned off. Remanent magnetism, as the term is used in geophysics, is the residual magnetism left in a rock when it crystallizes from a magma, or when it is recrystallized or cools in a magnetic field. The field referred to here is that of the Earth, but remanent magnetism has been found in lunar rocks (Hood *et al.*, 1981; Lin *et al.*, 1998), on Mars, and even in meteorites (Stacey, 1976). These surprising findings will be discussed later. Viscous remanent magnetism is that which increases during the time magnetism is applied, losing it at about the same rate when the impressed magnetic field is removed.

A few more elementary principles should be briefly reviewed, in particular the relation between electric currents and magnetism. Strange as it may seem, as late as 1800 a science text by Thomas Young could say that "there is no reason to imagine any immediate connection between magnetism and electricity" (Ratcliffe, 1951). But within a few years, the work of Oersted, Ampère, and Faraday showed that although electrostatic fields can exist without magnetism, electricity and magnetism are intimately interrelated. An authoritative review (Lanzerotti *et al.*, 1993) of the Earth's magnetic field in fact bears the title "Geoelectromagnetism".

Electric currents can be produced several ways, as in the chemical reactions of a battery or the discharge of a capacitor in an electronic flash. The mechanism that must be understood if one is to appreciate the nature of geomagnetism is *electromagnetic induction*, the production of electric currents by moving magnetic fields. Put most simply, an induced current is produced by movement of a magnetic field in relation to an electrical conductor, such as a wire. (It doesn't matter which is moving; that is, there is no such thing as absolute motion. This superficially obvious observation was in part the basis for Einstein's (1905) theory of special relativity.) In a transformer, magnetic lines of force from the primary coil move through the secondary coil as the current in the primary reverses, which incidentally is why transformers work only on alternating current. In a generator, wires in the rotating armature cut across magnetic lines of force of a stationary field. Things start getting complicated at this point, for the currents generated either way in turn generate magnetic fields themselves.

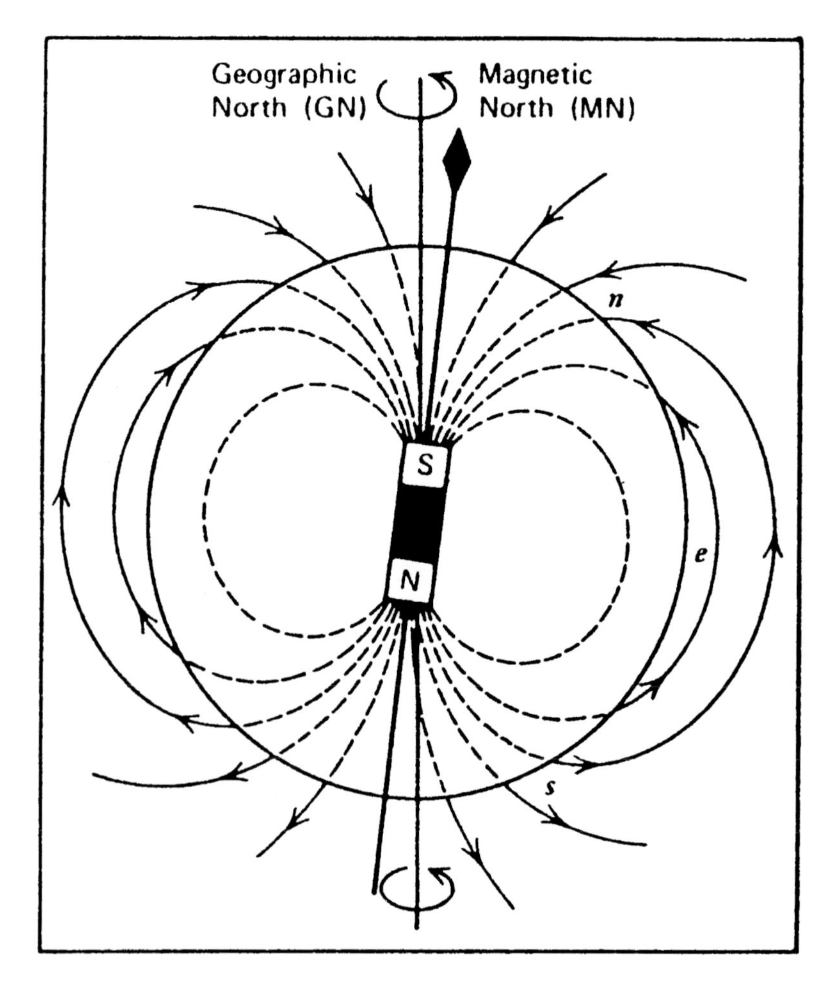

Fig. 3.1 Diagrammatic representation of the Earth's main field, which is analogous to that of a large bar magnet in the interior. Note orientation of field lines; roughly horizontal near magnetic equator and vertical near magnetic poles. From Wyllie (1976).

This elementary review may help the reader to appreciate the extraordinarily complex and dynamic nature of the Earth's magnetism. The general form of the main field, and the Earth's internal structure, are shown in Figs. 3.1 and 3.2. The magnetic field measured at any one point on or near the Earth's surface is the resultant of the main (core), crustal, and external magnetic fields (Cain, 1975; Langel, 1982, 1985, 1993). The external fields are those produced by currents outside the Earth, in the magnetosphere or ionosphere; useful reviews of this topic are those by Stern (1989), Van Allen (1991), and Russell (1991). The crustal fields detectable from orbit are those formed in magnetized materials (such as magnetite) in the

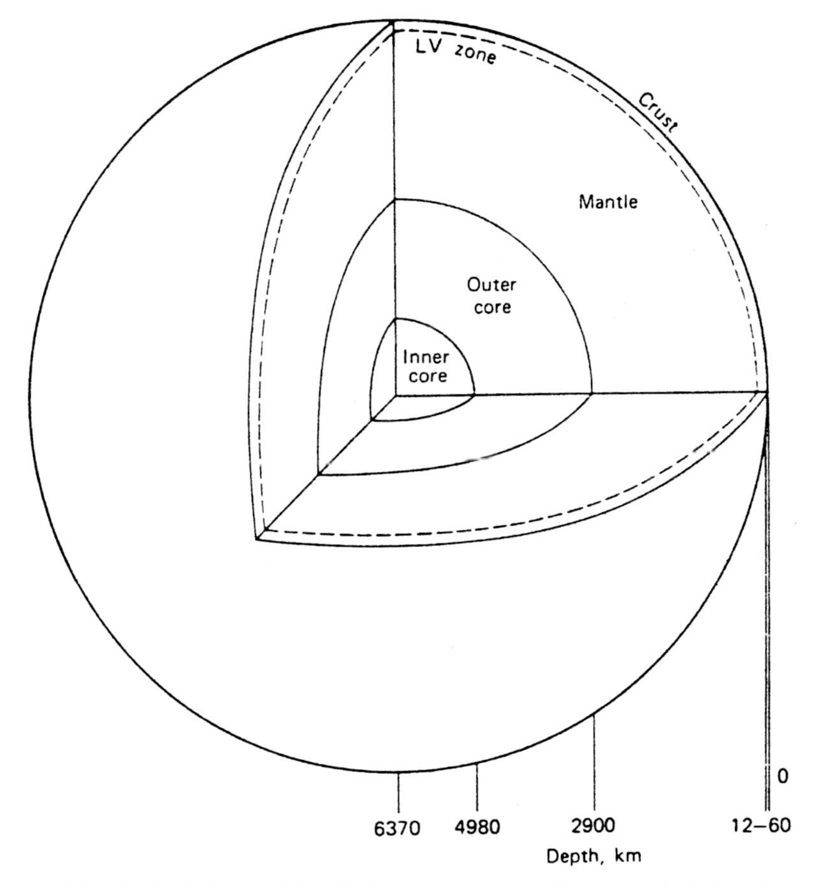

Fig. 3.2 Main division of Earth's internal structure. Outer core is deduced to be liquid from non-transmission of seismic shear waves. "LV" zone is a thin zone of low shear strength in upper mantle, found from low seismic velocities under ocean basins and tectonically-active continental areas. From Wyllie (1976).

crust roughly down to the Mohorovičić discontinuity. The Moho (crust–mantle boundary) appears to be a magnetic as well as a lithologic boundary (Wasilewski *et al.*, 1979) although it has been argued (Haggerty and Toft, 1985) that there are magnetic materials in the upper mantle as well. Both external and crustal fields are relatively weak, not more than a few thousand nanoteslas (nT) and generally much less. The main field is far stronger, typically 30,000–70,000 nT at the surface, depending on geomagnetic latitude.

The origin of the Earth's main field was called one of the great problems of physics by Einstein many years ago, although its general nature had been known since publication of Gilbert's *De Magnete* in 1600. The gross behaviour of the field, as summarized

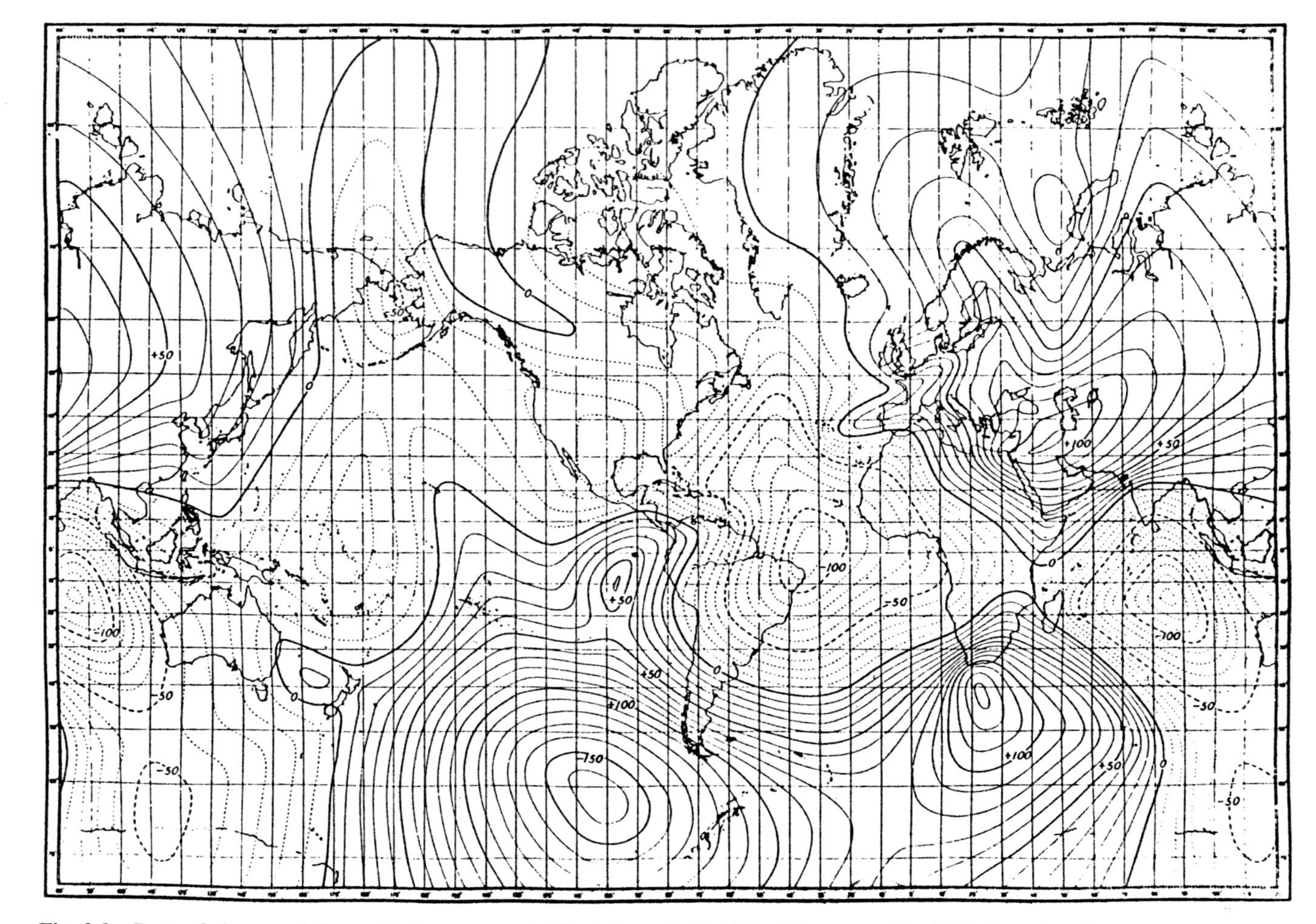

Fig. 3.3 Rate of change of the vertical component (Z) of the main field, in nT per year, for 1942. From Vestine *et al.* (1947). Compare with Fig. 3.5.

by Bullard (1954), gives us some indication of where and how it is generated. The most important aspect of this behaviour is the field's secular variation, i.e., its changes in strength and orientation on a time-scale of months to centuries. The map in Fig. 3.3, from Vestine *et al.* (1947), shows these changes for only one year. As commented by Bullard (1954), such maps superficially resemble weather maps. Collectively, the main field's behaviour clearly points to an origin related in some way to fluid motion deep in the Earth. At this point, seismology comes to our aid, for the general structure of the deep interior has been inferred from the propagation of seismic waves. As shown in Fig.3.2, the core of the Earth has a liquid outer part and a solid inner part. The bulk density of the Earth, and the cosmic abundance of elements, among other things, point convincingly to iron as the main constituent of the core, probably alloyed with nickel and containing an unknown fraction of lighter elements such as sulfur, oxygen, and potassium (Jacobs, 1992).

It was originally thought in the 17th century that the interior of the Earth might act as a giant permanent magnet, but temperatures in the core and mantle are far too high for that, to say nothing of the dynamic behaviour of the main field. The generally-accepted theory today for the origin of the main field stems from a concept proposed by Larmor in 1919 to explain the origin of the Sun's magnetic fields. As applied to the Earth by Elsasser and by Bullard, this can be termed the "self-exciting dynamo" theory. A good contemporary discussion of this theory has been published by Jacobs (1992); the reader may find it helpful to consult a text on elementary electricity such as that by Marcus (1968). The mechanism is roughly this.

The liquid iron of the outer core, cutting across a weak initial magnetic field – perhaps the interplanetary field – would generate electrical currents in the core. These electrical currents in turn produce a magnetic field which, being crossed by the molten iron, produces a still stronger current, and thus a stronger field, until a steady value is reached. This scheme, in which the field regenerates itself, may sound like perpetual motion, but thermal energy is generated in the system by some mechanism(s) – perhaps radioactivity and latent heat from solidification of the inner core – to keep the iron moving. This theory accounts in a general way for the dynamic behaviour of the main field discussed previously, such as the slow movement of the magnetic poles and the subsequent variation of magnetic declination from year to year.

It is well known that the Earth has a north and a south magnetic pole, and the main field is generally described as a dipole. However,

Table 3.1 *Magnetometer-carrying satellites and spacecraft in Earth orbit, 1958–1980. From Langel (1987).*

Satellite	Inclination	Altitude range (km)	Dates	Instrument	Approximate accuracy (nT)	Coverage
Sputnik 3	65°	226–1881	5/58–6/58	Fluxgates	100	USSR
Vanguard 3	33°	510–3750	9/59–12/59	Proton	10	Near ground station
1963–38C	Polar	1100	9/63–1/74	Fluxgate	30–55	Near ground station
Cosmos 26	49°	270–403	3/64	Proton	Unknown	Whole orbit
Cosmos 49	50°	261–488	10/64–11/64	Proton	22	Whole orbit
1964–83C	90°	1040–1089	12/64–6/65	Rubidium	22	Near ground station
OGO-2	87°	413–1510	10/65–9/67	Rubidium	6	Whole orbit
OGO-4	86°	412–908	7/67–1/69	Rubidium	6	Whole orbit
OGO-6	82°	397–1098	6/69–7/71	Rubidium	6	Whole orbit
Cosmos 321	72°	270–403	1/70–3/70	Cesium	Unknown	Whole orbit
Azur	103°	384–3145	11/69–6/70	Fluxgate (2-axis)	Unknown	Near ground station
Triad	Polar	750–832	9/72–present	Fluxgate	Unknown	Near ground station
S3–2	97°	230–900	10/72–present	Fluxgate	300λ (components)	Whole orbit
Magsat	97°	325–550	11/79–5/80	Fluxgate	6	Whole orbit
				and Cesium	3	

this applies only to the field as it is measured at the surface, and is an approximation even there (Sugiura and Heppner, 1968). The field in the core itself is much more irregular (Jacobs, 1992), and the main field at the surface may not have been dipolar through geologic time.

3.2 Satellite investigations of the Earth's magnetic field

It will be apparent, from the above generalized discussion, that the Earth's magnetic "field" is actually a composite of many fields that are continually changing in magnitude and direction, changes that must be taken account of even on an hourly basis for geophysical surveys. When it became possible to put magnetometers in artificial satellites a new era in the study of geomagnetism began (Table 3.1). Orbital measurements made it possible to monitor the Earth's field globally, frequently, and with identical instruments at similar altitudes.

Systematic magnetic measurements began with *Vanguard 3* and *Sputnik 3*, which produced scalar (non-directional) data useful chiefly for the study of external fields. Later satellites, in particular *Cosmos 49*, the *Polar Orbiting Geophysical Observatories* (*POGO*), and especially *Magsat* produced far better data on the main and external fields. A remarkable achievement (Zietz *et al.*, 1970; Regan *et al.*, 1975) was the extraction of extremely weak crustal anomalies from the main field, a feat analogous to photographing the stars by day. Further analyses (Langel, 1990a, b) produced more comprehensive and detailed maps from the *POGO 2*, *4*, and *6* data which, although obtained from high altitude, covered several years and all local times. One of these maps has already been presented in Fig. 1.6. The achievement of Regan and his colleagues led to development of *Magsat*, the first such satellite flown (in 1979) primarily for study of the main and crustal fields rather than the external ones.

Magsat was an unusually successful project in several aspects, including an on-time and within budget launch (Langel, 1982). Every proposal to NASA for a *Magsat* follow-on mission has to date been rejected. Fortunately, the European Space Agency's (ESA's) *Oersted* satellite has now picked up the torch, and has already produced an initial field model (Olsen *et al.*, 2000). *Magsat* results fill two dedicated issues of *Geophysics Research Letters* (**9**, No. 4, 1982) and the *Journal of Geophysical Research* (**90**, B3, 1985), which alone gives some idea of its success. The major achievements related to the solid earth are summarized in the following two sections.

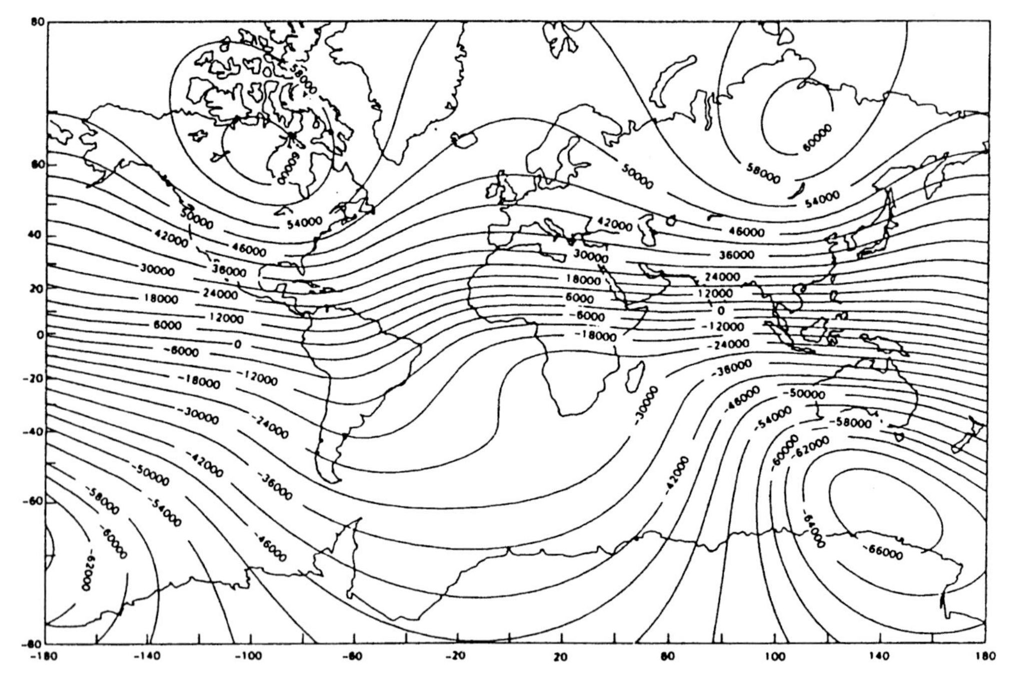

Fig. 3.4 World contour map, based on *Magsat* data, of the vertical component (Z) of the main field, in nT, for 1980. From Langel (1987). Note lack of relationship between crustal features (oceans, continents) and main field, because the latter's origin is in core.

3.3 The main field

Magsat carried both scalar and vector magnetometers (Langel *et al.*, 1982), thus permitting construction of the first global vector (as well as scalar) model (Fig. 3.4) of the core field, while accurately measuring the external fields as well. It shows that the main field has no relationship to crustal features, even to continents and ocean basins, because of its origin in the core. A comprehensive and accurate main-field model is vitally important for a variety of reasons. Aeromagnetic surveys for petroleum or mineral deposits must have a reference field to take out regional gradients, as well as a record of magnetic activity for a particular time. The review by Hood *et al.* (1985), though focussed on aeromagnetic maps, gives a good overview of how such maps are produced. Maps must have up-to-date corrections for changing declination; even with modern navigational methods, the magnetic compass is still a major navigational tool for ships, aircraft, and ground surveys. Scientific studies of the sort to be described below similarly require good reference fields. A World

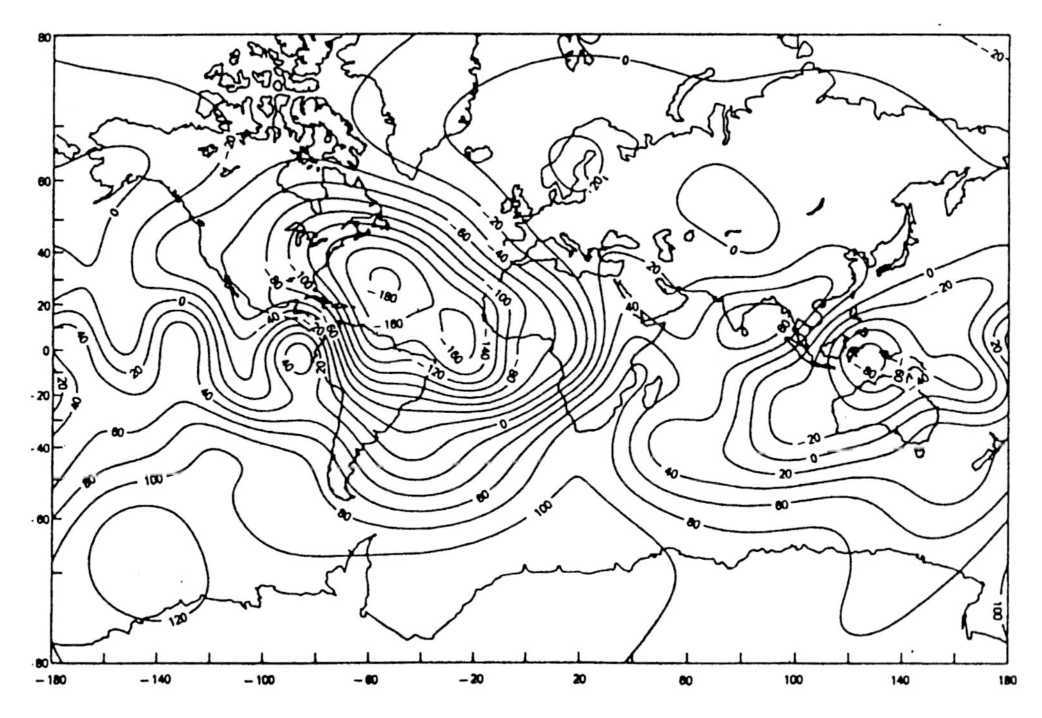

Fig. 3.5 World contour map, based on *Magsat* data, of the rate of change of the vertical component (Z) of the main field, in nT per year, for 1980. From Langel (1987). Compare with Figs. 3.3 and 3.4.

Magnetic Survey was proposed in 1954 by S. K. Runcorn (Langel, 1987) to the International Union of Geodesy and Geophysics to approach some of the needs for a reference field listed above.

An International Geomagnetic Reference Field (IGRF) for the 1955–1972 period was first adopted in 1968 (Langel, 1992). A second IGRF was adopted in 1975. However, it became increasingly outdated. The launch and operation for several months, in 1979–80, of *Magsat* permitted compilation of a third field, IGRF 1980, based mainly on data from *Magsat*. IGRF 1980 was a major improvement on previous fields, in accuracy and coverage, although even it was shortly improved.

Although *Magsat* operated for only seven months before re-entering the atmosphere, it permitted compilation of global maps showing not only the main-field components but also their rate of change in nT/year (Fig. 3.5).

New knowledge of the main field has been applied to a number of scientific problems. Voorhies and Benton (1982) used *Magsat* models to estimate the radius of the Earth's outer core, finding a value agreeing within 2% of that determined from seismic data. They also confirmed that the core is an electrical conductor and the mantle, in general, an insulator, a customary but unverified assumption. Other studies used field models to study fluid motion in the outer core, structure at the core–mantle boundary (Bloxham and Jackson, 1992), and similar problems of core structure and behavior (Langel, 1985). The greatly improved model of crustal magnetism is useful for studies of the main field in that crustal anomalies are essentially noise in main-field measurements.

The importance of satellite data should not be exaggerated. The contributions of fixed ground observatories, of historical studies, and of surface surveys are still the backbone of geomagnetic research. Bloxham (1992), for example, used surface measurements made as far back as 1690 to study the secular variation of the main field, and to map flow at the core–mantle boundary. Nevertheless, even the few adequate orbital surveys made to date have provided a solid global background for other studies.

3.4 The crustal field

The crustal anomaly field has been studied locally with surface, aeromagnetic, and marine methods for many years. Development of modern magnetometers for submarine detection during World War II (Bates *et al.*, 1982) led to great progress in aeromagnetic surveys, which have been extensively used for mineral and hydrocarbon searches since then. One of the problems with using aeromagnetic surveys for regional or continental studies is the difficulty of tying together local surveys, often carried out years apart and at different altitudes. In addition, there are enormous areas of land and sea over which adequate conventional surveys are difficult or impossible. The contribution of satellite-derived crustal anomaly maps has been, first, to provide truly global coverage, and second, to provide coverage with similar altitudes and instruments. In addition, the speed of orbital surveys makes it possible to allow for time variations in the field. The main disadvantage of orbital magnetic data is their relatively coarse spatial resolution, generally several hundred kilometers. Since magnetic field strength follows an inverse cube law, magnetic anomalies decrease rapidly with altitude, and it was thus a major achievment simply to detect crustal anomalies from space. Most of the continental anomalies were thought to result largely from induced rather than remanent magnetism, although the marine anomalies may be dominantly remanent. The discovery of what are probably remanent magnetic anomalies on Mars, to be discussed in

Section 3.5, indicates that terrestrial anomalies are largely remanent (Purucker *et al.*, 2000), not induced – a striking example of comparative planetary geophysics.

Dozen of papers presenting and interpreting satellite magnetic maps have been published, the vast majority based on *Magsat* data because of its spatial resolution and the fact that these data are vector as well as scalar. *Magsat* and *POGO* data have been combined by Arkani-Hamed *et al.* (1994) to produce a new crustal anomaly map (Fig. 3.6). General reviews of satellite magnetic anomaly studies have been presented by Mayhew *et al.* (1985) and Schnetzler (1989). The definitive treatment of this subject is that of Langel and Hinze (1998).

The first feature to be identified in satellite magnetic data was the large east–west-trending anomaly over central Africa (Fig. 3.6), since termed the Bangui anomaly (Regan and Marsh, 1982). Apart from its historic significance, the investigation of the Bangui anomaly illustrates the peculiarities of satellite data, the methods of investigating them, and their geological interpretation. Accordingly, we shall discuss the work of Regan and Marsh in some detail.

The first satellite data studied were those from the *POGO* series, which produced measurements from various altitudes over many months cumulative time. The first step in extracting crustal anomalies from these measurements was selection of those acquired from low altitudes during magnetically quiet periods. An immediate problem faced by investigators working with low-latitude data is allowing for the equatorial electrojet; for a recent treatment of this phenomenon, see Cohen and Achache (1990). After intensive mathematical analysis, it was possible to delineate a broad east-trending magnetic low with a maximum amplitude of 12 nT (contrasted with the main-field strength in that area of 30,000 nT).

Previously acquired gravity surveys in the area showed a Bouguer anomaly partly coinciding with the magnetic anomaly, further confirming its reality and general outline. The geologic field reconnaissance carried out by Regan and Marsh, in combination with published geologic maps, showed that although magnetic iron formations occur in the area, they are insufficient to cause the satellite anomaly. They made the important observation that the anomaly is caused by a rock type not exposed on the surface (Fig. 3.7). This, and the broad extent of the anomaly, point clearly to an origin in the lower continental crust, probably a large dense mafic intrusion.

As first demonstrated by the Bangui investigation, the most general discoveries from continental anomalies concern the structure and

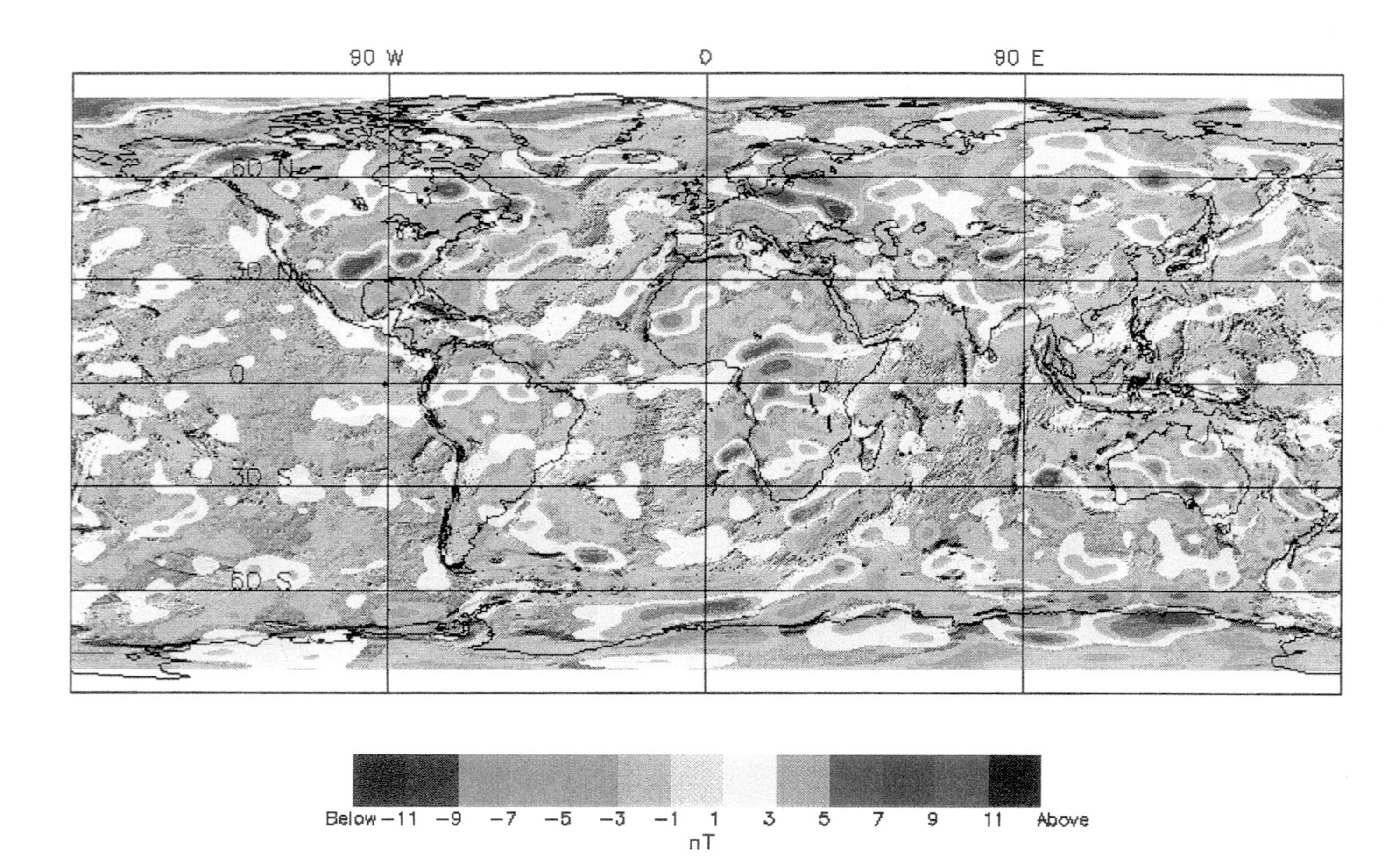

Fig. 3.6 (See also Plate X) World scalar map of crustal magnetic anomalies, from *Magsat* and *POGO* data. From Arkani-Hamed *et al.* (1994).

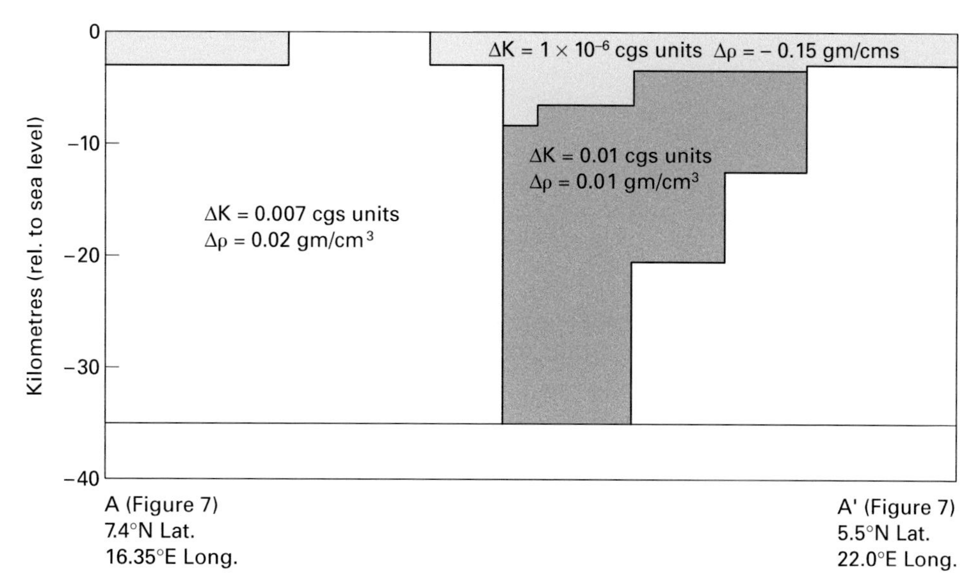

Fig. 3.7 Crustal model proposed by Regan and Marsh (1982) to account for magnetic and gravity anomalies shown in previous maps.

composition of the deep continental crust (Fig. 3.8) (Schnetzler, 1985). The lower crust, exposed in very few places, has until recently been poorly understood. Studies of crustal anomalies, coordinated with laboratory and field investigations of rock magnetism, have strongly supported the view that the lower continental crust is more mafic (iron- and magnesium-rich) and more highly metamorphosed (granulite grade) than the exposed upper crust (Wasilewski and Fountain, 1982; Coles, 1985; Schnetzler, 1985). Such interpretations, added to new knowledge of the lower continental crust from seismic reflection profiling and other sources, is rapidly opening up this important part of the lithosphere. This will be discussed further in Section 3.6.

A related finding from satellite magnetic data is that a number of large anomalies are the expression of intrusions of mafic rock possibly related to rifts, in the lower crust. Such interpretations have been made of anomalies over the Mississippi Embayment (Thomas, 1984), and Kentucky (Mayhew *et al.*, 1982). The significance of these findings is becoming more apparent in the light of recent studies of crustal evolution, indicating that much of the mantle differentiation since the Archean has involved separation of basalt from the mantle, with subsequent underplating of the continental crust. This will be discussed in Chapter 6. An interesting possibility suggested by Girdler *et al.* (1992) is that some *Magsat* anomalies, in particular the Bangui anomaly, may mark large Precambrian impact structures, or impact-triggered mafic intrusions.

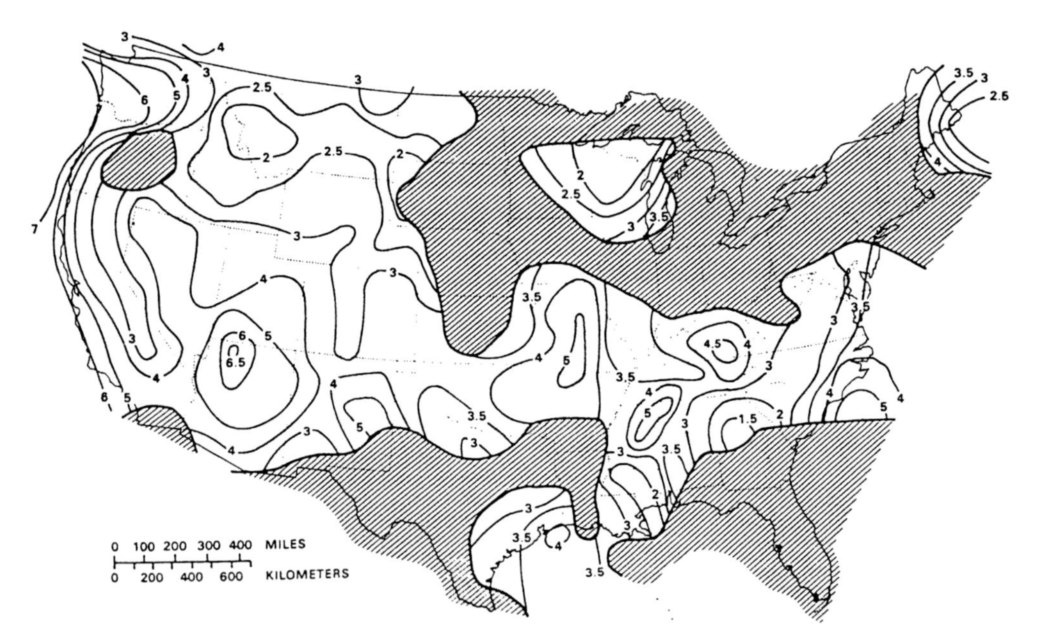

Fig. 3.8 Magnetization in the lower crust, as determined from seismic data, based on *Magsat*-derived anomaly field; units amperes per meter. Shaded areas are more than 225 km deep from measurement of either the Moho or Conrad discontinuities. From Schnetzler (1985).

The *Magsat* crustal anomaly maps have been compared with global and regional tectonic structure by several investigators. Mayhew and Galliher (1982) produced maps from *Magsat* data whose main features (Fig. 3.9) express the physiographic or tectonic provinces of the coterminous US surprisingly well. The Basin and Range Province, for example, is distinct from the Interior Plateaus. The Gulf of California, generally agreed to represent incipient ocean basin formation, also shows up distinctly. Frey (1982) found that many aulacogens (failed rifts) in central Asia had magnetic expression even on the scalar anomaly maps, as did several other major structures. Hinze *et al.* (1982) made similar interpretations for South America (Fig. 3.10). An interesting feature of the Hinze *et al.* map is that the continental margins show little expression in the scalar anomaly values at 350 km altitude. Given the great contrast in composition, one would expect this crustal boundary to be conspicuous. Newer studies (see Fig. 3.21) (Purucker *et al.*, 1998), incorporating crustal thickness and susceptibility values, have delineated the boundary.

Hall *et al.* (1985) studied *Magsat* data from several passes over the boundary between the Churchill and Superior Provinces of the

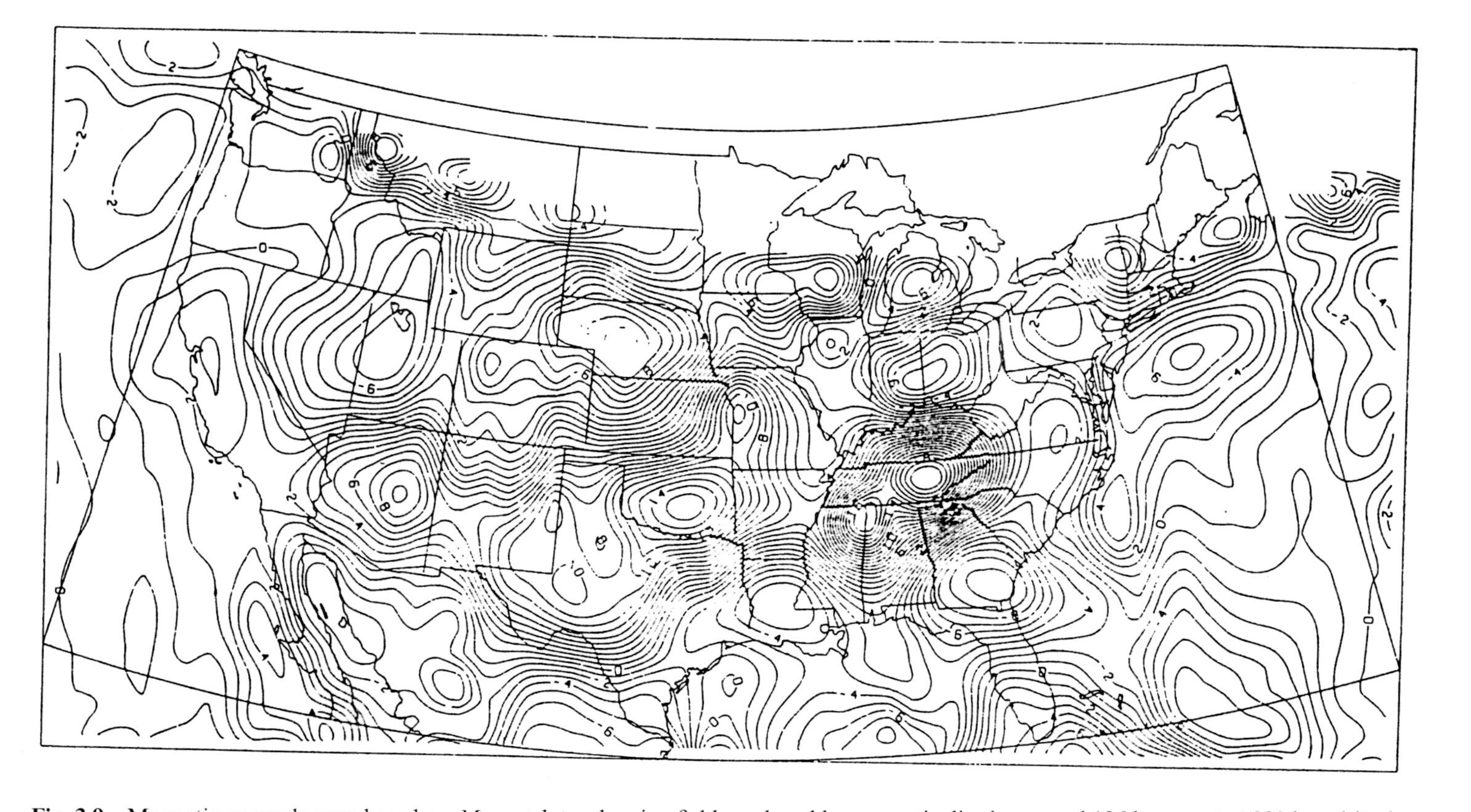

Fig. 3.9 Magnetic anomaly map based on *Magsat* data, showing field produced by magnetic dipoles spaced 136 km apart at 320 km altitude (equivalent source representation). Contour interval 1 nT. From Mayhew and Galliher (1982).

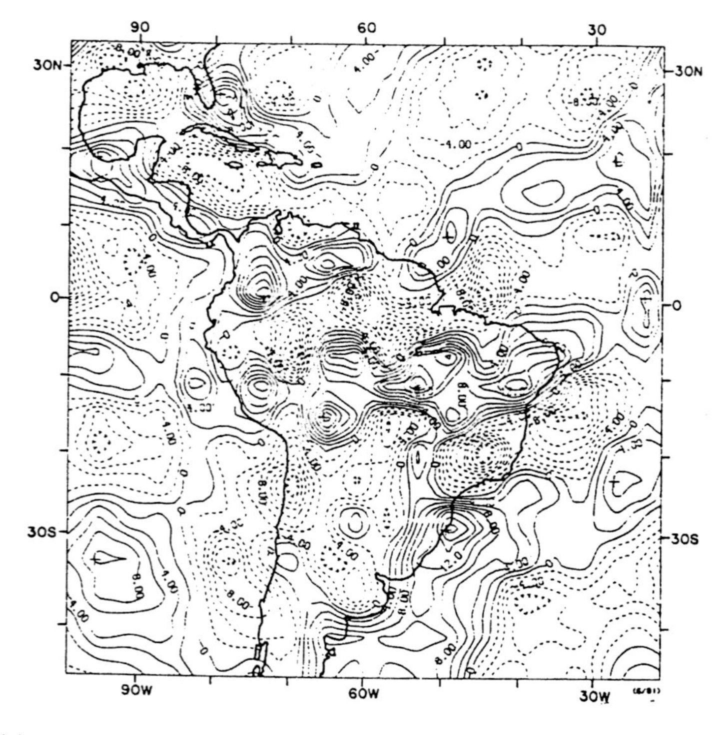

(a)

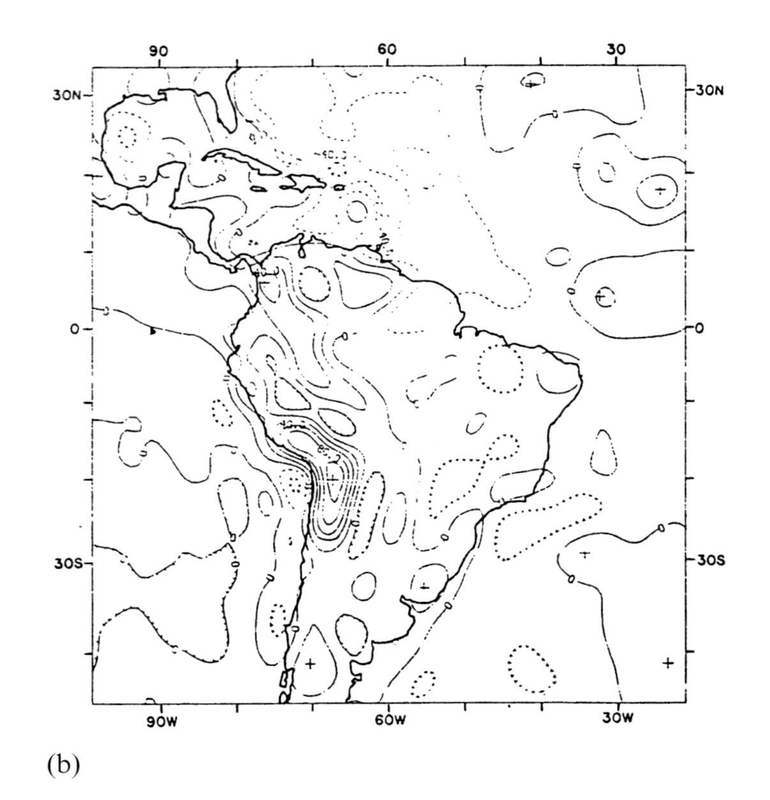

(b)

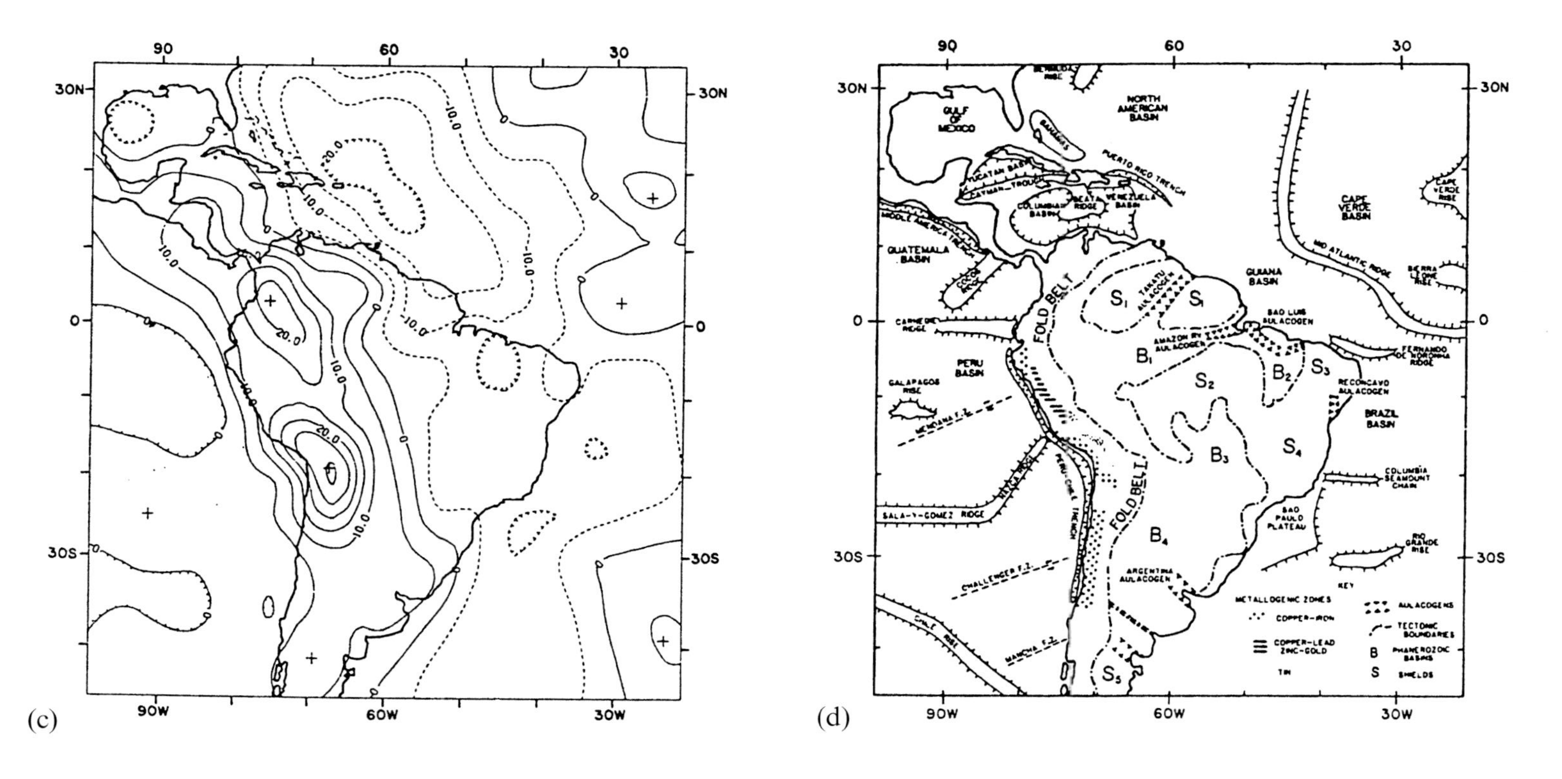

Fig. 3.10 (a) South American crustal anomaly field, based on *Magsat* data, showing equivalent point source scalar field at 350 km altitude. Contour interval 2 nT. (b) South American surface free-air gravity anomaly field, long-wavelength-pass filtered. Contour interval 20 milligals. (c) South American free-air gravity anomaly field, upward continued to *Magsat* altitude (350 km). Contour interval 5 milligals. (d) Generalized tectonic features. S1: Guiana Shield; S2: Central Brazilian Shield; S3: Sao Luiz Craton; S4: Sao Francisco Craton; S5: Patagonia Platform; B1: Amazon River Basin; B2: Parnaiba Basin; B3: Parana Basin; B4: Chaco Basin. All from Hinze *et al.*115 (1982). Lambert conformal projection.

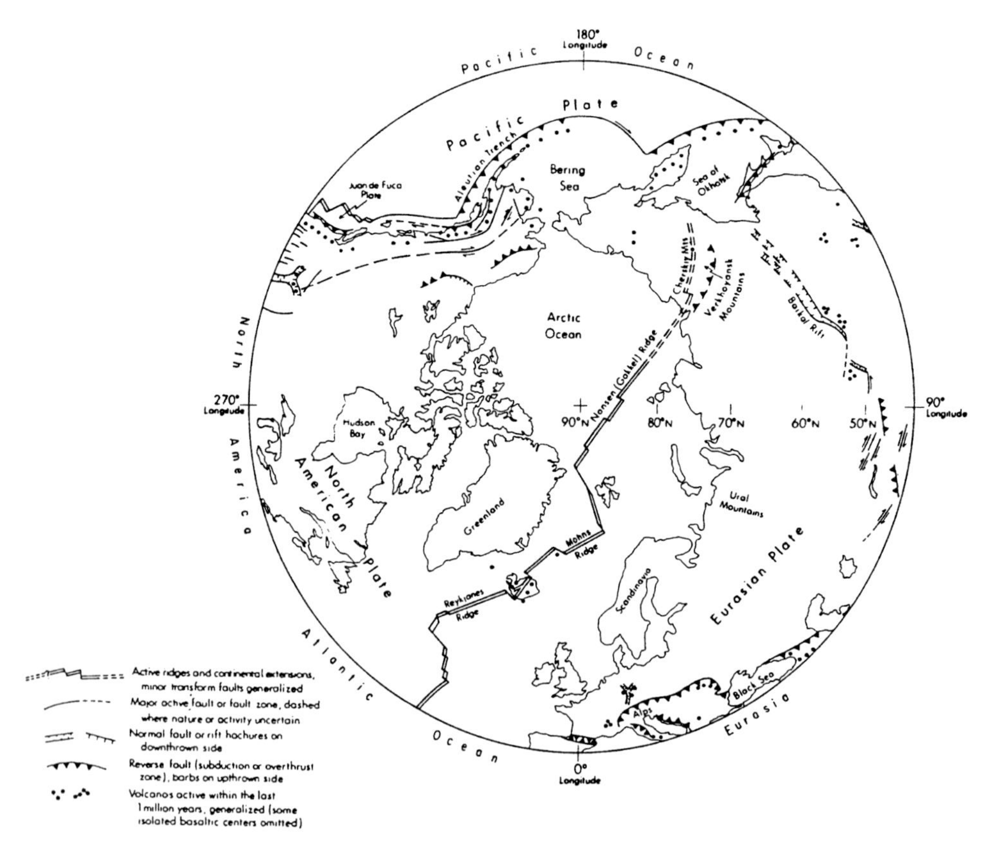

Fig. 3.11 Tectonically and volcanically active (within the last one million years) features of the Arctic Regions, north of 40th parallel. From Lowman (1984). Orthographic projection.

Canadian Shield, or Nelson Front, hypothesized to be a suture formed by terrane accretion (Gibb and Thomas, 1976). They found a definite magnetic signature over the boundary that was compatible with models derived from seismic data, thus providing an example possibly applicable to supposed sutures on other shields. Since terrane accretion is thought by many to be the major mechanism for formation of the continents, this local study is more significant than it might appear. (Field investigations along the Nelson Front later showed the suture concept to be incorrect, the magnetic anomaly noted by Gibb and Thomas resulting from interbedded serpentinites, not ophiolites (Lowman *et al.*, 1987).)

Improvements in analytical methods for *Magsat* in the decade after the satellite was launched are illustrated by the study by Ravat *et al.* (1993) who studied tectonic structures of Europe. They found a wide variety of features expressed in the *Magsat* data, including

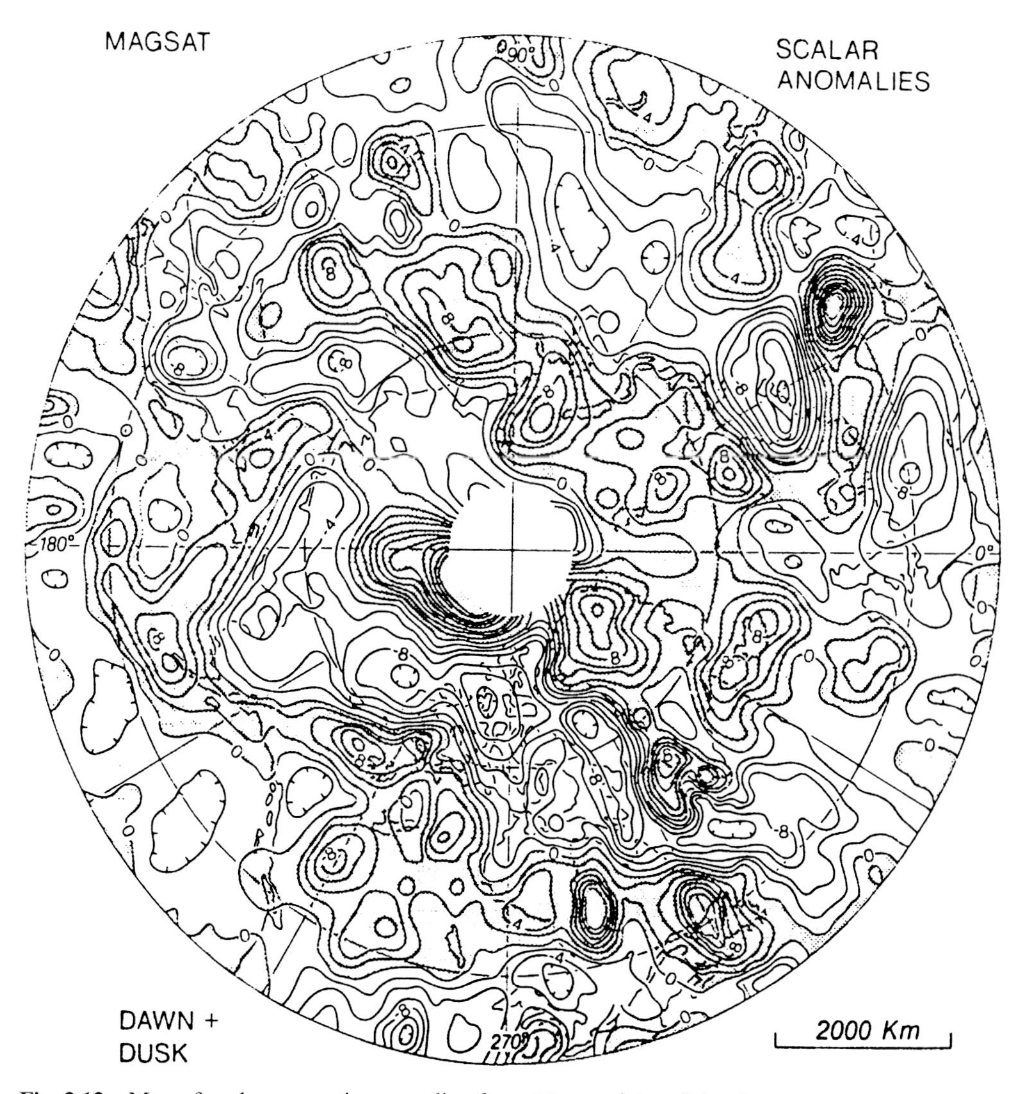

Fig. 3.12 Map of scalar magnetic anomalies, from *Magsat* data, of Arctic Regions. Average altitude 415 km; average of dawn and dusk passes. Contour interval 2 nT. From Coles (1985).

geologic provinces, regions of high heat flow and thin crust, the Kursk and Kiruna iron deposits, and others. The Ravat *et al.* study is a notable demonstration of the wide range of crustal features that can affect the magnetic field measured at satellite altitudes, and the great amount of information needed for a valid interpretation.

A comprehensive *Magsat* investigation of northern hemisphere high-latitude anomalies was carried out by Coles (1985) (Figs. 3.11, 3.12). On the one hand, these areas are inherently hard to study magnetically because of the high intensity and rapid variability of the main field near the magnetic poles. On the other hand, the nature of the *Magsat* orbit helps compensate for these problems by crossing over any given area many more times than at low latitudes. As shown in Fig. 3.11, many of the scalar anomalies around the Arctic Ocean correspond to tectonically active features such as subduction zones and spreading centers. The large anomaly in Canada, just east of Hudson Bay, is a good example of the distinctive magnetic signatures of high-grade metamorphic terrains. This area is part of the Superior Province, in which lower-crust granulites are exposed over large regions.

An interesting study of a tectonically important area was carried out by Langel and Thorning (1982). The Nares Strait region, between Greenland and Ellesmere Island, Canada (Fig. 3.11), must be an area of major horizontal displacement if Greenland has drifted away from North America as the result of sea-floor spreading in Baffin Bay. However, a number of geologic markers appear not to be offset in the Strait (Kerr, 1982; Lowman, 1985). Langel and Thorning found that the magnetic contours from the *POGO* satellite data paralleled the Innuitian fold belt, and the Strait, suggesting that this feature is an extremely old and fundamental crustal boundary. This finding, supported by the map of Coles, neither proves nor disproves continental drift in the area, but suggests the application of satellite data to another fundamental problem.

Another category of new knowledge from the satellite data, namely the Earth's internal temperatures, has been demonstrated by the work of Mayhew (1985). It has been known since the time of Gilbert (1600) that magnetic materials become less magnetic with increasing temperatures, and totally non-magnetic above what is now called the Curie temperature in honor of Pierre Curie's 19th-century studies. For the lower crust, this temperature is generally estimated to be about 550 °C (823 K). The mantle is generally hotter than this, one reason it is considered non-magnetic. The temperature dependence of magnetic susceptibility (degree to which a substance can be magnetized) has been applied by Mayhew (1985) to a study of the depth of the subcontinental Curie isotherm. By comparing *Magsat* anomalies to heat-flow data for the western US, Mayhew showed that these anomalies often reflect the depth of the isotherm. This is not generally true for cratonic areas, but for tectonically and magmatically active areas this discovery points the way to a new method of studying the Earth's heat flow.

It is generally known that the "revolution in the earth sciences," or plate tectonics, was triggered largely by the discovery of system-

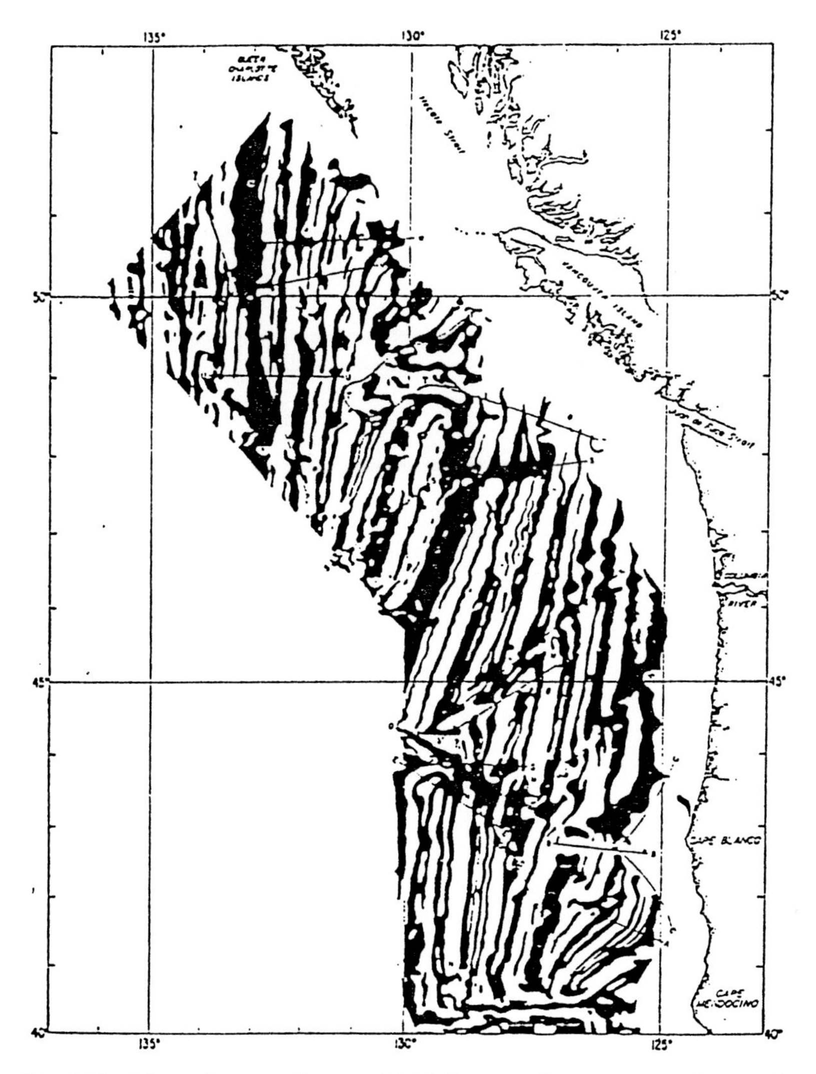

Fig. 3.13 Magnetic anomalies, total field, from marine surveys southwest of Vancouver Island, on Juan de Fuca Plate. Positive anomalies in black. From Raff and Mason (1961).

atic linear magnetic anomalies in the ocean basins. These were explained by Vine and Matthews (1963) (and independently, by Morley and Larochelle, 1964) as having been formed by successive reversals of the main field, impressing thermal remanent magnetism bands of alternating polarity on the moving oceanic crust (Figs. 3.13, 3.14). Vine's description of the sea floor as "a conveyor belt and a tape recorder" is classic. It is natural to expect that satellite magnetometers would produce new information on the oceanic

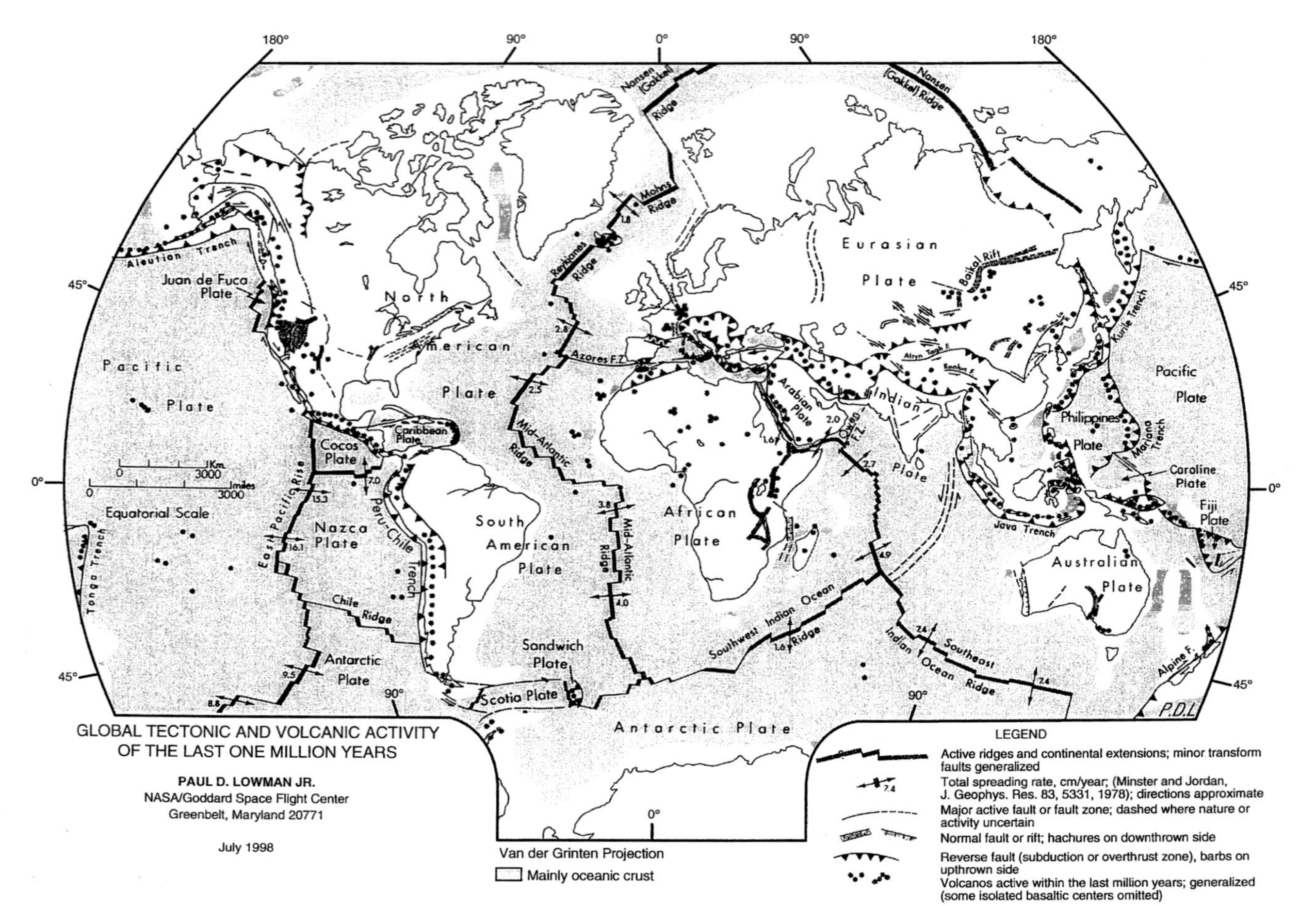

Fig. 3.14 Global tectonic and volcanic activity map for the last one million years, showing regional setting of Juan de Fuca Plate. From Lowman (1979).

"tape recorder." However, there are very few distinct oceanic anomalies visible on the *Magsat* and *POGO* maps. The reason for this is the relatively low spatial resolution of the orbital data, on the order of 200 kilometers, and for some ocean areas the north–south inclination of the satellite orbits (Purucker and Dyment, 2000). The marine anomalies in most areas cancel out, resulting in the relatively featureless oceanic areas on the map (Thomas, 1987). Similarly, the continent–ocean boundaries, which should have magnetic expression, are not distinct (e.g., Taylor, 1991). One possible explanation suggested by Heirtzler (1985) is that, especially on active margins, the magnetic layer may be heated above the Curie temperature as it descends. Another explanation, by Meyer *et al.* (1985), is dominance of the main field over very-long-wavelength crustal anomalies. Development of the standard Earth magnetization model by Purucker *et al.* (1998, Chapter 1, Fig. 6) overcomes this problem by allowing for crustal thickness and susceptibility. Despite the difficulties, interpretations have been made of the magnetic properties of the ocean basins as seen from space.

Although the dominant linear anomalies found in most ocean areas are not resolved by the satellite data, some features expressing sea-floor spreading can be identified. LaBrecque and Raymond (1985) and Purucker *et al.* (1998) have shown (Figs. 3.15, 3.16), that the broad northeast-trending magnetic low in the Atlantic east of North America results from the Jurassic and Cretaceous "quiet zones." These are broad belts of crust, roughly parallel to the spreading centers, formed during long periods of constant polarity of the main field and with consequent uniform crustal polarity. Similar features were identified in the North Pacific by LaBrecque *et al.* (1985) and Cohen and Achache (1990), and in the South Atlantic by Fullerton *et al.* (1989) and Purucker and Dyment (2000). They appear to be the only features on the satellite anomaly maps dominated by remanent rather than induced magnetism (Thomas, 1987).

Some tectonic features in the ocean basins can be readily identified on the satellite magnetic maps. The Aleutian Island subduction zone (see Fig. 3.14), a well-studied classic Benioff–Wadati zone of descending oceanic lithosphere, has been studied by Clark *et al.* (1985). They found that the *Magsat* anomaly along the Aleutian chain (see Fig. 3.6) could be interpreted in terms of magnetization contrast between the relatively cold down-going slab and the hotter (and thus non-magnetic) surrounding mantle. However, an additional slab of material from a former subduction zone was required to account for the anomaly north of the Aleutian chain in the Bering Sea, an interesting example of the tectonic value of *Magsat* data.

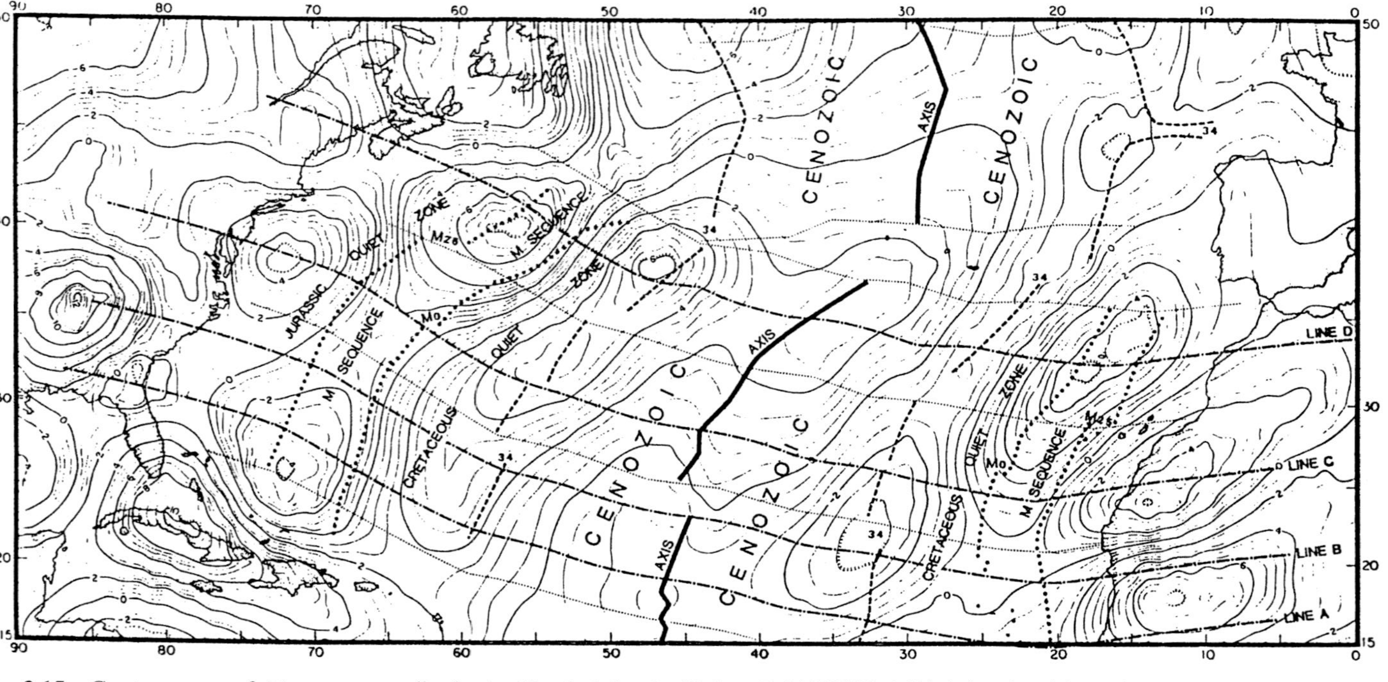

Fig. 3.15 Contour map of *Magsat* anomalies in the North Atlantic. Units nT. "AXIS" is Mid-Atlantic Ridge (Fig. 3.14), from which sea-floor spreading is thought to originate. Major isochrons derived from polarity reversal time-scale: AXIS, 0 Ma; 34, ca 84 Ma; M25, ca 150 Ma; ocean–continent boundary, 200 Ma. From LaBrecque and Raymond (1985).

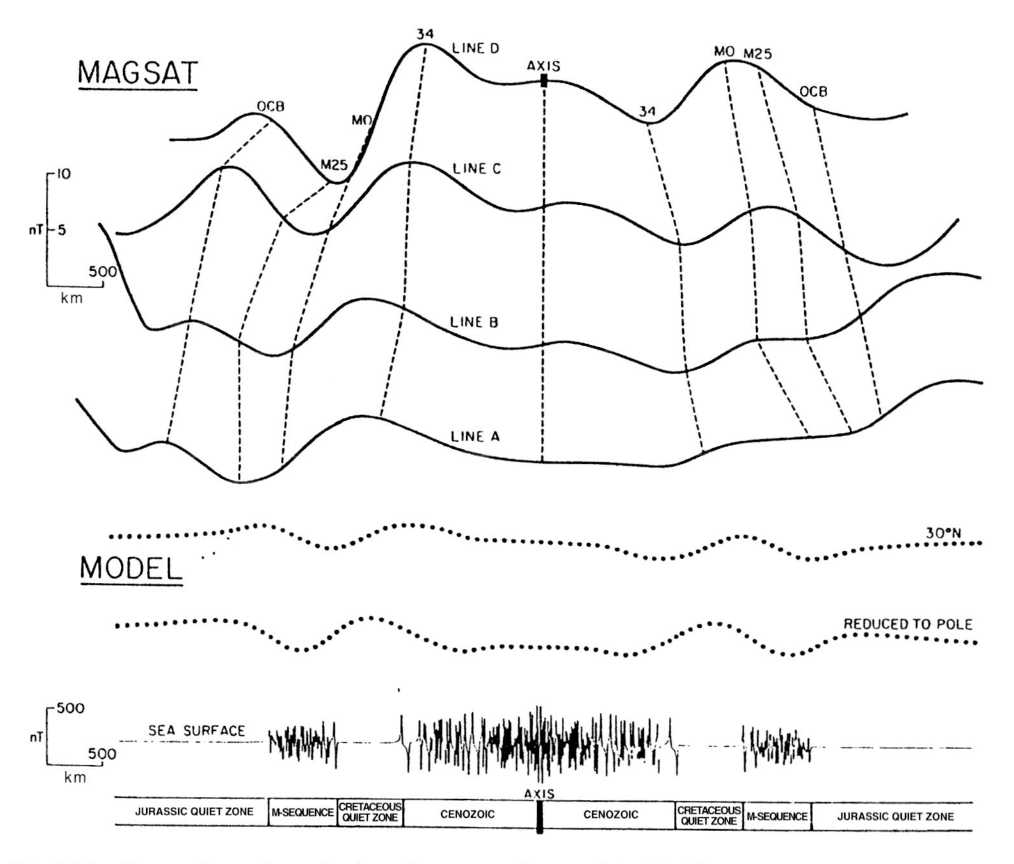

Fig. 3.16 Comparison of standard sea-floor spreading model with *Magsat* profiles A, B, C, and D on Fig. 3.15. Reading from bottom up, the profiles are: anomalies measured at sea surface; same anomalies extrapolated upward to satellite altitude and reduced to pole (i.e., corrected for magnetic latitude); and same anomalies at 30 deg. N, line C on Fig. 3.15 (top profile). From LaBrecque and Raymond (1985).

Another study of subduction zones as seen on *Magsat* data was carried out by Arkani-Hamed and Strangway (1987). On comparing the magnetic signatures of known subduction zones around the Pacific Ocean, they found that age of the subducted oceanic crust, as inferred from the magnetic time-scale, had strong influence on the magnetic anomalies. Older crust, such as that in the northwest Pacific, produced distinct anomalies, whereas younger (and warmer) crust, such as that of the Nazca Plate, produced none. A laboratory study of island arc xenoliths by Warner and Wasileski (1997) showed that mafic xenoliths might account for the magnetic anomalies detected over areas such as the Aleutians and Japanese islands.

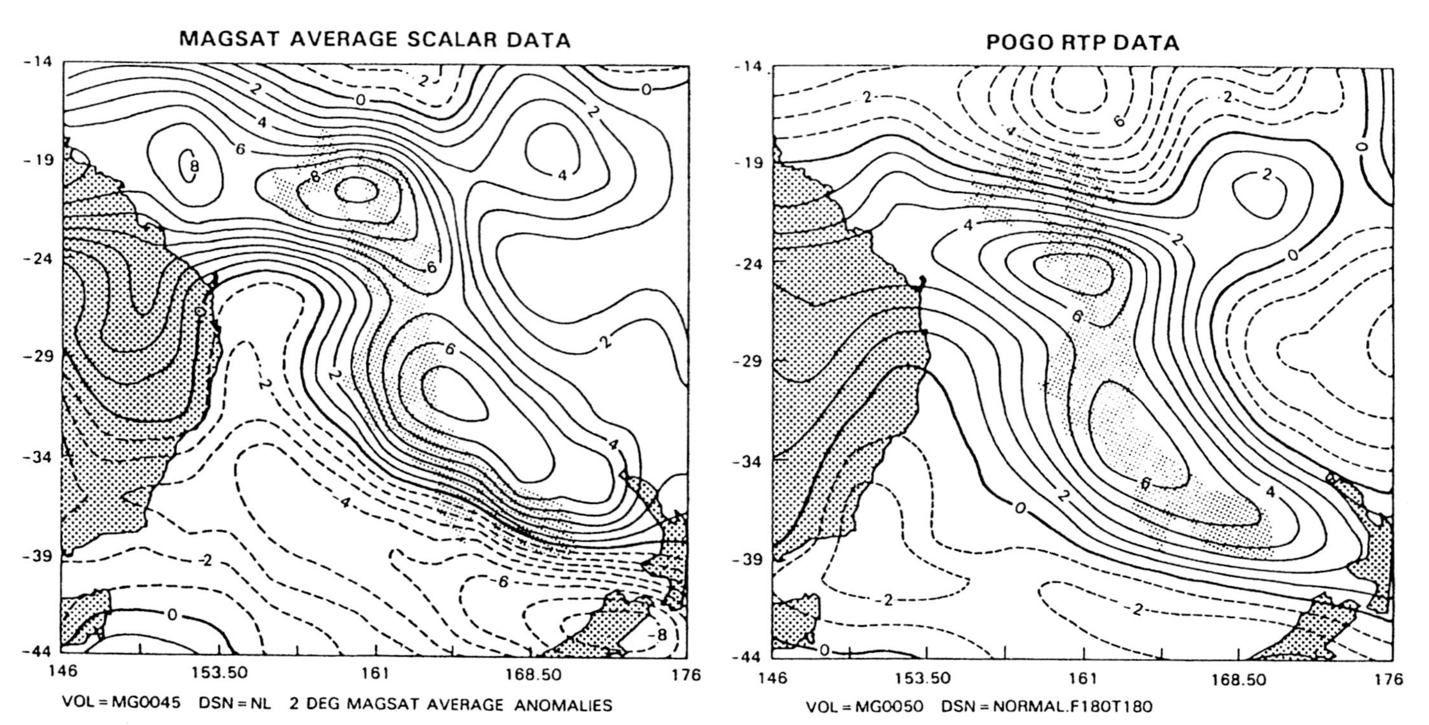

Fig. 3.17 *Magsat* and *POGO* magnetic anomalies over the Lord Howe Rise, between Australia (left) and New Zealand (lower right). Lord Howe Rise shaded, outlined by 2000 meter isobath. *POGO* anomalies reduced to pole. Contour interval 1 nT, scaled to constant 50,000 nT field throughout the map. From Frey (1985).

It has been proposed by several workers that, geophysically impossible as it may seem, continental crust can be converted to oceanic crust. A *Magsat/POGO* study (Fig. 3.17) by Frey (1985) showed what may be an actual example of such conversion, the Lord Howe Rise between Australia and New Zealand. This a deeply-submerged marine plateau whose lithology (e.g., ignimbrites) shows that it was at one time emergent; the fact that it is now at oceanic depths alone suggests that it is not normal continental crust. Frey showed that the magnetic susceptibility implied by the positive satellite magnetic anomaly indicates possible conversion of the lower crust to a more mafic rock type. Although further data would be needed to confirm this interpretation, its novelty suggests the value of the *Magsat/POGO* data for an extremely fundamental problem, the origin of ocean basins. The anomaly maps of Frey, like those of Hinze *et al.* (1982), show little expression of the continental margin, a problem to which we shall now turn.

The apparent failure of *Magsat* to delineate the continental/oceanic crust boundary, demonstrated in Indonesia by Taylor (1991), has been addressed by Purucker *et al.* (1998). They point out that the apparent lack of contrast at the boundary may be due to removal of these features by the spherical harmonic separation of the main field, or by the lack of spatial resolution in the satellite data. They therefore have developed an inverse technique to include a-priori information on crustal magnetization. The block diagram of this approach will be helpful (Fig. 3.18) in discussing it.

The technique, in brief, is an iterative one, starting with construction of a simple Earth magnetization model (SEMM) by assuming values of crustal susceptibility and thickness. The field such a crust (SEMM-0) would produce at 400 km altitude is then calculated, followed by other steps shown on the diagram until a new model field is produced. This field is then compared with observed anomalies. If it does not reproduce the anomalies, the procedure is repeated after dipole corrections, a new SEMM calculated, then its field constructed as before. Some idea of the number of geophysical parameters involved may be gained from Fig. 3.19. The result of this approach is a global magnetization model, SEMM-1 (Fig. 3.20) giving a much more understandable and realistic picture of the crustal anomaly field. Purucker *et al.* show how this model can be used to refine interpretations of crustal structure, composition, or heat flow in the Gulf of Mexico and over the Kentucky anomaly. Other factors are involved, such as the strength of the main field at different latitudes, which affects the induction process.

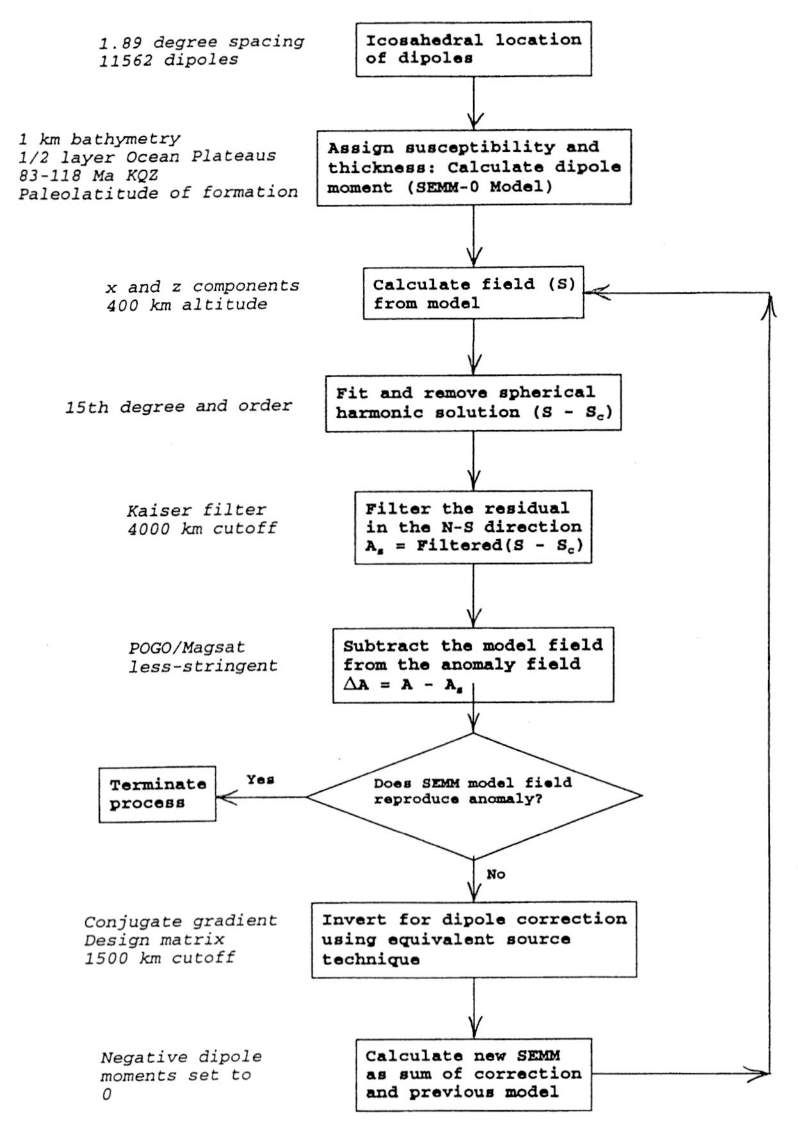

Fig. 3.18 Block diagram of the derivation of the global magnetic model. From Purucker *et al.* (1998).

3.5 Extraterrestrial magnetic fields

As we have seen with gravity fields, we can gain a broader perspective from the magnetism of other planets (Ness, 1979; Dyal, 1992; Kivelson, 1995). Working outward from the Sun – which, being composed of intensely hot turbulent plasma, is strongly magnetic on all scales – we come first to Mercury. Given its resemblance to the Moon – an inactive and largely non-magnetic body – we might

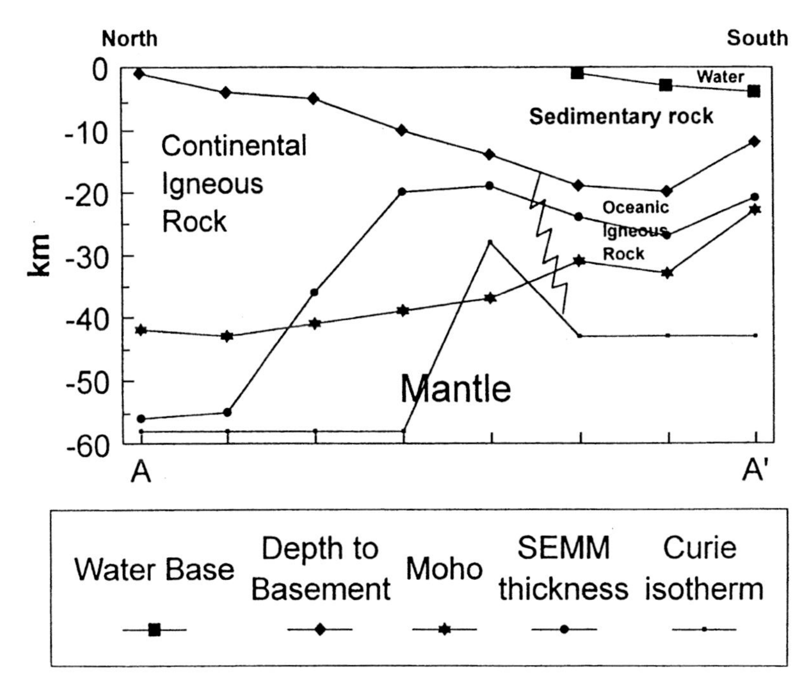

Fig. 3.19 North–south cross section through the crust of the Mississippi River embayment and adjacent Gulf of Mexico. From Purucker *et al.* (1998).

expect little or no main field, but it was found (Ness *et al.*, 1974) that Mercury has a surprisingly strong one, several hundred nanoteslas. Although only 1% of the Earth's field, this is much more than the interplanetary field, and has led to intensive study for an explanation. As summarized by Ness (1979), the consensus is that Mercury is differentiated into a liquid core and mantle, and has an active "dynamo" more or less analogous to that of the Earth. However, remanent magnetism may play a role, especially in view of the relatively large iron core.

Venus is in many ways similar to the Earth, and it might be expected to have a comparable magnetic field. Early Soviet and American missions failed to detect a planetary field. The magnetometer on *Pioneer Venus*, flown in 1978, did detect nightside fields of 20 to 30 nT (Russell *et al.*, 1979), but these were horizontally oriented and variable from one orbit to another. Generally consistent results were obtained by the *Galileo* mission, which went by Venus in 1991 as part of a gravity-assist maneuver (Kivelson, 1995). The ionosphere of Venus is electrically conducting, and thus interacts with the magnetic field of the solar wind ("moving electrical fields"). The weak fields detected were accordingly interpreted as resulting from interaction of the venusian ionosphere with the

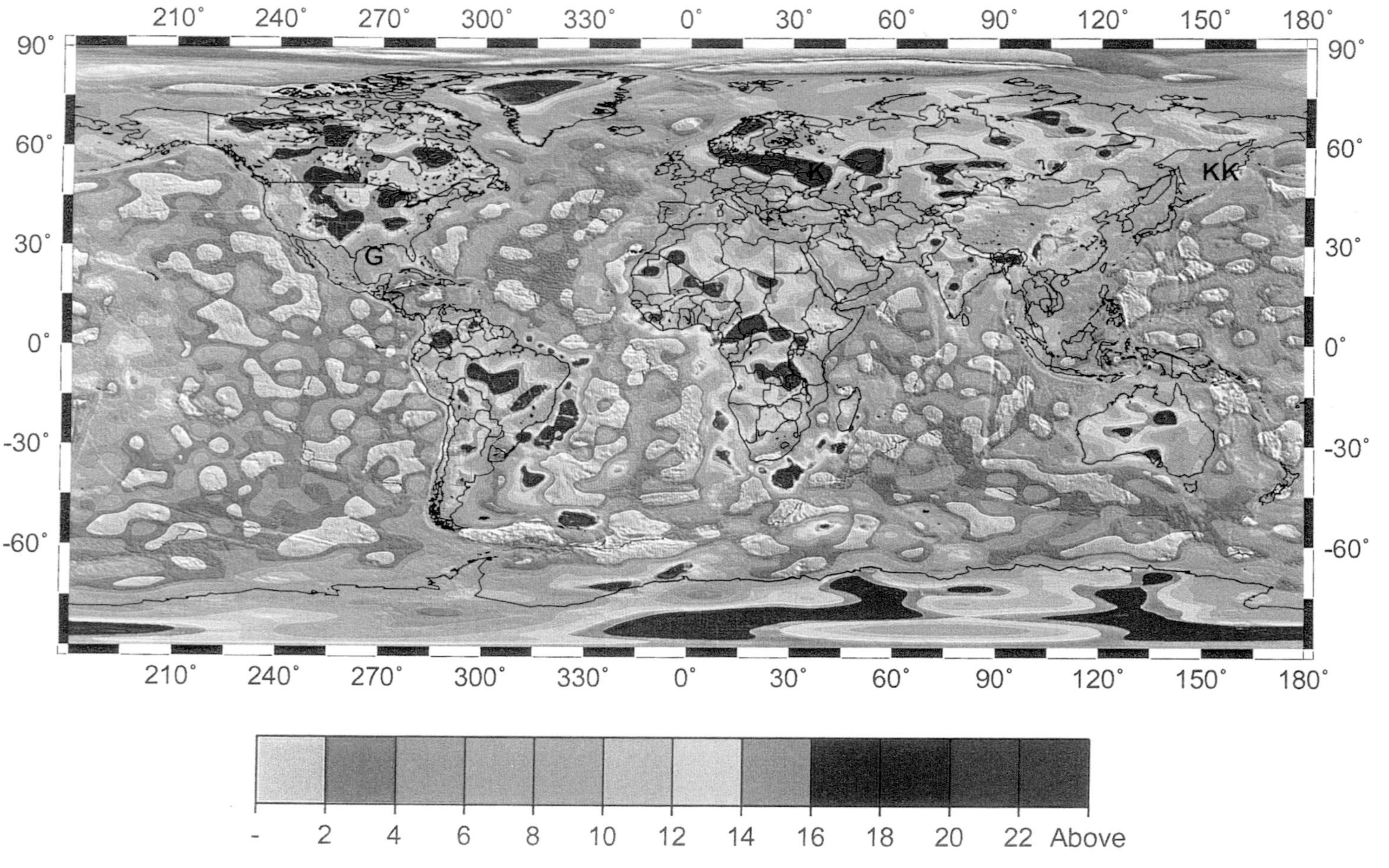

Fig. 3.20 (See also Plate XI) Global map of susceptibility (SI) times thickness times 10 of the SEMM-1 model shaded by surface topography for correlation with major bathymetric and topographic features. Negative values in gray. Units are SI × km × 10. From Purucker *et al.* (1998).

solar wind, not from an internal dynamo. Venus must have a hot interior and a liquid core, but its very slow rotation rate (about 243 days) apparently inhibits generation of a core field (Russell and Luhmann, 1992).

The Moon has no bipolar main field (Ness, 1979), a discovery made by Soviet spacecraft in the early-1960s. However, more detailed investigations, notably those by subsatellites launched from the *Apollo* spacecraft, showed that there are local magnetic anomalies on the lunar surface (Figs. 3.21, 3.22) as strong as a few hundred nanoteslas (Hood *et al.*, 1981). Reiner Gamma (Fig. 3.23), the strongest of these, is particularly puzzling, having no visible relief, consisting apparently of high-albedo "swirls." These anomalies in general have stimulated much theorizing. One explanation for their origin is that of Runcorn (1967), who suggested that the lunar anomalies are caused by remanent magnetism produced when the Moon had a strong main field, or was exposed to a strong field (that of the Earth or the Sun). Runcorn actually carried out paleomagnetic studies for the Moon. Another possible cause for the lunar anomalies is magnetization by impact, or shock remanent magnetization (Pilkington and Grieve, 1992). First suggested for terrestrial rocks by Pohl (1971) and Cisowski and Fuller (1978), the obvious importance of impact on the Moon suggested that this possibility be further investigated. Such investigation became possible with the *Lunar Prospector* mission (Lin *et al.*, 1998).

The *Lunar Prospector* (*LP*) spacecraft carried a magnetometer and electron reflectometer experiment in a near-polar low-altitude (ca. 100 km) orbit. Electron reflectometer magnetometry depends on the fact that magnetic fields will deflect or reflect charged particles (electrons), a technique only possible over airless bodies. The *LP* magnetic survey carried out confirmed previous findings, but permitted their application to almost the entire lunar surface. Perhaps the most interesting result of the *LP* magnetic survey was confirmation of the existence of strong magnetic anomalies antipodal to the large near-side impact basins, such as Mare Imbrium. Although the exact mechanism responsible for these anomalies is not understood, they are probably related to shock remanent magnetization, mentioned above. For reasons discussed by Lin *et al.* (1998), such remanent magnetization implies that when the mare basins were formed, around 3.9 to 3.6 billion years ago, the Moon had an internally-generated main field. This field has long since disappeared, presumably as the result of the Moon's cooling and solidification to depths of several hundred kilometers, as discussed in later chapters of this book.

One of the most surprising disoveries related to extraterrestrial

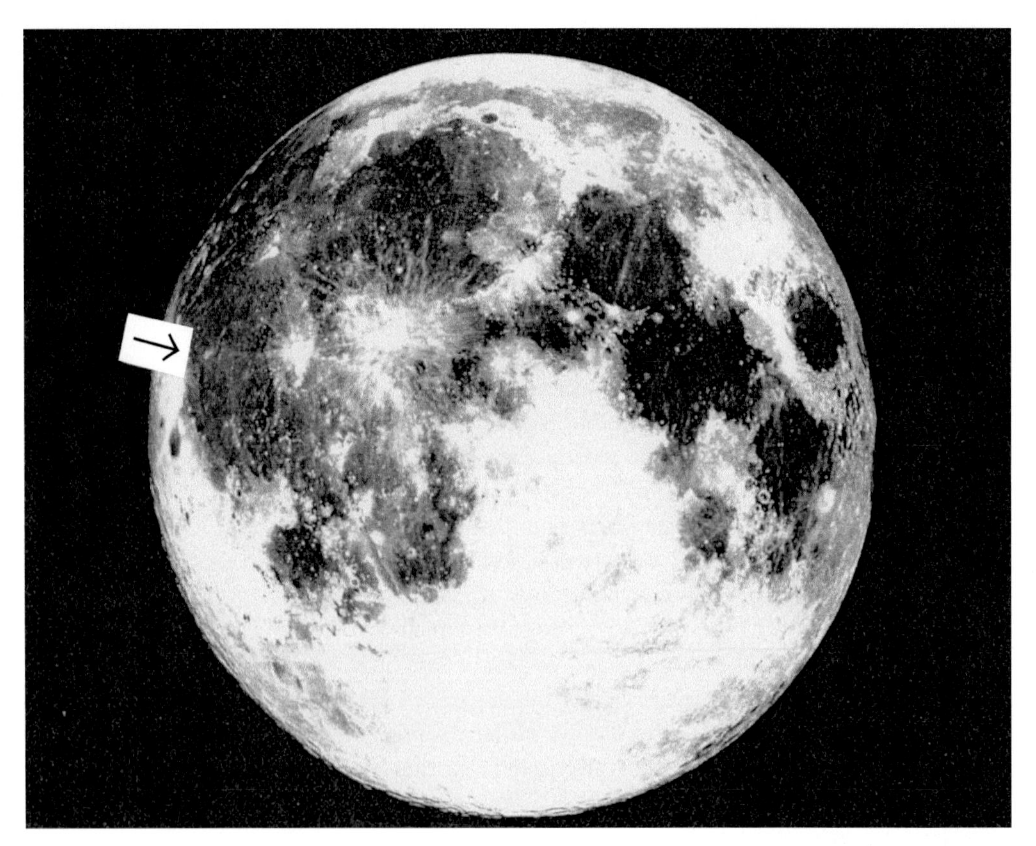

Fig. 3.21 Lick Observatory composite photograph of first and last quarter Moon. Reiner Gamma indicated with arrow: white, tadpole-shaped feature.

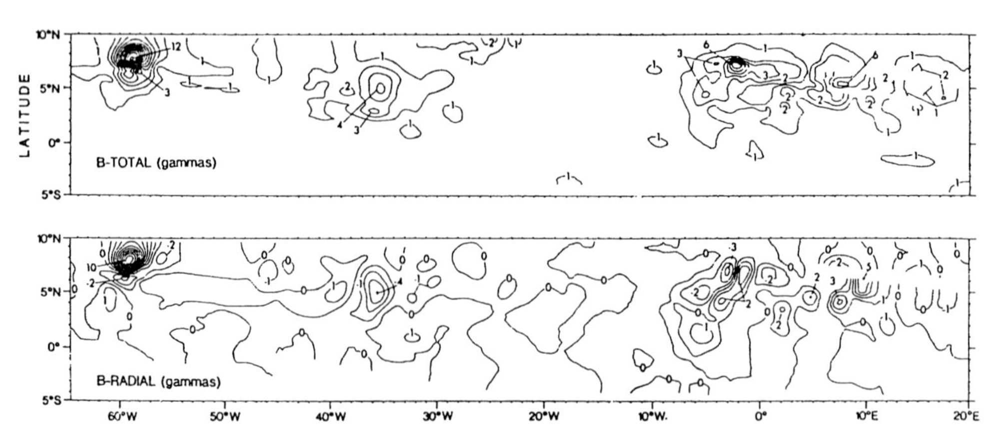

Fig. 3.22 Magnetic anomaly maps from *Apollo* subsatellite measurements: (*top*) total field intensity in nT ("gammas"); (*bottom*) vertical intensity. Reiner Gamma is anomaly at upper left. From Hood *et al.* (1981).

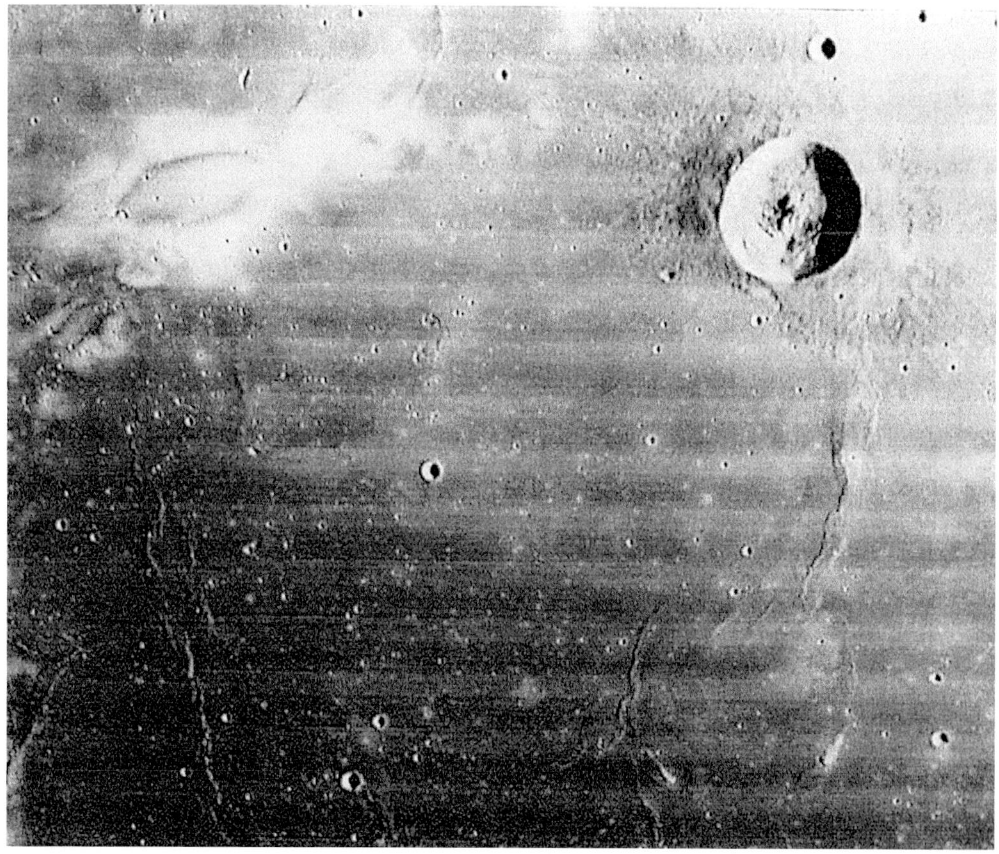

Fig. 3.23 *Lunar Orbiter IV* photograph 157H1 (north at top) showing crater Reiner at upper right, 30 km in diameter, and Reiner Gamma at upper left.

magnetic fields comes from Mars, specifically from the *Mars Global Surveyor* (*MGS*) (Acuña *et al.* 1998; Connerney *et al.*, 1999). It has been known for decades that Mars has little if any global magnetic field. But just as the Moon's lack of air and water makes it invaluable for comparative planetology, the absence of a global martian magnetic field has turned out to be extremely revealing. The *MGS* measurements have shown that there is no significant main field corresponding to that of the Earth. However, there are extremely strong linear magnetic anomalies detectable at altitudes of several hundred kilometers (Purucker and Clark, 2000; Purucker *et al.*, 2000) (Fig. 3.24). These are much stronger than comparable terrestrial anomalies (see Fig. 3.6). In the absence of a strong main field, the martian anomalies must be produced by some form of remanent magnetism (Kletetschka *et al.*, 2000). Connerney *et al.* (1999) interpreted the anomalies that they discovered as being essentially analogous to

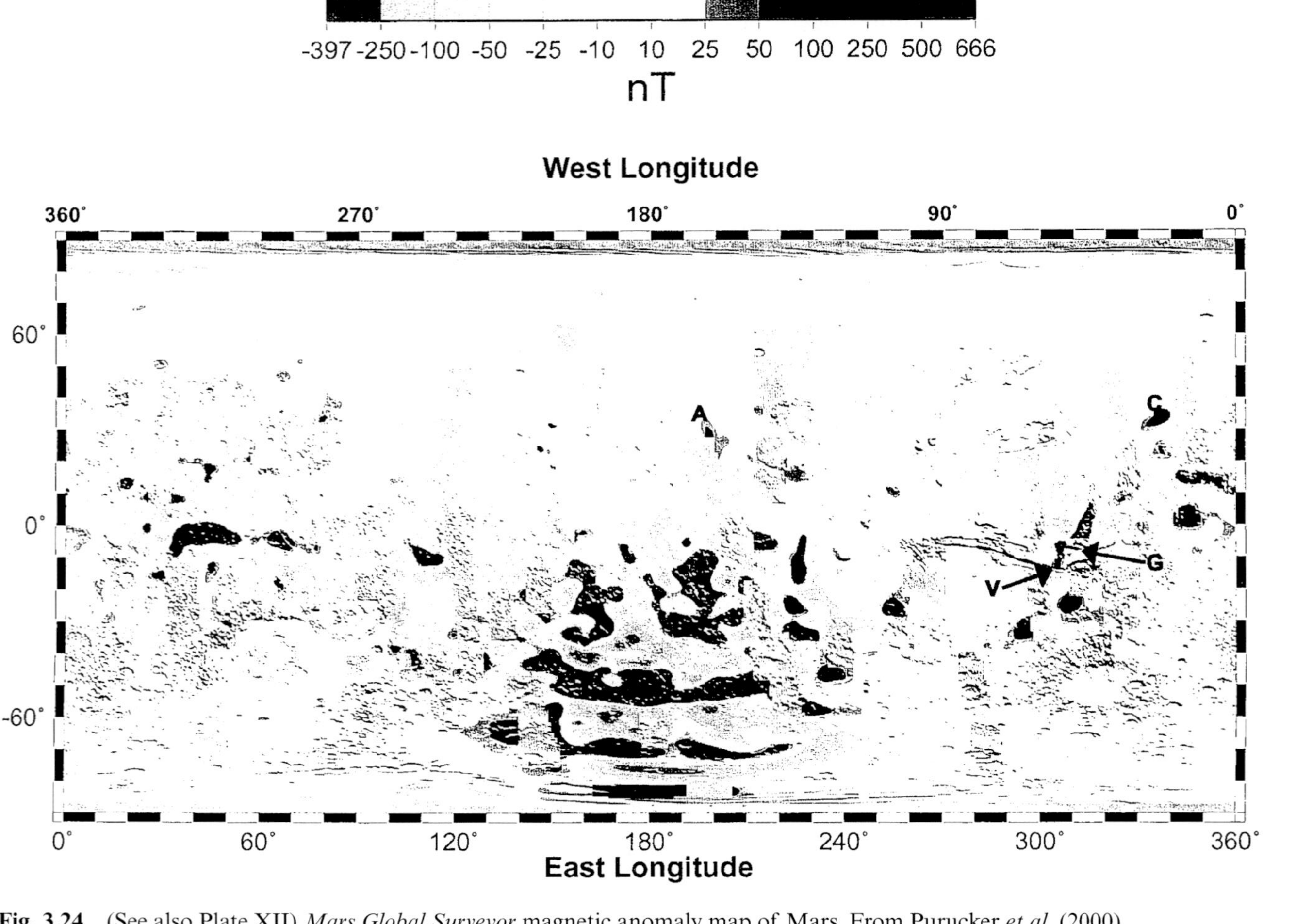

Fig. 3.24 (See also Plate XII) *Mars Global Surveyor* magnetic anomaly map of Mars. From Purucker *et al.* (2000).

those of the terrestrial oceanic crust and, like them, to have resulted from sea-floor spreading in the presence of a reversing primordial martian magnetic field. The *MGS* data then might imply that Mars, like the present Earth, has undergone a period of sea-floor spreading, the highland crust being the reworked "remnants" of an ancient oceanic crust.

Regardless of interpretation, the *MGS* discoveries and the interpretation by Connerney *et al.* are extraordinarily interesting, furnishing a striking example of the value of comparative planetary geophysics. The visible geology of Mars gives no support at all to the plate tectonic hypothesis, as in fact Connerney *et al.* note. More generally, Mars appears to be a planet that although still internally active never reached the stage of true plate tectonics. One specific implication of the *MGS* anomalies is the insight they may give to interpretation of terrestrial anomalies.

It has been uncertain whether the crustal anomalies detected from space are produced by induced or by remanent magnetism. The Earth's main field is, as we have seen, very strong, quite strong enough to induce magnetic anomalies. However, Mars has no comparable field, nor does the Moon, at the present time. It thus seems likely that the anomalies detected on each body must be remanent, not induced, formed when there were strong inducing fields, of whatever origin.

The magnetic fields of the terrestrial (inner planet) bodies, in summary, appear to reflect several of their characteristics: rotational rate, size, and internal temperature. The Earth's strong and relatively well-organized (dipolar) main field results from its unique combination of rapid rotation and a large, partly liquid, iron core.

The giant (outer) planets have all been visited, by *Pioneer*, *Voyager*, *Galileo*, or *Ulysses* spacecraft, and each has a magnetic field (Connerney, 1987; Kivelson, 1995). That of Jupiter is the strongest and most variable, as we might expect from Jupiter's near-stellar composition and structure. The jovian magnetic field is thought to be produced by a dynamo mechanism, but one involving currents of liquid hydrogen (perhaps metallic) and comparably exotic conditions. An interesting complication in the jovian magnetosphere is the magnetic connection to its nearest large satellite, Io, through a torus of sulfur and oxygen ions produced by the erupting volcanos on Io. This connection is not a trivial effect, involving a million megawatts of power (Bagenal, 1998).

The years-long *Galileo* and *Ulysses* missions have produced a cornucopia of magnetic field results that can be summarized only briefly here; a good review has been published by Kivelson (1995).

The *Galileo* results are particularly relevant to the theme of this book.

Perhaps the most surprising *Galileo* findings, from the viewpoint of terrestrial geophysics, are that bodies that might be expected, from their size, composition, or temperature to be non-magnetic apparently do have significant magnetic fields. Ganymede and Callisto, both ice-covered satellites of Jupiter, have been found to perturb the solar wind, strongly implying that they both have magnetic fields. The most favored explanation (Kivelson *et al.*, 1998; Khurana *et al.*, 1998) is that both satellites have, under their icy crusts, oceans of salt water. Being a moving electrical conductor, such water could produce magnetic fields as they cut the lines of force of Jupiter's immense and strong field. The possible implications of this discovery for extraterrestrial life are extremely interesting, raising the possiblity that at least simple life-forms may exist in these superficially hostile bodies.

The *Galileo* mission carried out a long and complex series of gravity-assist maneuvers. The spacecraft was diverted toward two asteroids, Gaspra and Ida (Kivelson, 1995), and found that even these small rocky bodies also perturbed the solar wind. Although the trajectory did not go close enough to detect magnetic fields directly, the existence of such fields can be inferred with some confidence.

Further discussion of other planetary magnetic fields would lead us too far astray, but it is worth noting that strange and unpredicted as they were, these phenomena have so far all proved understandable in terms of conventional electromagnetic theory. Quantum mechanics, quarks, and "strange" particles have not been invoked to explain them. Faraday, Oersted, and Maxwell would be able to follow discussions of magnetism in these unimaginably alien planets, indeed, to contribute to them.

3.6 Summary

How can we summarize the impact of space flight on studies of the Earth's magnetism? We should begin by pointing out that the first satellite dedicated to study of the global and crustal fields, *Magsat*, was only launched in 1978, and there was, until the 1999 launch of *Oersted*, no comparable successor. The situation is analogous to that of satellite meteorology if there had been no additional weather satellites launched after *Tiros 1*. NASA has been criticized by Dyson (1979) for its common practice of apparently considering a program completed with one successful mission. For *Magsat*, the criticism can not be easily dismissed.

What has been accomplished with the space data so far acquired? The most general and unarguable benefit is simply the broader perspective acquired from spacecraft that have already left the solar system after visiting every planet but Pluto. Geophysicists now have magnetic field data from all these planets to compare with that of the Earth. In addition, we now have a reasonably complete picture of the interaction of the Earth's magnetic field with the interplanetary field.

For the main field, the value of orbital data is obvious. This field is a dynamic and complex feature, affected by factors as diverse as shifting currents in the core and variations in the solar wind, changing on time-scales from millions of years to minutes. The first dedicated magnetic field satellite, *Magsat*, provided the primary data for the first adequate International Geomagnetic Reference Field. These data have provided the first synoptic vector data on the main field, giving the beginning of real understanding of the liquid core, starting with confirmation of its size and gross electrical properties.

The application of crustal anomaly orbital data can be summarized very generally as follows. First, the global nature of satellite orbits has provided a consistent picture of crustal magnetism in remote areas on land and sea that had been geomagnetically almost unexplored. A second result of satellite data has been a new and independent source of information on the lower continental crust, until very recently almost entirely unknown. Satellite magnetic measurements, revealing extremely broad and deep features, have thus proven a valuable complement to aeromagnetic and ground surveys that reveal much smaller and shallower features such as mafic dikes, banded iron formations, and greenstone belts. Many different structures and lithologies of the lower crust may be expressed in satellite data. However, interpretations of these data are beginning to converge with those from other lines of investigation, such as seismic reflection profiling. The lower crust in most areas appears to consist largely of high-grade metamorphic rocks, significantly more mafic than those of the exposed basement (Lowman, 1984; Rudnick, 1992). Several broad anomalies detected on satellite data have been interpreted as very large mafic intrusions, supporting petrologic and geophysical studies indicating that basaltic underplating, or intrusion, has been a major factor in evolution of the continental crust. We will return to this subject in Chapter 6. These developments typify the application of orbital magnetic data to geology. By themselves, they have produced few discrete discoveries, but combined with other lines of evidence they are contributing to a solid if still imprecise understanding of the structure, composition, and evolution of the continental crust.

The flood of global magnetic data from *POGO* and *Magsat* has had a stimulating effect on laboratory and field studies of what has been termed magnetic petrology, the study of magnetic materials in the crust. The importance of the satellite data in magnetic petrology stems from the nature of the anomalies these data reveal, arising largely from the deep and inaccessible crust. The magnetic properties of major rock types have been studied before, for interpretation of aeromagnetic and surface studies. But the satellite data have stimulated investigations of uncommon and poorly-exposed rocks, granulites in particular. Results of these studies are of course fed back into interpretations of satellite data, with mutual benefit.

An unexpected application of satellite magnetic data to geophysics has been in mapping the depth of the Curie isotherm, the surface below which the crust or mantle is too hot to be magnetic. Since the Curie isotherm reflects heat flow, the satellite data are giving us a new source of information on the Earth's thermal behavior. Because most tectonic activity in the Earth is fundamentally thermal in origin, this application of satellite magnetic field measurements has great potential value (no intentional pun).

Perhaps the most important long-term result of satellite magnetic field studies to date has been the discovery or at least the redefinition of geologic problems. For example, the Bangui anomaly has been suggested to be the possible expression of a major impact that triggered basaltic magmatism. This may stimulate study of the relations between terrestrial magmatism and impact, already a lively subject because of the flood of new data on such relations from the Moon, Mercury, Mars, and Venus. Another line of inquiry stimulated by satellite data concerns the distribution and origin of aulacogens, failed rifts that sometimes localize mafic igneous intrusions, expressed as satellite magnetic anomalies. Satellite data focus attention on the problem and may contribute to its solution.

CHAPTER 4

Remote sensing: the view from space

4.1 Introduction

Geophysics is sometimes distinguished from geology as being the study of the *inside* of the Earth, geology being the study of the *outside*: the surface of the Earth and structures expressed at the surface. These light-hearted definitions are actually a convenient introduction to what has become the most pervasive and important direct effect of space flight on geology: **remote sensing.**

This is a new term for an old technology, briefly defined as **the study of objects at a distance by means of electromagnetic radiation, reflected or emitted**. Vision obviously fits this definition, and in fact "remote sensing" can be thought of as the extension of human vision to far wider parts of the spectrum, and to far greater distances, than the unaided eye can reach. Astronomers have been doing "remote sensing" for centuries. However, as currently used, remote sensing refers to the acquisition, processing, and interpretation of images from satellites and aircraft (Sabins, 1998), images formed by electromagnetic radiation, as distinguished from potential fields (gravity and magnetic). This definition is to a degree artificial, in that electromagnetic radiation itself consists of transversely oriented electric and magnetic fields traveling through space at 300,000 km/s (in vacuum). Furthermore, as we will see, remote sensing and geophysics are increasingly used together, and many data-processing techniques and formats such as shaded relief maps are common to both.

The definition cited above, by Sabins, is notable for the priority given to "satellites." Remote sensing from aircraft goes back to the first applied aerial photography in World War I, and for the first few years after the term was coined (in 1958, by Virginia Prewitt) it referred only to airborne sensors. Aerial methods continue to be widely used, but since the 1980s "remote sensing" has increasingly meant *orbital* sensing. For convenience, this usage will be followed here; unless specified otherwise, "orbital" is implied by "remote sensing." This chapter will cover only remote sensing of the Earth, with treatment of other planets reserved for Chapters 5 and 6.

It was suggested by Naisbitt (1984) that the achievement of orbital flight had been a major contributor to the emergence of the "information" or "post-industrial" society after World War II, the specific cause being development of communications satellites. Remote sensing from space has made a comparable and growing contribution to this "megatrend," as discussed by Cary (1997). The field of remote sensing has consequently become an extremely large one, filling many library shelves, and this chapter does not pretend to cover the subject in general. The objective is to summarize the impact of remote sensing on geology. Accordingly, only the most important principles will be outlined, as well as the main events in the development of geologic remote sensing.

The electromagnetic spectrum is illustrated in Fig. 4.1, differing from diagrams in physics textbooks in the emphasis on atmospheric transmission as a function of wavelength. This aspect of the diagram can be thought of as window shades (black), pulled down over certain parts of the spectrum. It is extremely important in remote sensing as defined here, since orbital techniques as used around the Earth depend on radiation that must get through the entire thickness (actually a double thickness, if we include the Sun's radiation) of the planet's atmosphere. Another important aspect of the diagram may not be so obvious: the dependence of wavelength on size of the radiation's source. Gamma rays come from the nucleus, X-rays from inner electron shells, visible light from the outer shells, thermal infrared from molecular motion, magnetron-generated microwaves from resonant cavities a few millimeters across, and broadcast band radio, with wavelengths of a few hundred meters, from large antennas on towers. The dependence of wavelength on antenna length is well known in radio engineering, and given by an equation that the reader will be spared. The more general dependence on radiation source, however, is fundamental to understanding the physics of remote sensing: **the longer the wavelength, the bigger the source.**

There are two general categories of remote sensing, depending on the source of the electromagnetic radiation: *passive* and *active*. Passive methods use only radiation reflected or emitted by the object being studied; the target is also the radiation source. Vision is the most obvious example of passive remote sensing, usually depending on visible light coming from the source. Active methods involve generation of electromagnetic radiation beamed at the object to be studied. Radar is the best-known active method, generally using

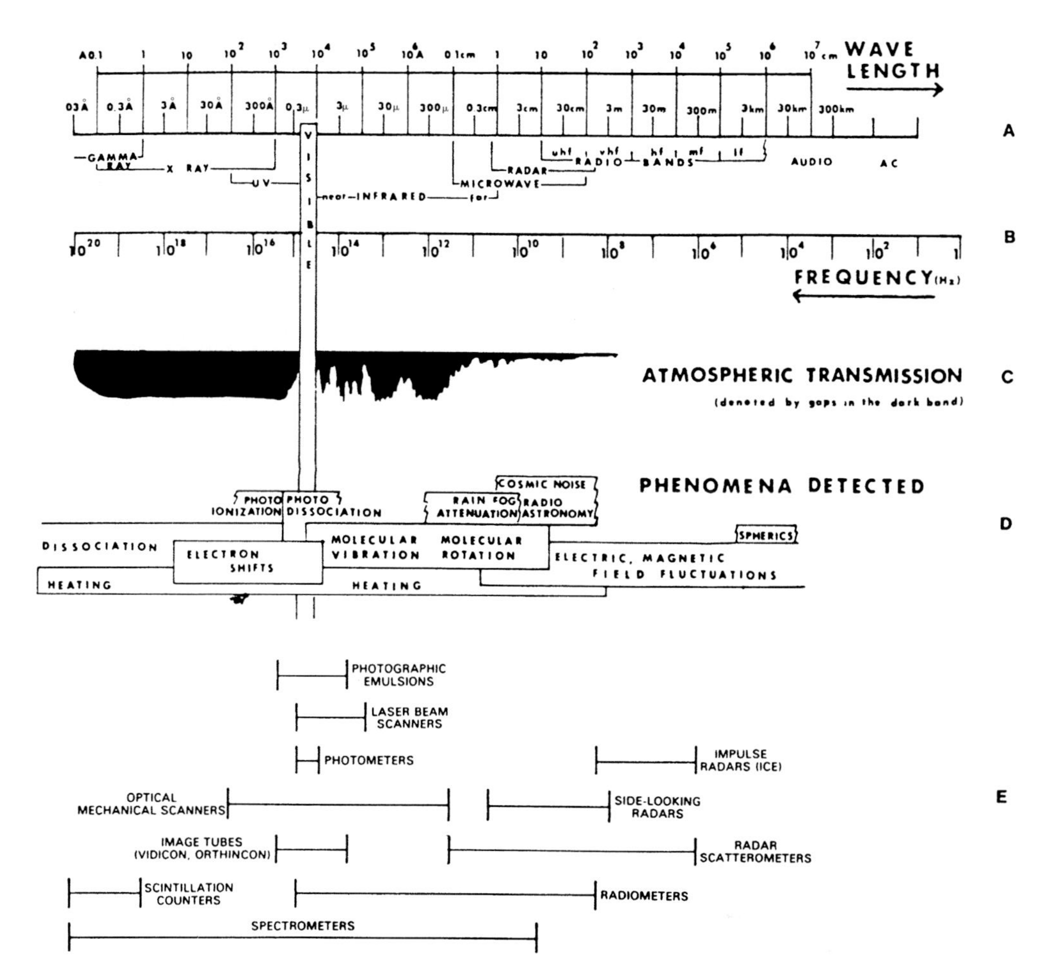

Fig. 4.1 Electromagnetic spectrum, atmospheric transmission, radiation sources, and instruments used.

microwave radiation reflected from the target. Optical radar, or lidar, using laser-generated visible or near-visible radiation, is coming into wide use. One category of active remote sensing has already been covered in Chapter 2, under space geodesy: radar and laser altimetry, techniques that are increasingly being included, correctly, in "remote sensing." Altimetric data, showing marine or land topography, is fundamental to understanding of both satellite geophysical and geological methods.

The use of electromagnetic radiation in remote sensing involves a wide range of data-analysis techniques, most generally categorized as *analog* and *digital*. Analog methods, in this context, are illustrated by human vision, film photography, and older types of radar as displayed on an air-traffic control screen. Digital methods, treated at length by Vincent (1997), are now by far the dominant method of data analysis used in remote sensing, depending not only on digital primary data but also on initially analog data such as photographic film that has been digitally scanned. A remarkable example of this latter approach is the spectacular collection of orbital photography by Apt *et al.* (1996), in which returned film images have been digitized and enhanced for publication. "Colorization" of old movies is a better-known example of digital reprocessing of originally analog data.

With this unavoidably brief sketch of the principles of remote sensing data acquisition and handling, let us turn to the application of remote sensing from space to geology. This chapter will be confined to orbital methods only, but the first space photographs used for geology were taken from sounding rockets, such as the *Viking* series, at altitudes of up to 200 km (Lowman, 1965). Many excellent photos were taken on lunar missions by American astronauts, as shown in the collection edited by Schick and Van Haften (1988).

4.2 Orbital remote sensing in geology: a brief history

The advantages of orbital remote sensing for meteorology were recognized well before artificial satellites were actually launched: global coverage, systematic repetition of coverage (or continual coverage from geosynchronous orbits), wide field of view, and capability for thermal measurements. The value of orbital images for topographic mapping and military reconnaissance was also realized at an early stage, but there was virtually no appreciation of the value of orbital methods for geology until the mid-1960s, after hundreds of satellites had been launched. The reasons for this are an interesting aspect of the history of space flight in general, and will therefore be discussed at some length; a detailed account has been given elsewhere (Lowman, 1999).

President Kennedy's 1961 challenge to land a man on the Moon by the end of the decade led to a large and rapid increase in the civilian space program, i.e., NASA. Although focussed on the lunar landing, NASA planning in the early-1960s included several potential applications of our growing space capability. One of these was the use of manned space stations for earth-oriented remote sensing, primarily for earth resource studies, as distinguished from satellite meteorology. "Earth resources" of course included geology, and NASA studies in this field were managed by geologist Peter Badgley in cooperation with the US Geological Survey (USGS), the

Department of Agriculture, and the Office of Naval Research among others. These studies involved a wide range of airborne remote sensing experiments, carried out by the Manned Spacecraft Center (now the Lyndon B. Johnson Space Center) as well as by many other governmental and non-governmental organizations (Vincent, 1997). Many of the techniques developed were actually used successfully on the first American space station, *Skylab*, in 1973–4. However, geologic remote sensing was given a jump start, so to speak, by the *Mercury* and especially the *Gemini* programs.

Project *Mercury*, the first American manned space effort, began in 1958, with the first successful orbital flight by John Glenn in 1962. Beginning with the second flight, by Scott Carpenter, the *Mercury* pilots carried out terrain photography for geologic purposes, stimulated by the suggestion of Paul Merifield on the basis of his work with rocket photographs. The last *Mercury* flight, a 22-orbit one flown by Gordon Cooper, permitted acquisition of twenty-nine 70 mm color photographs, chiefly of southern Asia (O'Keefe *et al.*, 1963). In combination with Cooper's remarkable visual sightings, the *Mercury* terrain photographs generated the beginnings of wide appreciation of the geologic value of orbital photography (Lowman, 1965). They gave rise to a much more extensive photographic effort, the S005 Synoptic Terrain Photography Experiment, carried on the *Gemini* missions (Gill and Gerathewohl, 1964). By the time the 10 *Gemini* flights were over, some eleven hundred 70 mm color photographs suitable for geology, geography, or oceangraphy study had been acquired (Lowman, 1969). Published widely, these spectacular pictures generated world-wide interest among public and scientists alike (Merifield *et al.*, 1969). Their most important result for geology was the stimulus to the US Geological Survey's *Earth Resources Observation Satellite* (*EROS*) proposal of 1966, triggered primarily by the "demonstrated utility of the *Mercury* and *Gemini* photographs" (Pecora, 1969).

The *EROS* concept was based on use of a television system, proposed by the Radio Corporation of America (RCA), rather than returned film. The story becomes extremely complicated at this point, with *EROS* getting entangled in interagency conflicts. An authoritative account of these conflicts has been published by Mack (1990) and need not be recounted here. However, when the intrabeltway dust had settled, *EROS* had evolved into the *Earth Resources Technology Satellite* (*ERTS*), a NASA program developed with the cooperation of the USGS, Department of Agriculture, and other government agencies. The first *ERTS* was launched in July 1972, and shortly thereafter re-named *Landsat*.

From a historical viewpoint, the most important aspect of *Landsat* was its derivation from the *Apollo* program. *Gemini* was solely technological preparation for a lunar landing; the remote sensing efforts sponsored by NASA Headquarters were explicitly part of the *Apollo* program; and the Manned Spacecraft Center was built for *Apollo*. *Landsat*, and the extensive remote sensing research that supported it, can thus be considered part of the *Apollo* legacy (Lowman, 1996, 1999).

The first *Landsat* was followed by two essentially identical satellites and then by the much more advanced "D" version, *Landsats 4* and *5*, carrying the Thematic Mapper (Salomonson and Stuart, 1989). However, the success of the first three, coupled with progress in remote sensing technology, led to development of remote sensing satellites by other countries, including France, the former Soviet Union, Japan, and India. Several countries also had been operating classified reconnaissance satellites since the early-1960s, but although their existence was common knowledge, they had little impact on non-military remote sensing. In 1995, the formerly-secret American *Corona* program was declassified (McDonald, 1995), and its thousands of photographs will no doubt find scientific application. By the mid-1990s, satellites for study of the Earth's surface, using reflected visible or near-infrared radiation, were in wide use. Leadership in orbital remote sensing has largely passed from the United States to other countries, notably France, Canada, Japan, and western Europe in general. Private industry is taking an increasingly greater role in the field.

Returned film orbital photography, which as we have seen was a major stimulus to *Landsat* and its cousins, has been carried out more or less continuously since the end of the *Apollo* program in 1975, by cosmonauts and, when the American Shuttle program began in 1981, again by astronauts. The Space Shuttle Earth Observation Program, SSEOP (Wood, 1989; Lulla *et al.*, 1993; Apt *et al.*, 1996) has produced thousands of high-quality color photographs, a valuable low-cost public domain supplement to *Landsat*, and one area in which the United States still leads. The quality of the photographs can be enhanced by digitization (Apt *et al.*, 1996), a technique applicable to older pictures as well. The ability of the astronauts to pick out specific features has made the SSEOP pictures valuable for studies of surface changes and transient phenomena (Lulla and Helfert, 1989; Strain and Engle, 1993). The journal *Geocarto International* has been the main outlet for SSEOP photography.

An important development in geologic remote sensing has been orbital imaging radar (Settle and Taranik, 1982). The first civilian radar satellite, *Seasat*, demonstrated the geologic value of orbital radar in its short (3-month) lifetime in 1978 (Ford, 1980). Further imaging radar experiments were carried out in 1981 by the *Shuttle* Imaging Radar-A (SIR-A), followed by SIR-B and SIR-C in later years. The success of these short experimental missions stimulated the development of applied orbital radar by the European Space Agency, the Canada Centre for Remote Sensing, and the Japanese space agency NASDA (Singhroy, 1992a,b). The 1995 launch of Canada's *Radarsat-1* marked the beginning of operational radar from orbit (Mahmood *et al.*, 1998), producing by 1998 the first complete coverage of the entire land area (including the polar caps) of the Earth, in addition to many ocean areas.

Specific geologic applications of orbital radar will be presented later.

To summarize this brief historical section, remote sensing of the Earth's surface has evolved, from localized sounding rocket photography and the 70 mm pictures taken by *Mercury* and *Gemini* astronauts, into a major field of space applications, with global coverage of the Earth being returned regularly by dozens of satellites using passive and active methods. Geologists were initially slow to realize the advantages of orbital imagery, and the leaders in applications of such imagery have been physical geographers (Estes and Senger, 1974). However, this situation has finally changed, and remote sensing from space has begun to have fundamental impact on geologic research and geologic applications. Thousands of papers and reports on geologic remote sensing have been published, most primarily on technique development. This immense mass of material can be summarized only by covering the main areas of geology in which the data from space have been used. A basic reference for all types of geology is the compilation by Short and Blair (1986).

4.3 Tectonics and structural geology

4.3.1 Global tectonic activity map

The first and most important direct geologic application of orbital remote sensing has been in the study of crustal structure, both regional (tectonics) and local (structural geology). Taking the broad view first, one result of orbital remote sensing is the global tectonic activity map (GTAM) presented in previous chapters, but repeated here for convenience (Fig. 4.2). This is the first map to show global tectonic and volcanic activity of the geologic present, "present" being defined as the past one million years. It has been reproduced

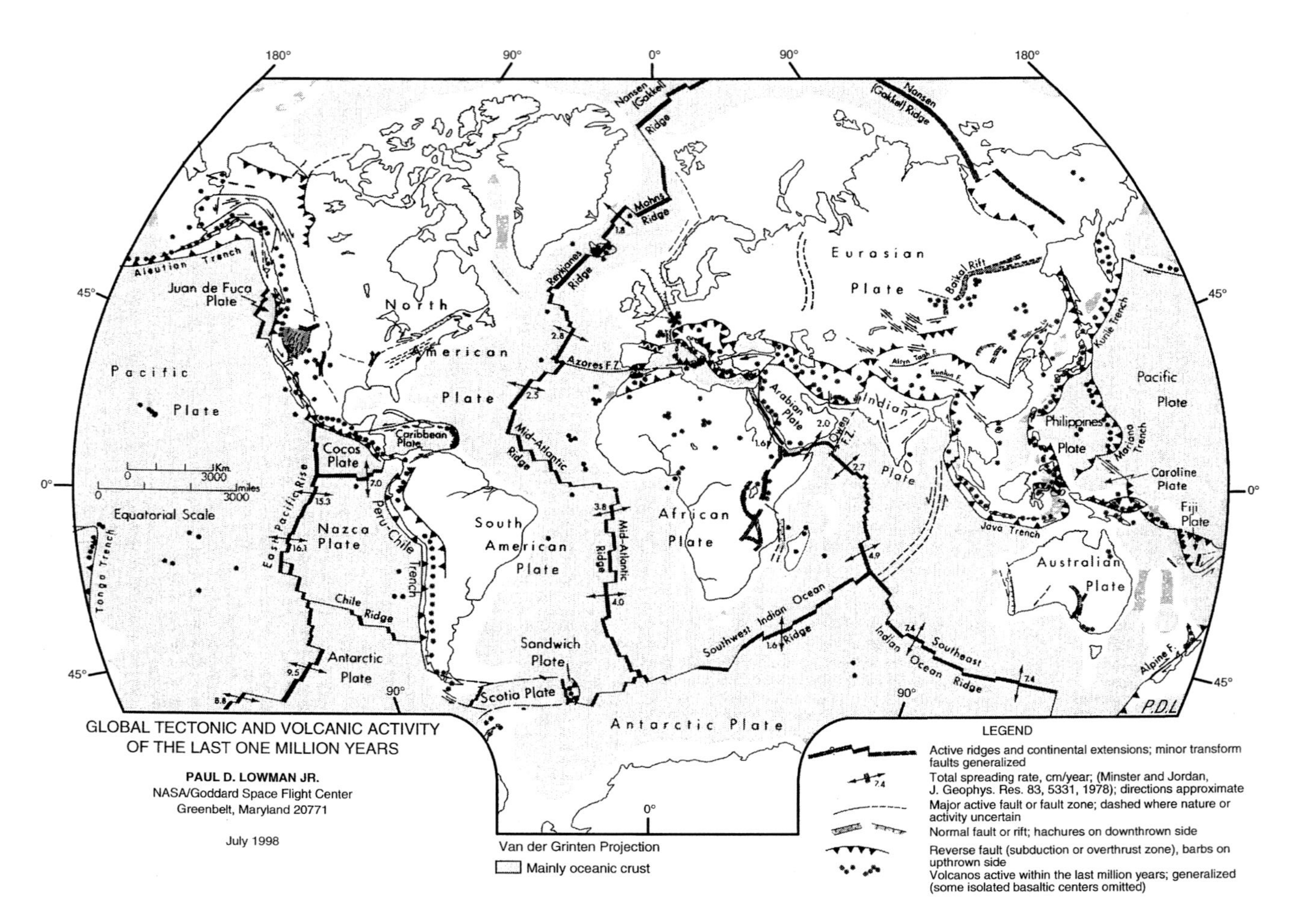

Fig. 4.2 Tectonic and volcanic activity of the last one million years.

in more than a dozen textbooks (e.g., Davis, 1984; Best, 1982; Peltier, 1989) and many scientific papers (e.g., Rubincam, 1982; McKelvey, 1986), and is currently available on the World Wide Web as discussed in Chapter 1. It was the hand-drawn precursor to the Digital Tectonic Activity Map presented in the first chapter, but has been updated in parallel with the compilation of the DTAM. The GTAM is based on remote sensing directly and indirectly: directly in that many tectonically active features were first mapped with *Landsat* imagery, in areas such as southern Asia, and indirectly in that a single compiler with a small travel budget could never have acquired enough knowledge of the planet's geology without the background provided by orbital photography, both film and digital. Data sources and compilation methods of this map have been described elsewhere (Lowman, 1981, 1982), and only the aspects relevant to remote sensing need be briefly summarized.

The primary value of remote sensing in compilation of the GTAM was the global coverage of large sparsely-settled areas such as North Africa, the Canadian Shield, and the Tibetan Plateau. For example, the Haruj al Aswad in Libya, a Quaternary volcanic field, is conspicuous from space (Fig. 4.3), but being historically inactive is not shown on maps of global volcanism. Once aware of its existence, the author was easily able to document the feature. Furthermore, its geomorphic freshness, apparent from orbital photographs, implied an age of the last activity of less than a million years. This example demonstrates the value of remote sensing for initial reconnaissance of an area, and subsequent focussing of attention on specific geologic features.

Like all global maps, the GTAM is a compilation from previously-published maps, many of which are themselves compilations. Remote sensing has been essential to the GTAM in that many of its source maps, especially in Asia, had been drawn from *Landsat*, *SPOT* (*S*ystème *P*our l'*O*bservation de la *T*erre), or other orbital imagery. The impact of remote sensing on tectonics can be illustrated first by studies of the structure of southern Asia with *Landsat* images.

4.3.2 Tectonics of southern Asia

It was the 70 mm photographs of Tibet by Gordon Cooper on the final *Mercury* mission in 1963 (O'Keefe *et al.*, 1963) that broke the ground, so to speak, for geology, but it was not until *Landsat* provided systematic vertical coverage of Asia that serious tectonic studies with orbital imagery were undertaken. Tibet and adjacent

Fig. 4.3 *Gemini 11* photograph over North Africa, looking northeast. Tibesti Mountains at upper right; Marzuk sand sea bottom center, over *Agena* transponder antenna; Haruj al Aswad is largest dark area at left edge.

areas were for centuries isolated both physically and politically, and the regional geology was accordingly little known until the *Landsat* era. Molnar and Tapponnier (1975) acquired early *Landsat* MSS (MultiSpectral Scanner) imagery, whose quality is illustrated by Fig. 4.4. Similar work was done by Ni and York (1978), who compiled a monumental *Landsat* mosaic of all China and adjacent areas. Chinese geologists themselves began using *Landsat* images as soon as they became available, as demonstrated by maps in the volume

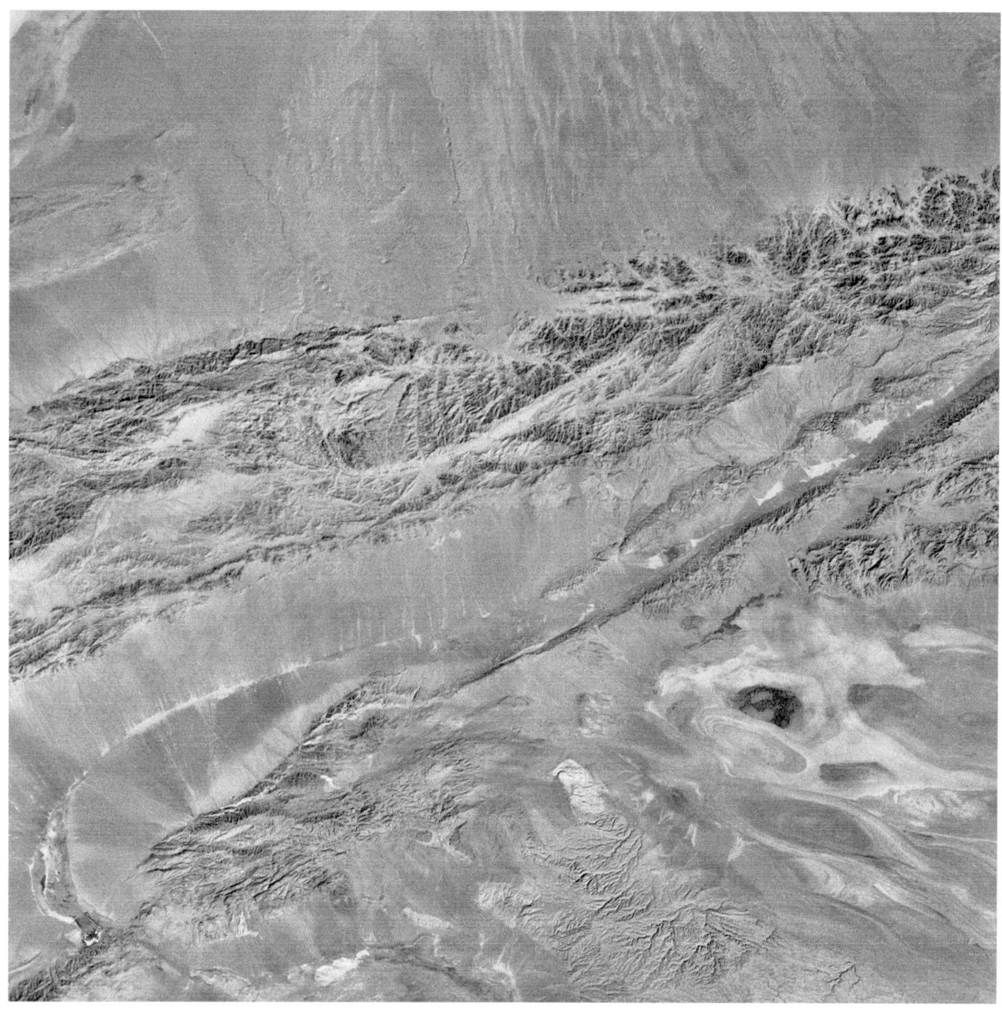

Fig. 4.4 *Landsat* picture of Altyn Tagh fault, western China, active left-lateral strike-slip fault with estimated 400 km offset. From Molnar and Tapponier (1975). *Landsat* scene 1449-04062, 15 October 1973. Width of view ~185 km.

Geotectonic Evolution of China (Huang, 1987). Dozens of papers have been based on the use of *Landsat* and more recently *SPOT*. An especially interesting use of *SPOT* images has been made by Zhang *et al.* (1995), who showed that the grabens of the Ordos Plateau (Fig. 4.5) and the faults just to the south could be explained by the extrusion models of Molnar and Tapponier.

The result of these efforts has been nothing less than a revolution in our knowledge of the tectonics and seismicity of southern Asia. Dozens of little-known active faults have been mapped and

Fig. 4.5 *Landsat* picture of grabens in the Ordos Plateau, Inner Mongolia, about 500 km west of Beijing. Width of view 185 km.

their sense of motion determined, both from satellite imagery and from correlated seismic data. The main tectonic implications of the various studies are the following.

The validity of plate tectonic theory for southern Asia has been both confirmed and in a sense contradicted. Taking the contradictions first, the *Landsat* images and maps derived therefrom show that, contrary to early publications (e.g., Dewey and Bird, 1970), there is no definite plate boundary in southern Asia, in particular between the Eurasian and adjacent plates such as the Indian Plate (see Fig. 4.2). A north–south region more than 2500 km wide, from the front of the Himalayas to Siberia, is actively deforming, as shown both by the distribution of seismic activity and the geomorphology of faults (see Figs. 4.4, 4.5). This area has been termed "intraplate" by York and co-workers in their *Landsat* study, and this term has been applied by many other authors to similar broad deformation zones elsewhere such as the Basin and Range Province (Fig. 4.2). However, if a "plate" is defined as a relatively rigid and inactive segment of lithosphere bounded by some combination of ridge, trench, or transform fault, the term is not at all valid for such areas. This contradiction gets to the very heart of modern tectonic theory: Can continental deformation be validly described in terms of plate interaction? It has been recognized for many years (e.g., McKenzie, 1968; Atwater, 1970; Burchfiel, 1980) that plate boundaries in continents tend to be diffuse ones. The tectonic studies of Asia with *Landsat* have shown clearly that simple application of plate tectonic concepts to such areas is unrealistic (Lowman *et al.*, 1999).

Taking a more positive view, the *Landsat* studies of Molnar and Tapponier, and the many subsequent ones, have shown that tectonic and seismic activity in southern Asia can be interpreted by a coherent model of rigid indentation (Molnar and Tapponier, 1975). The Indian subcontinent is treated as a rigid indenter, deforming a plastic medium, the resulting slip lines being expressed as faults. The resulting regional motion is one in which Tibet and western China are being extruded to the east. Similar interpretations have been applied to other continental areas. Whether these models are correct can not be discussed here at any length. For example, another *Landsat* study of western Tibet by Searle (1996) found that the Karakorum fault has apparently undergone only limited right-lateral offset, far less than the extrusion model requires. Searle thus argued that this fault was a relatively young feature, not related to the supposed early-Cenozoic of India with Asia. Allen *et al.* (1993) had previously used *Landsat* imagery of the Turfan Basin, far north but at roughly the same longitude as the Karakorum Fault, to infer thrusting, possibly along Paleozoic trends. The main point in the present context is that the use of remote sensing imagery of Asia has not only transformed our knowledge of this area, it has produced testable theories of the very nature of continental deformation.

The examples just discussed are from a large, remote, and until recently poorly-mapped region, and the advantages offered by satellite coverage were obvious immediately. However, it can be asked if remote sensing has much to show in areas already covered by modern, large-scale geologic maps. When *Landsat* imagery was first acquired, L. W. Morley commented, referring to the Canadian Shield, that this imagery would have been useful 20 years earlier, but that the Shield had by 1973 already been mapped at a reconnaissance scale. Our next remote sensing example will therefore be taken from a nominally well-mapped area, southern California.

4.3.3 Elsinore Fault

Because of its low latitude, the part of California from Los Angeles south was covered by astronaut photographs beginning with the first *Gemini* mission (GT-3) in 1965. These photographs eventually produced an excellent example of remote sensing photointerpretation of easily accessible and well-mapped regions (Singhroy and Lowman, 1997).

The example begins with an early oblique 70 mm view of the Salton Sea (Fig. 4.6) and adjacent areas taken by astronauts Conrad and Cooper in 1965. Although notable for the conspicuous gyre in the Salton Sea (probably suspended sediment, circulated by the usually strong southerly winds from San Gorgonio Pass), this photograph also shows a number of conspicuous lineaments in the Peninsular Range, accentuated because they parallel the camera axis. It was noticed by the writer (Lowman, 1980) that several of these lineaments intersected the Elsinore Fault without horizontal offset. Since the Elsinore Fault is active, and parallels the other members of the horizontally-moving San Andreas fault family in southern California, this was a critical anomaly. Further orbital photography (Figs. 4.7, 4.8), this time from the *Apollo 9* mission flown by astronauts McDivitt, Scott, and Schweikart, produced systematic multispectral coverage as part of the S065 experiment (Lowman, 1969), essentially a returned film simulation of *Landsat*. The Peninsular Range photograph was used to make a much more detailed map, and showed the most crucial areas for field checking and low-altitude reconnaissance flights (Figs. 4.9, 4.10).

This field checking showed that a bedrock ridge, the extension of a lineament visible on the original GT-5 photograph, was cut by the Elsinore Fault, but with no horizontal offset. Close-range examination of the fault plane on the ridge showed slickensides (Fig. 4.11) indicating pure dip slip, at least for the last movement. It was concluded that, contrary to all previous interpretations, the Elsinore Fault in this area is a dip-slip fault, not a strike-slip one (Lowman, 1980). Independent field mapping by others (Todd and Hoggatt, 1989), and an orbital radar study by Schultejann (1985) confirmed this interpretation.

The sequence of events in this example is worth summarizing to

Fig. 4.6 (See also Plate XIII) *Gemini 5* photograph over Salton Sea, looking northeast. Note linear valleys at lower left, crossing Elsinore Fault without offset. Gyre in Salton Sea is formed by winds through San Gorgonio Pass (left). Width of view ~150 km at center of photograph.

illustrate the geologic use of remote sensing even for well-mapped areas. The first photographs, from GT-5, revealed an apparent anomaly: bedrock features cut by an active supposed strike-slip fault but without horizontal offset. Subsequent photographs, from *Apollo 9*, revealed critical areas on which field work should be focussed, guided by standard air photos and low-altitude reconnaissance obliques. This field work confirmed the photogeologic interpretation, one directly contradicting all previous views. Furthermore, the field work occupied less than three weeks total outcrop time, and was done by an eastern geologist initially unfamiliar with the area. The

Fig. 4.7 *Apollo 9* view of Peninsular Ranges, California. The original (one of a four-frame series) was a false color infrared image. See Fig. 4.8 for interpretation of features.

point of this example is not the cleverness of the interpretation, which could have been done – given the photographs, and having the problem defined – by a third-year geology student with strong legs. Rather, the Elsinore Fault study is presented as a typical example of the methodology permitted by remote sensing, starting with the most crucial part of any scientific research: recognition of the problem (Singhroy and Lowman, 1997).

A much more advanced remote sensing investigation of southern California structure has been carried out by Ford *et al.* (1990), using *Landsat* Thematic Mapper (TM) imagery to map faults in the Mojave Desert. This example is notable in that Ford *et al.* used

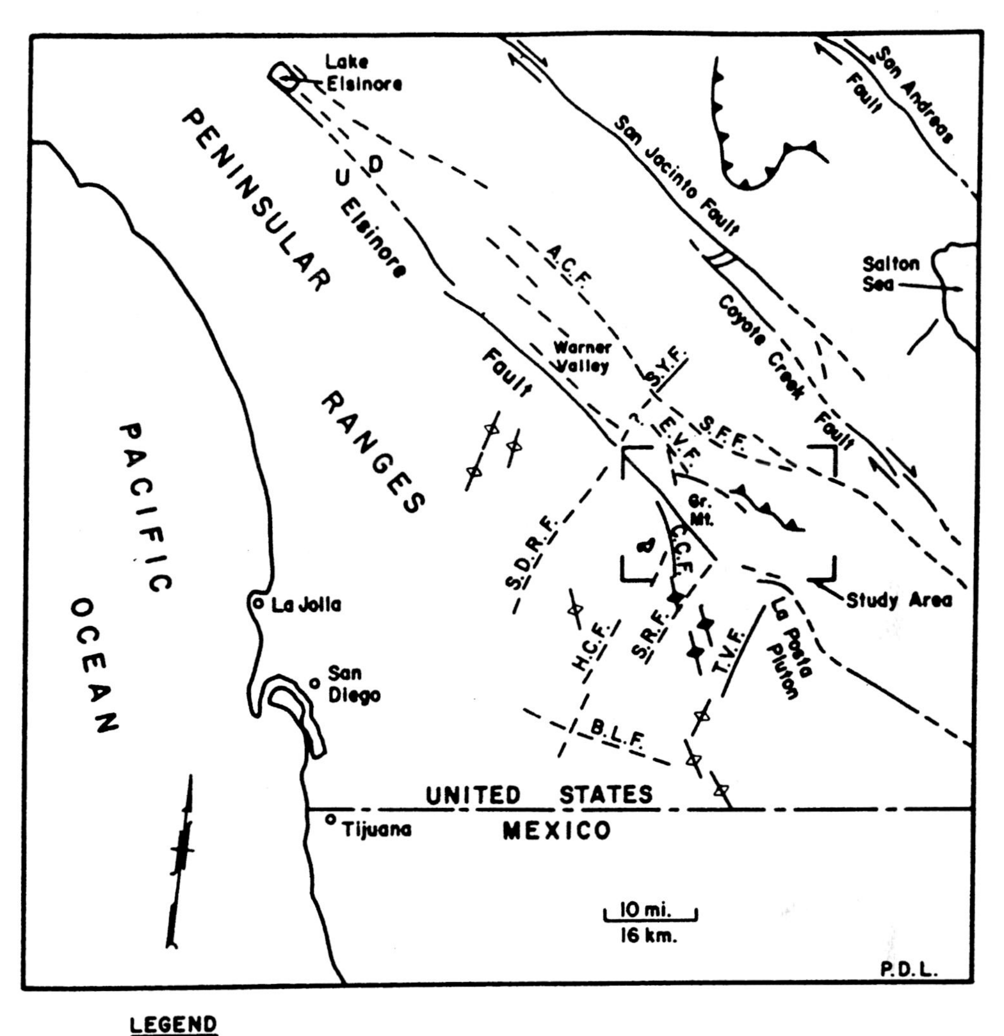

LEGEND

Major fault:

A.C.F.	Aqua Caliente
B.L.F.	Barrett Lake
C.C.F.	Chariot Canyon
E.V.F.	Earthquake Valley
H.C.F.	Horse-thief Canyon
S.F.F.	San Felipe
S.D.R.F.	San Diego River
S.R.F..	Sawtooth Range
T.V.F.	Thing Valley

Thrust fault; barbs on upper block

Generalized foliation in steeply dipping metamorphic or igneous rocks

Fig. 4.8 Sketch map of Fig. 4.7. From Lowman (1980).

Fig. 4.9 High-altitude aerial photograph of east edge of Peninsular Ranges; note angular ridge at center (Sawtooth Range). Width of view 10 km.

lithologic discrimination based on TM infrared bands, rather than simple photogeology based on landforms. The example also demonstrates again the value of remote sensing in an accessible and supposedly well-mapped area, in this case just 2 hours drive from Los Angeles.

Radar interferometry, to be discussed in Subsection 4.5.1 on volcanism, has produced an entirely new type of tectonic activity map of the Mojave Desert (Massonet *et al.*, 1993a). This technique permits continuous mapping of crustal displacement as small as a few centimeters over hundreds of square kilometers.

Fig. 4.10 Low-altitude oblique photograph from 2000 feet above ground, looking to northeast, over Sawtooth Range. From Lowman (1980). Elsinore Fault crosses ridge where highway makes a U-curve; exposed in road cut at Campbell Grade. Photograph taken 1970, by author. Width of view ~5 km at center of photograph.

4.3.4 Lineament tectonics

It has been known for many years that the continental crust in stable areas, i.e., cratons, is pervaded by a network of lineaments: straight or nearly straight topographic features of regional extent that express some sort of bedrock structure, commonly fractures (Fig. 4.12). It was proposed by Hobbs (1911) that these features form a relatively simple pattern of roughly orthogonal fractures, global or nearly so in extent (Hodgson, 1976). Aerial photographs were used to map these features for many years, but the sudden availability of global coverage from *Landsat*, beginning in 1972, triggered a surge of interest in what became known as "basement tectonics"

Fig. 4.11 Outcrop view of fault plane of Elsinore Fault, showing slickensides indicating dip slip for last movement. Pen gives scale.

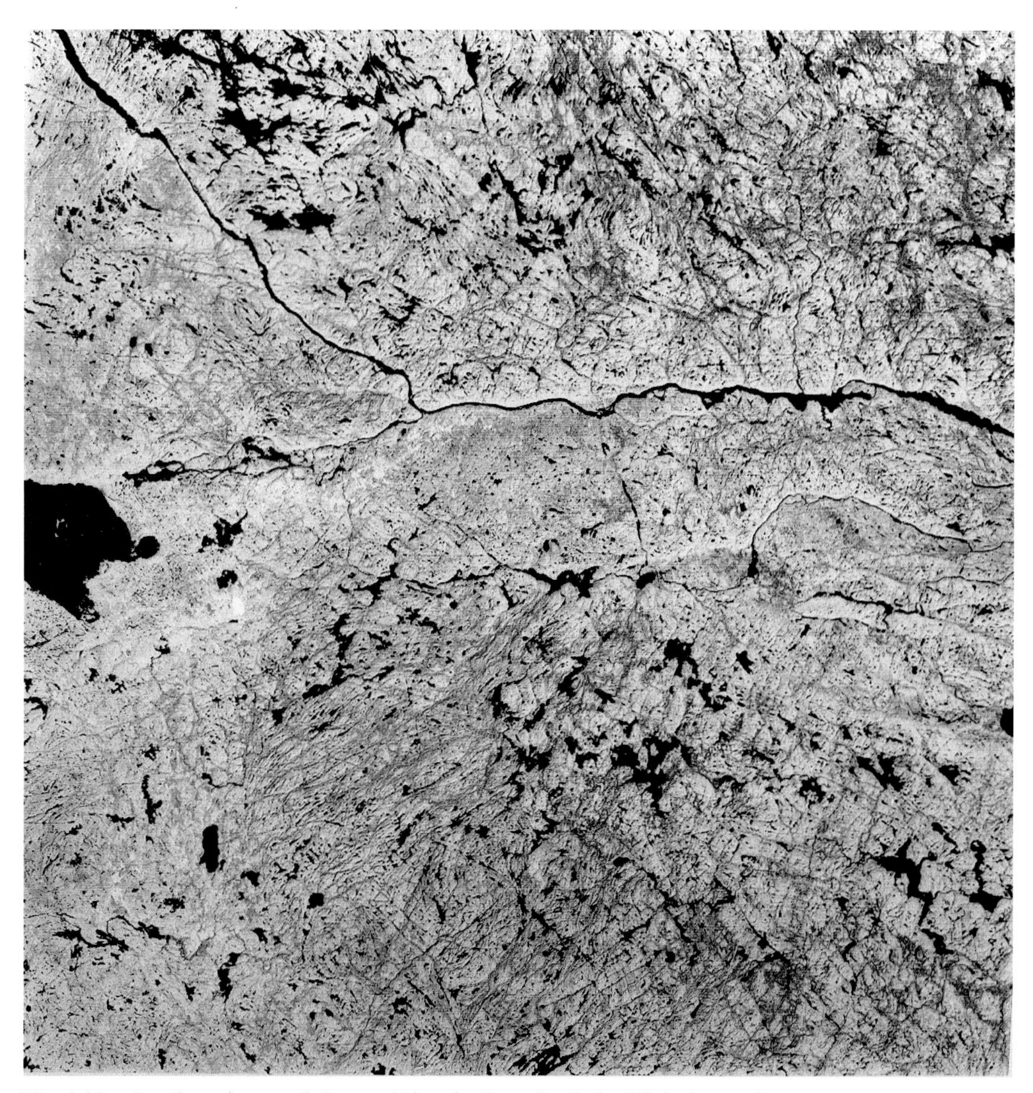

Fig. 4.12 *Landsat* picture of Ottawa River in Ontario; Lake Nipissing and North Bay at extreme left. Scene acquired October 1973; accentuates topography and fracture systems. Valley considered part of Ottawa–Bonnechere graben system, suggested by Kumarapelli and Saull (1966) to be a branch of the world rift system through the St. Lawrence River. Width of view 185 km.

(Nickelsen, 1975; Hodgson *et al.*, 1976). Several symposia with that title were held, and scores of papers on lineaments were published in other places. However, the topic remains controversial, with little agreement on the nature, origin, and to some extent even the existence of lineaments. As used here, "lineament" will refer to well-defined, relatively narrow linear topographic features, although the term is sometimes applied to broad zones of faults, dikes, igneous intrusions, and volcanic features. The term will also exclude lines expressing the strike of sedimentary strata, igneous flow structure, or volcanic layering. This exclusion, following conventional photogeologic techniques, is necessary to prevent lineament maps from becoming simply pen-and-ink sketches of the aerial or space photographs.

The reality of lineaments, as the term is used here, is actually unarguable for many areas such as the Canadian Shield (Fig. 4.12). Furthermore, they are generally agreed to express brittle fractures of some sort, either faults or joints, or dikes intruding such fractures. Lineaments are of far more than scientific interest, being known in many areas to localize mineral deposits (Kutina and Hildenbrand, 1987), oil or gas reservoirs, or ground water. In addition, as basement fractures, they may represent local geologic hazards, such as mine roof falls or slope failure in road cuts, and may localize seismic activity (Mollard, 1988). Kusaka *et al.* (1997) have applied orbital radar and optical imagery to lineament mapping in Japan, finding the orbital data a valuable aid in estimating landslide risk. Remote sensing has thus found immediate and wide application in lineament tectonics. Geophysical studies, aeromagnetic and gravity field in particular, have also been applied to lineament problems (Quershy and Hinze, 1989).

The primary questions about lineaments are, first, whether they do in fact form a global unified network of more or less orthogonal fractures, and, second, whether they are tensile or compressional (i.e., vertical shear) fractures. Given the purpose of this chapter – to review the impact of space remote sensing on geology – treatment will be restricted to a few examples of how remotely-sensed data are being applied to lineament tectonics, with some of the conclusions reached.

The Canadian Shield is an ideal area in which to study lineaments by remote sensing. It has been tectonically quiet for at least one billion years, providing ample time for differential erosion to etch out lineaments. By definition, the basement, i.e., the Precambrian crystalline rock, is generally well exposed. Finally, the Shield lies in two countries with close scientific ties and strong remote sensing capabilities. For these reasons, a major study of lineaments on the Shield was undertaken by the writer and his colleagues (Lowman *et al.*, 1992), with the general objective of testing the theory that lineaments form a unified global or at least continental network, or "regmatic shear pattern" (Sonder, 1947). Standard

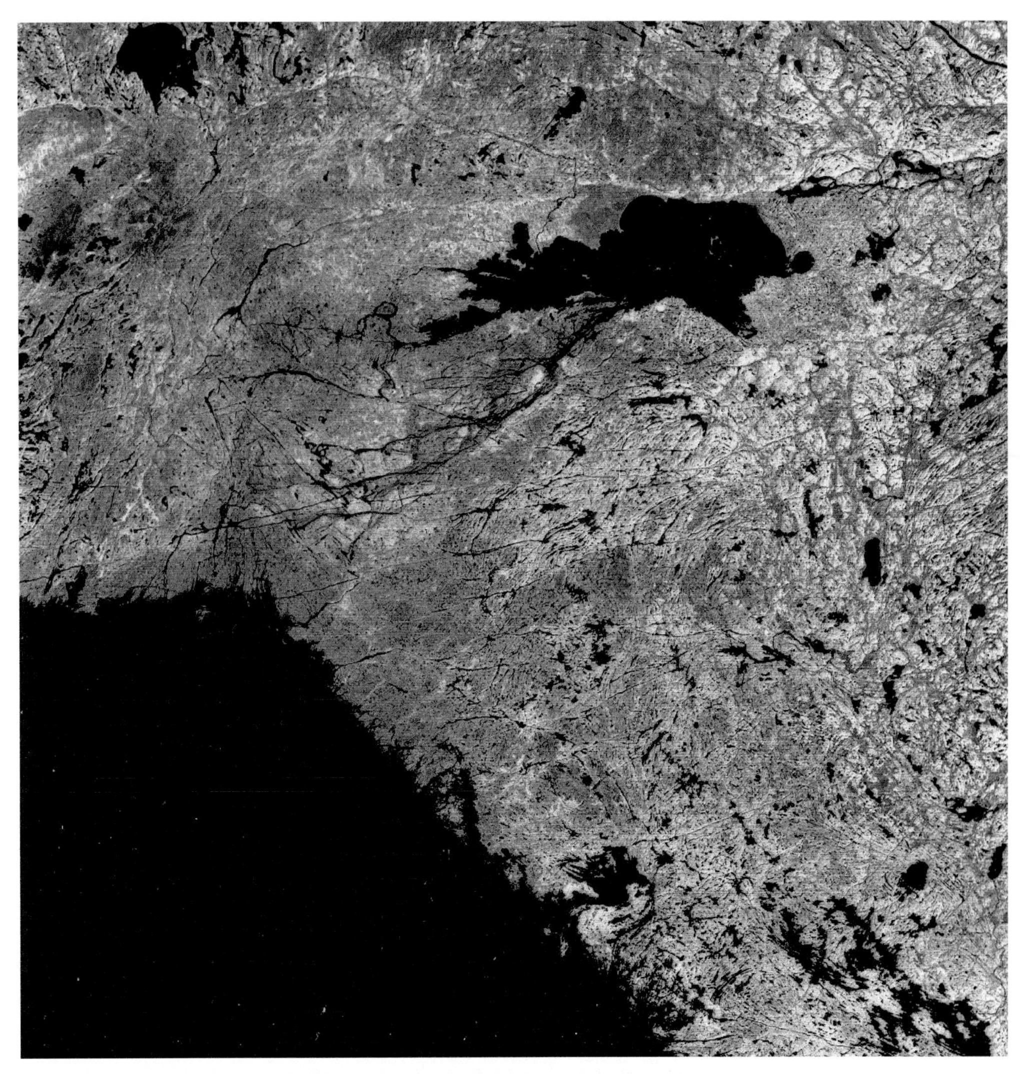

Fig. 4.13 *Landsat* picture of Georgian Bay, Ontario, and adjacent Grenville Province. North Bay at top center. From Lowman *et al.* (1992). Width of view 185 km.

photogeologic methods were applied to 60 low-Sun-angle *Landsat* scenes covering parts of all structural provinces of the Shield (e.g. Fig. 4.13). The resulting lineament maps (Fig. 4.14 (a–c)) were digitized and rose diagrams (azimuth-frequency plots) drawn by computer. Some field checking was done, although the enormous area covered by 60 *Landsat* scenes obviously made anything like thorough field work impossible. Orbital radar imagery was available for

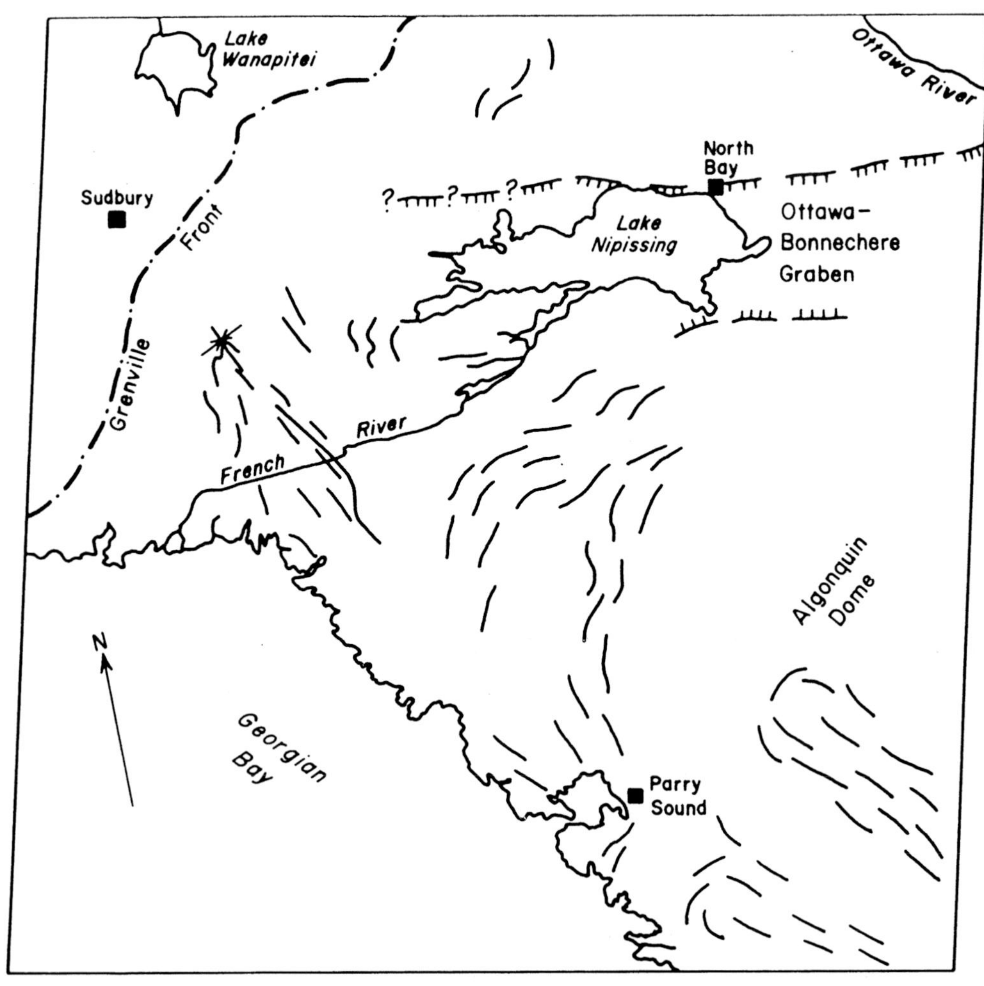

GEOLOGIC SKETCH MAP

Landsat 2 MSS Scene, 2620–15192–7
Scene Center N45°54′, W80°10′; Width 185 km.

LEGEND

Strike of foliation, lithologic contacts, or flow structure

Normal fault; teeth on down-thrown side

Trace of Grenville Front

P.D. Lowman

Fig. 4.14(a) Sketch map of Fig. 4.13.

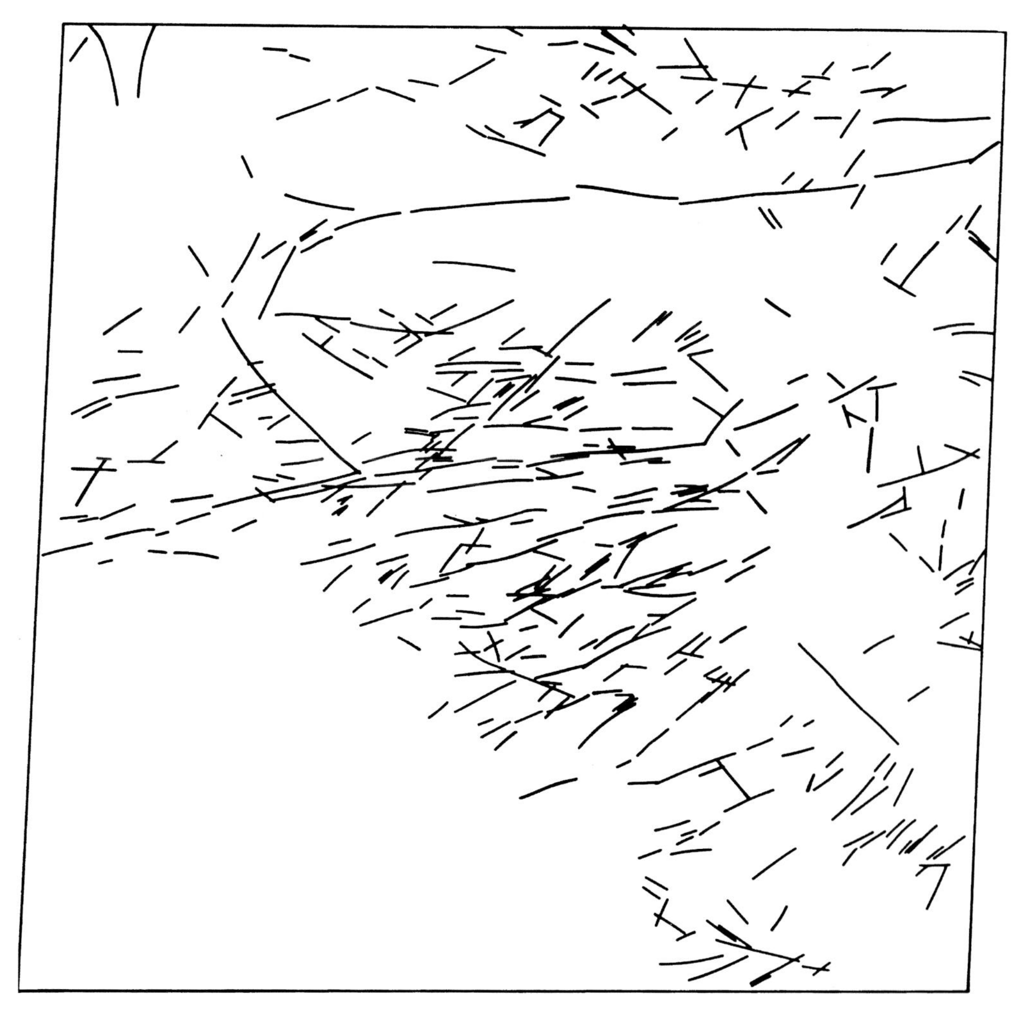

Fig. 4.14(b) Lineaments drawn from Fig. 4.13.

some areas (e.g. Fig. 4.15), providing an interesting comparison with *Landsat*. Radar has proven unusually valuable for structure mapping and lineament mapping in general (Singhroy *et al.*, 1992; Lowman, 1994; Kusky *et al.*, 1993), one reason being its sensitivity to look direction. Structures nearly perpendicular to the illumination are strongly highlighted, as Fig. 4.15 shows, bringing out subtle features not obvious on standard visual images.

Several conclusions were reached, some contradicting majority opinion on the nature of lineaments. The most general conclusion

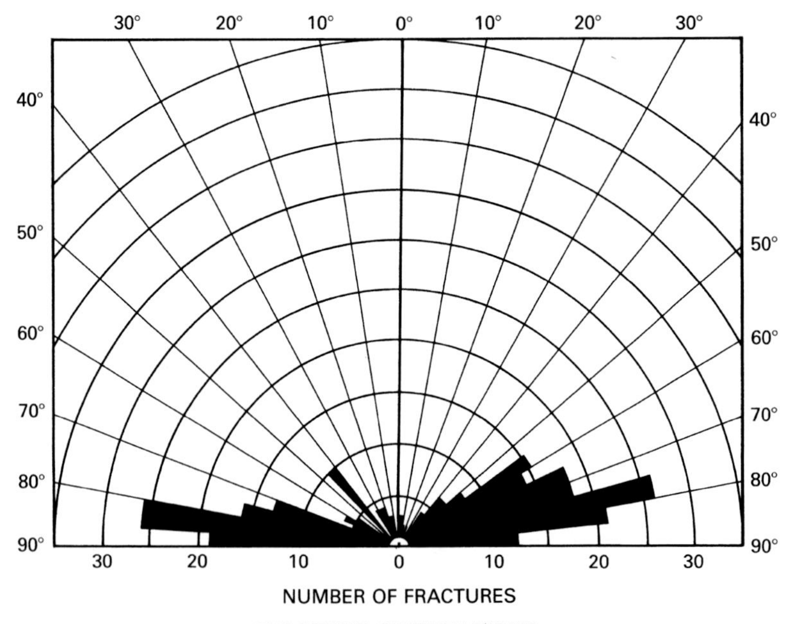

Fig. 4.14(c) Rose diagram of lineaments in Fig. 4.14(b).

was that if there is such a thing as a unified network of orthogonal fractures, it is not visible on *Landsat* images. The reality of the mapped lineaments is unquestionable, thanks to field checking and available geologic maps, and many of them turned out to be basaltic dikes of known dike swarms, or empty fractures parallel to such dikes (Fig. 4.16). The unfilled lineaments checked in the field proved to be extensional, either normal faults or joints, not shear fractures. A similar conclusion had been reached by Nur (1982) on the basis of *Landsat* images of Israel, where the desert climate has etched out lineaments.

An important conclusion reached by many lineament investigators is that lineaments in relatively young rock may be re-activated fractures of much greater age. The term "recurrent tectonics" has been applied by Onasch and Kahle (1991) to the Bowling Green fault in Ohio, which appears to represent movement along the one billion year old Grenville Front, whose southwest extension continues under the Paleozoic sediments of the mid-continent region. The Ottawa–Bonnechere graben, shown in Fig. 4.12, localizes low-level seismic activity, even though it is Paleozoic or older in age. The extent of such fracture re-activation is an important problem, since

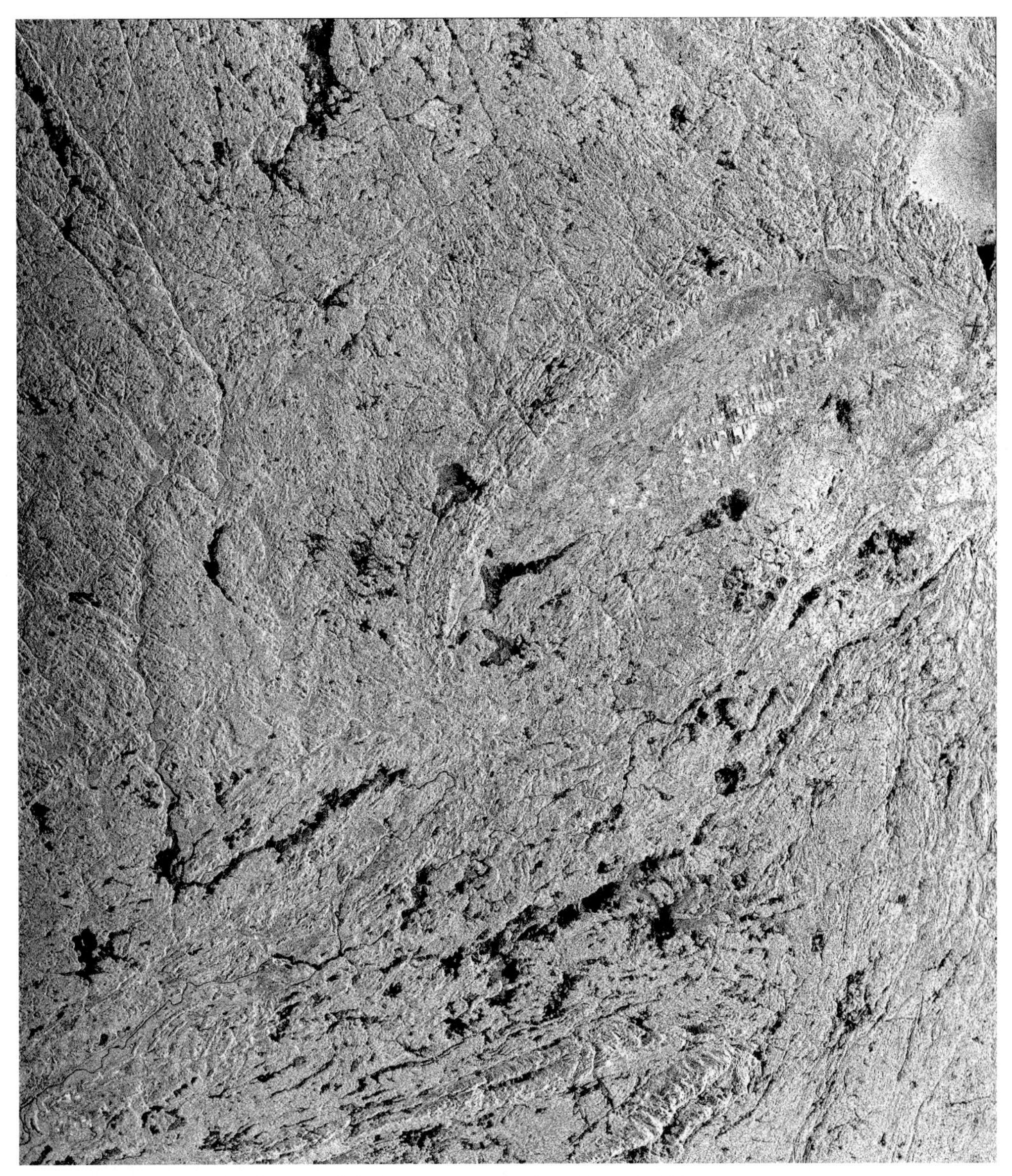

Fig. 4.15 *ERS-1* radar image of Sudbury area, northen Ontario. Linear contrast stretch applied at Goddard Space Flight Center. Image acquired July 1992. Illumination from right. Width of view ~100 km.

"neotectonic," i.e., newly-formed joints should parallel the principal horizontal stress direction (Engelder, 1982; Hancock, 1991) thus providing a good indication of regional stress fields.

One very interesting aspect of the Canadian Shield *Landsat* study was that the mapped lineaments were found to be fractal, i.e., scale-

CUMULATIVE FRACTURE
ORIENTATIONS
CANADIAN SHIELD

province boundary

shield boundary

Base: Tectonic Map of North America, P.B. King, 1969
Bipolar oblique conic conformal projection

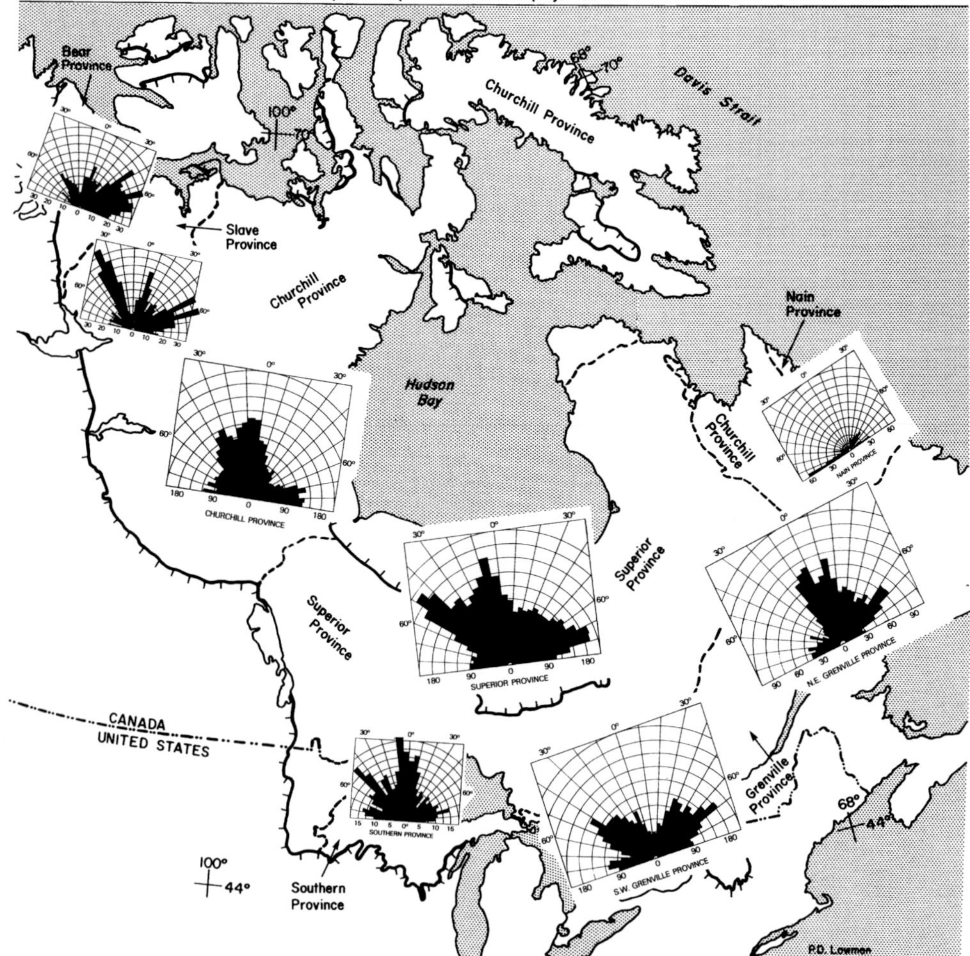

Fig. 4.16 Cumulative rose diagrams for provinces of Canadian Shield, drawn from 60 low-Sun-angle *Landsat* scenes. From Lowman *et al.* (1992 1994).

invariant. Those mapped along the Ottawa–Bonnechere graben in Ontario, for example, were found to be similar in orientation and origin to joints visible in outcrop (Figs. 4.17, 4.18). The Ottawa–Bonnechere graben has been shown by Kumarapeli and Saull (1966) to be possibly related to the world rift system through

Fig. 4.17 View to west from Highway 69 bridge over French River, Ontario; see Fig. 4.14(a) for location. Valley occupies site of a now-eroded diabase dike, part of the Grenville Swarm (Fahrig and West, 1986), controlled by Ottawa–Bonnechere graben.

the demonstrably tectonic St. Lawrence valley. If this interpretation is correct, it shows how tectonic features of global scale can be traced down to the outcrop – a striking example of the fractal concept.

The fractal concept has been familiar to structural geologists for a century, though not under that name, in that small structures such as joints often mirror regional joint networks. Small folds similarly parallel much larger ones. The fractal nature of geologic structures in general is important for the use of remote sensing data, by showing that structures visible from space may be repeated at increasingly smaller sizes. The fractal dimension of *Landsat*-mapped lineaments on the northern Great Plains was applied to tectonic interpretation by Shurr *et al.* (1994).

This summary has only touched on the now-enormous literature on lineament tectonics. The subject will be revisited below in relation to other geologic fields, such as mineral deposits and petroleum exploration. At this point, it can only be noted that despite the flood of remote sensing imagery, and the large number of studies using it, lineaments are in general still a controversial and poorly-understood feature of the Earth's crust.

Fig. 4.18 Joints in outcrop on south side of French River (left in Fig. 4.17). Joints parallel French River and other lineaments; main joint by Brunton compass strikes due west. Compass is 8 cm square.

4.4 Exploration geology

Exploration for oil, gas, and minerals has traditionally been the chief field of applied geology, and it is understandable that remote sensing has had perhaps its greatest geologic impact in these areas (Rowan, 1975; Goetz *et al.*, 1983). Potential oil traps in the then poorly-mapped Tibetan Plateau were found on the very earliest orbital photography, taken by Gordon Cooper on the MA-9 mission (O'Keefe *et al.*, 1963), one of the factors that, as mentioned previously, stimulated interest in remote sensing for earth resources. In the decades since, exploration geology has applied a wide range of remote sensing techniques, and orbital data have long since become an operational tool, not simply an experimental one. The following subsections and examples are intended to summarize this now-enormous field, and to guide interested readers to further information sources.

4.4.1 Petroleum exploration

The search for oil and gas, lumped hereafter for convenience as "petroleum," has for many years been one of the most fertile fields for application of new discoveries in geophysics and geology, and this is now true for remote sensing. As pointed out by Vincent (1997), the petroleum industry was initially sceptical about remote sensing because the resources it seeks are generally far below the surface. The easily-found on-shore reservoirs, i.e., those visible from above, have already been found, although Halbouty (1976) showed that many giant oil fields could have been located with *Landsat* had it been available earlier. However, remote sensing has turned out to be so valuable for petroleum exploration, directly and indirectly, that it is now a standard tool for discovery and development of new fields and a shining example of the ultimate value of space technology.

A brief review of the principles of petroleum exploration will be helpful. Most exploration techniques are not intended to find oil or gas directly; they are instead focussed on locating traps where hydrocarbons have accumulated. These traps may be structural, such as anticlines, or stratigraphic (concordant hydrocarbon-bearing strata sealed from the surface), or they may occur around salt domes or igneous bodies. However, the first oil fields were located by seeps, and even today seeps are useful targets, especially for offshore exploration where they may be visible from space, especially with radar (Estes *et al.*, 1985) as slicks. An increasingly useful indicator for petroleum has been developed primarily from *Landsat* (Donovan *et al.*, 1974) and other remote sensing methods: subtle discoloration of bedrock, soil, or vegetation caused by chemical reactions with escaping hydrocarbon gases.

Petroleum exploration has become an extremely complex process, generally involving a wide range of geophysical, field geology, and now remote sensing techniques, and it is only rarely that a successful exploration well is drilled on the basis of a survey with one technique. The most effective use of remote sensing data, such as *Landsat* imagery, is generally as a means of focussing surface and subsurface surveys. Seismic prospecting, for example, can be far more efficient if survey lines can be laid out on the basis of structure mapping with remote sensing techniques. At least one company has made *Landsat* reconnaissance a requirement before seismic surveys are begun in a new area (Sabins, 1998).

The obvious question to be answered sooner or later is: Has any oil or gas actually been discovered by remote sensing from space? A simple answer is not possible. One might as well ask: Did radar win World War II? The answer to both questions is essentially yes, BUT only in combination with many other technologies and as part of broad strategic approaches. Remote sensing has played a vital role in several petroleum discoveries, described by Sabins (1998), Vincent (1997), and many other authors. The use of remote sensing can best be summarized with two examples, from the work of Sabins (1998) of data from areas now in production.

The first of these examples, from Sabins (1997) is focussed on northwestern Colorado (Figs. 4.19, 4.20) shown on a *Landsat* MSS image that brings out the structure unusually well. The area shown includes several producing oil and gas fields, all localized by anticlines so well-expressed they are locally termed "sheep-herder structures." These fields long pre-date *Landsat*, but they illustrate the potential value of *Landsat* and similar imagery for regional reconnaissance of less-known areas.

The second example, from Sabins (1998), is in Saudi Arabia. It is an extremely useful one in that *Landsat* TM images were used to detect a structural trap more than 2 kilometers down. The area is on the Central Arabian Arch, a regional eastward-dipping homocline flanking the Precambrian Arabian Highlands, as shown first on a *Landsat* mosaic (Fig. 4.21) and then on a single *Landsat* frame (Fig. 4.22). As related by Sabins, the Saudi government directed Aramco to carry out petroleum exploration outside the already producing areas flanking the Persian Gulf. Using *Landsat* Thematic Mapper images (Figs. 4.23, 4.24), Sabins and his colleagues identified a geomorphic feature, the Raghib anomaly, a topographic depression

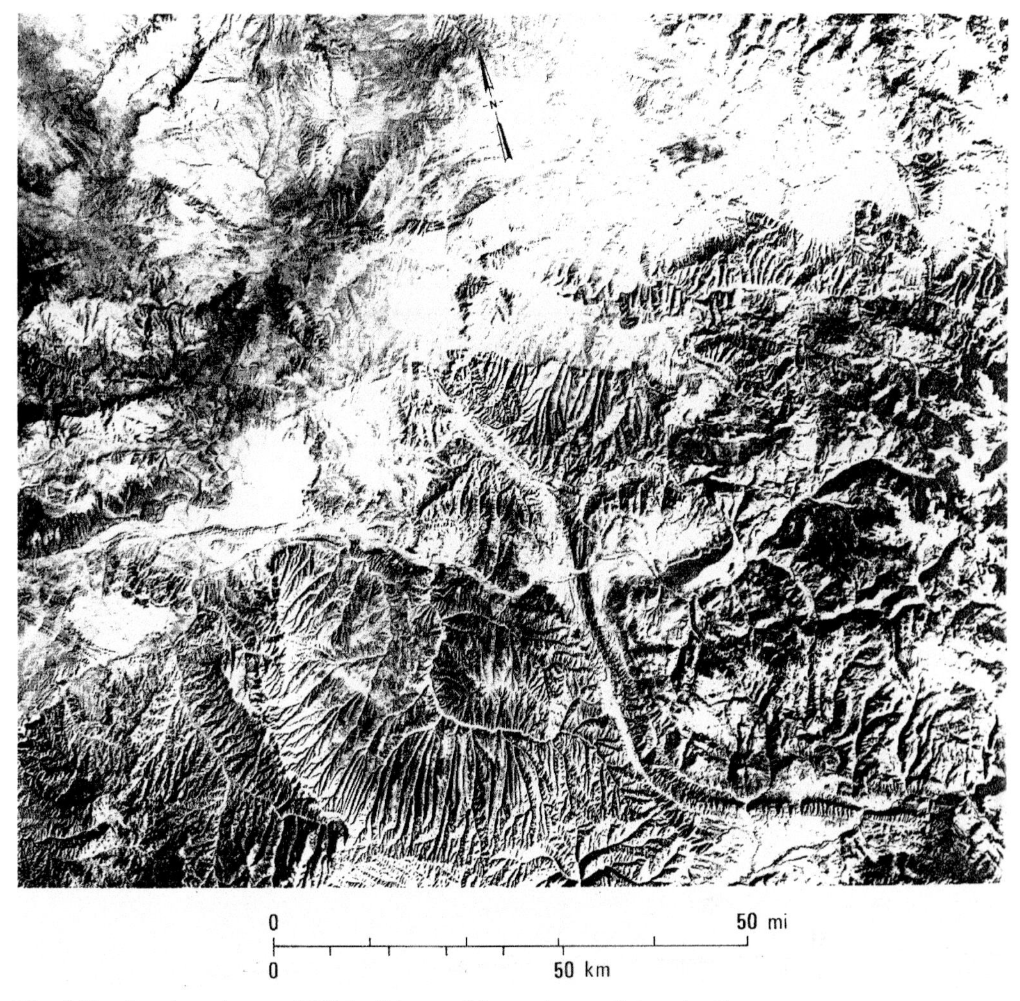

Fig. 4.19 *Landsat* picture of White River uplift, northwest Colorado. From Sabins (1997).

expressing a flattening of the regional dip caused, in turn, by an anticline in the Paleozoic strata about 2.5 km down. This anticline was formed by movement on steeply-dipping faults in the Precambrian basement, an incidental example of the importance of basement tectonics even though there are no visible lineaments in this area. Guided by the *Landsat* interpretations, seismic surveys and field checking were carried out. The Raghib anomaly was drilled in 1989, with commercial amounts of oil and gas found.

The Raghib case history should be a classic example of the use of remote sensing for petroleum exploration "to generate exploration targets" in the apt phrase of Agar and Villanueva (1997). In

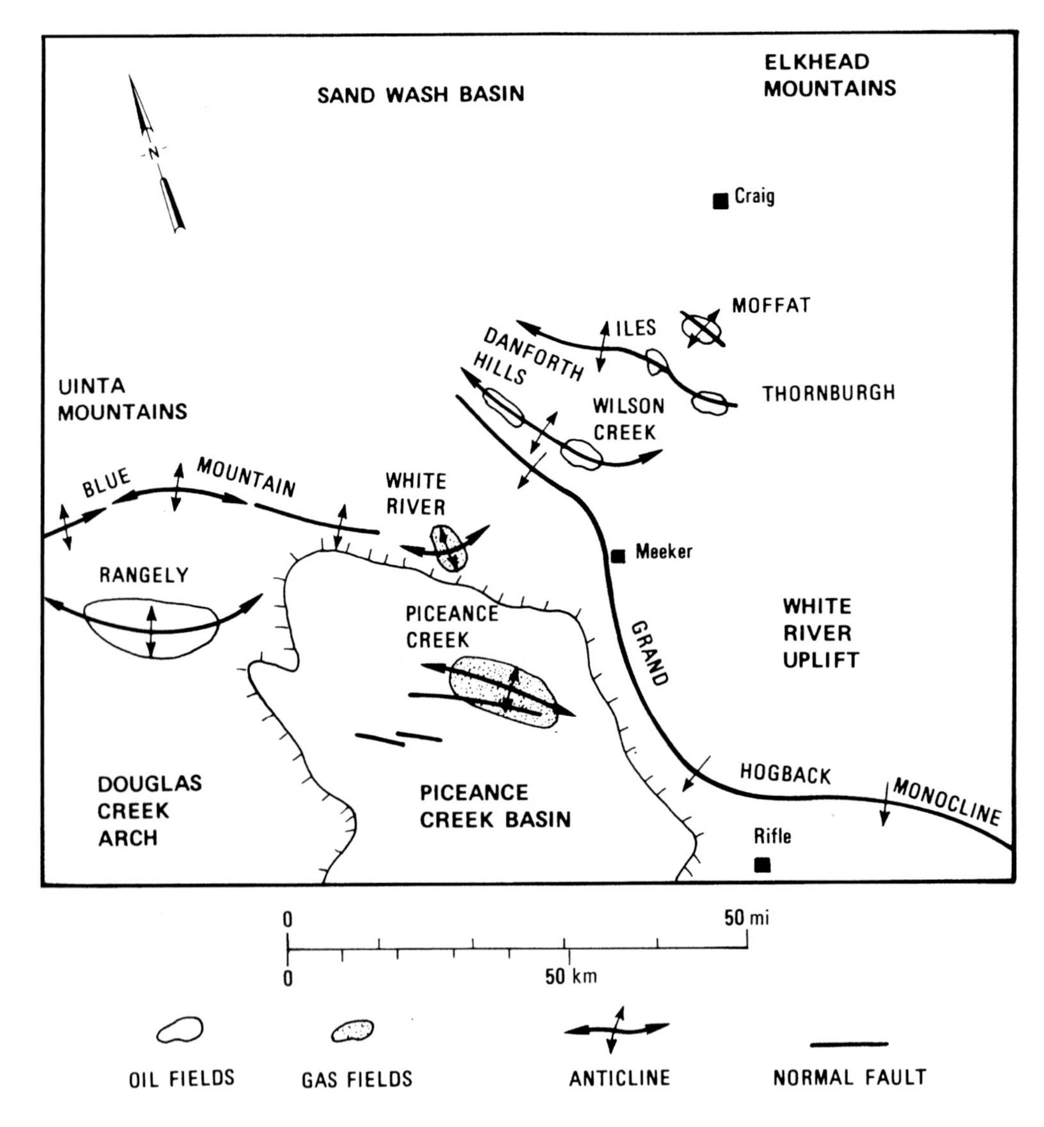

Fig. 4.20 Map of Fig. 4.19. From Sabins (1997).

some ways it parallels the Elsinore Fault study previously discussed, in that orbital imagery permitted recognition of a structural anomaly, and focussed subsequent field work on crucial areas. Furthermore, like the Elsinore Fault, the Raghib anomaly was found in an area that was already covered by excellent geologic maps. Finally, this example demonstrates the need for supplementing remote sensing data with conventional surface and subsurface methods, methods that can be used far more efficiently when they are combined with orbital imagery.

Fig. 4.21 (See also Plate XIV) *Landsat* MSS mosaic of Red Sea area.

Before leaving the subject of petroleum exploration, another application of remote sensing should be mentioned, one concerned with the final stage of petroleum production: minimizing the environmental impact *after* oil fields have been established. Groth and Rivera (1997) have presented an example of this from Ecuador,

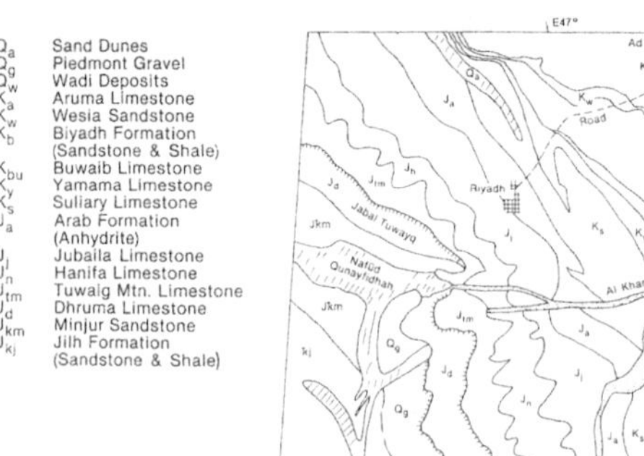

Fig. 4.22 (See also Plate XV) *Landsat* picture of Riyadh area, Saudi Arabia. From Short and Blair (1986). Area of scenes shown in Figs. 4.23 and 4.24 is at extreme lower right corner. Black circles (see higher-magnification view in Fig. 4.23) are irrigator patterns.

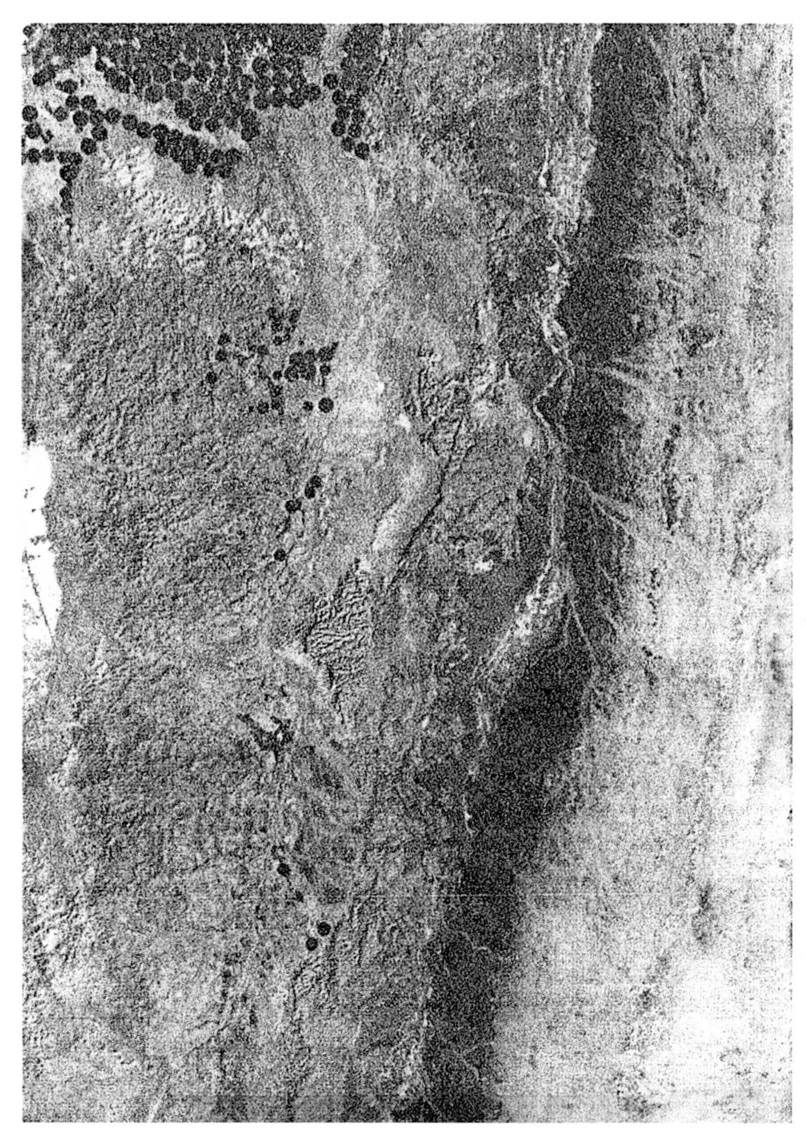

Fig. 4.23 Partial *Landsat* TM picture of Raghib anomaly area. From Sabins (1998).

where oil production from the Amazon basin rain forest, has begun. It was found that petroleum exploration inadvertently promoted settlement, with subsequent deforestation, of the region's rain forest, in that seismic profiling provided access to previously uninhabited areas. Using old and new *Landsat* imagery, Groth and Rivera were able to assess the patterns of such settlement. The Ecuadorian government subsequently imposed controls, such as check-points on former exploration trails, as a means of controlling destruction of

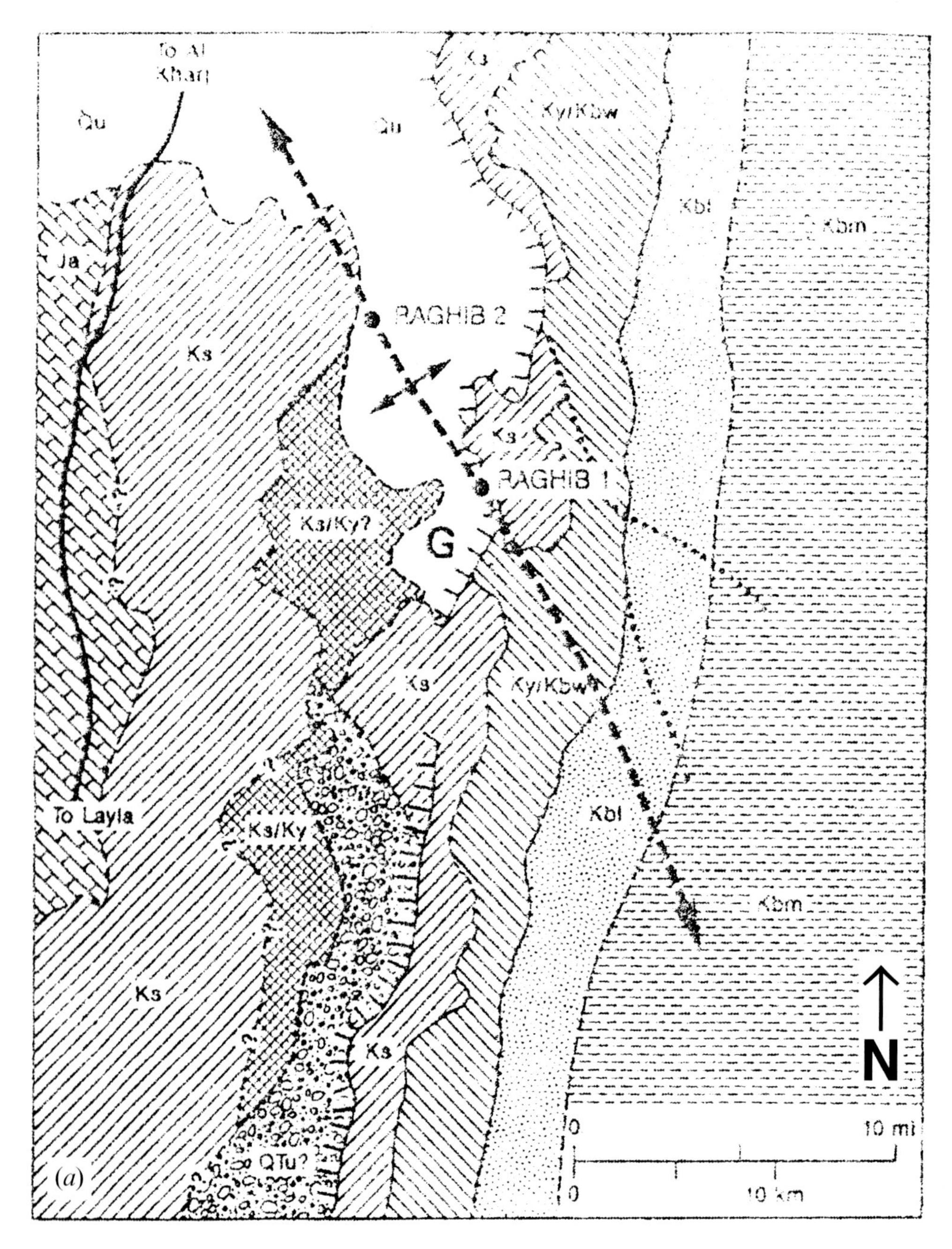

Fig. 4.24 (a) Sketch map of Fig. 4.23. (b) Structure map of Raghib anomaly. Both from Sabins (1998).

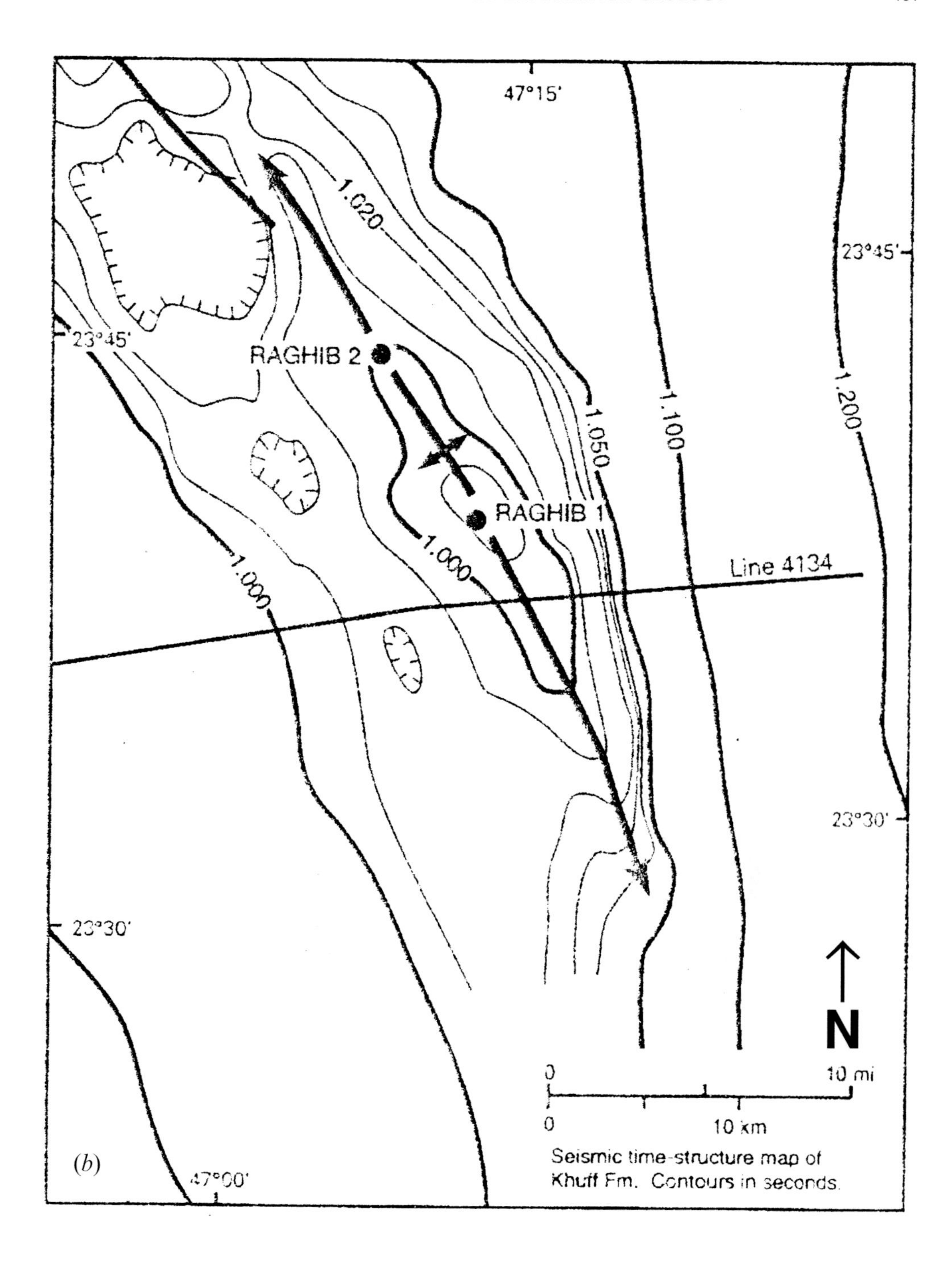

Seismic time-structure map of Khuff Fm. Contours in seconds.

the rain forest. Remote sensing can thus contribute not only to discovery and extraction of earth resources, but to protection of the environment after these resources have been extracted.

4.4.2. Mineral exploration

The term "mineral" as used in economic geology is a broad one, including not only ore deposits of metals such as iron, copper, and uranium, but minerals themselves, such as diamonds, and natural fuels such as coal (an organic rock, not a true mineral). In addition, a wide range of common materials such as sand, gravel, and limestone are generally categorized as "industrial minerals." Society has been dependent on minerals since the Bronze Age, and mineral exploration is thus a field dating back millennia. However, remote sensing has become an increasingly important tool for mineral exploration. Some helpful general references are those by Goetz *et al.* (1983) (and papers in the special issue introduced by this paper), Vincent (1997), and Sabins (1998). The *Proceedings of the Twelfth International Conference on Applied Geologic Remote Sensing* included a wide range of papers on mineral exploration.

Two general approaches have emerged for the use of remote sensing in mineral exploration, based primarily on *structural* or *compositional* methods although obviously the two aspects of geology are inseparable. These categories are a useful framework for discussion of the subject in general.

The "structural" approach is well illustrated by the study of mineral deposits in Nevada by Rowan and Wetlaufer (1979), using a *Landsat* mosaic. The example can be introduced by an aerial view of western Nevada (Fig. 4.25), showing the high relief and scarce vegetation of this huge desert, factors making Nevada an ideal test site for remote sensing research. The Basin and Range Province, as shown in Fig. 4.25, is a unique and still poorly-understood tectonic feature, best illustrated by a *Landsat* mosaic (Fig. 4.26). The area has since the 19th century been a rich source of gold, silver, copper, and other metals, to say nothing of non-metallic resources such as evaporites and even pumice (the "feather-rock" popular with landscapers).

Rowan and Wetlaufer carried out an extensive lineament mapping program on the *Landsat* imagery. Most of the lineaments proved to be previously-mapped faults, frequently the bedrock/alluvium contacts prominent in Fig. 4.25. However, the broad view provided by the *Landsat* mosaic also showed a group of much larger regional lineaments. One of these, the Walker Lane, goes through

Fig. 4.25 (See also Plate XVI) Aerial view from 37,000 feet, looking northeast in western Nevada near Lake Tahoe, showing typical Basin and Range topography. Photo taken July 1998, by author.

the area shown on the aerial photograph, but the restricted coverage even from 37,000 feet makes it impossible to see the unified nature of this lineament. Rowan and Wetlaufer showed (Fig. 4.27) that most metal production from Nevada had come from deposits localized by the lineaments visible on the *Landsat* mosaic, still another example of the importance of lineament tectonics.

The "compositional" approach to mineral exploration depends on a wide variety of remote sensing techniques to determine the lithology, mineralogy, or chemical make-up of the area under investigation. It will be obvious that composition mapping, especially from space, encounters major difficulties not faced in the "structural" approach (which depends largely on gross topography). Much of the Earth's land area is covered by vegetation, and even in apparently barren areas, such as the Canadian Arctic, outcrops will be found at close range to be coated with lichens. In extremely dry deserts, such as the Arabian Peninsula, the well-exposed bedrock will often have a weathered crust, and obviously large parts of most deserts are covered with wind-blown sand or dust, alluvium, or

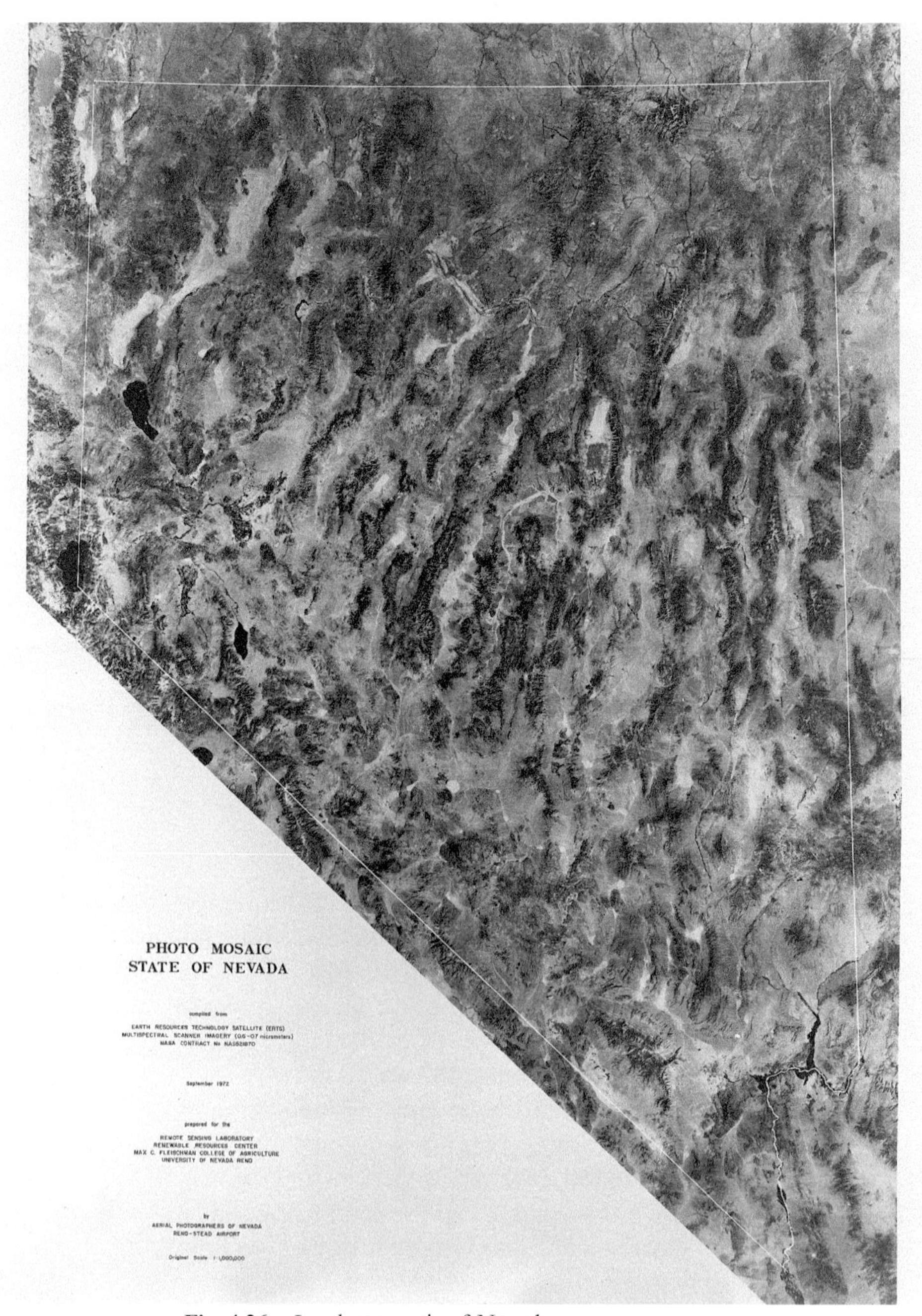

Fig. 4.26 *Landsat* mosaic of Nevada.

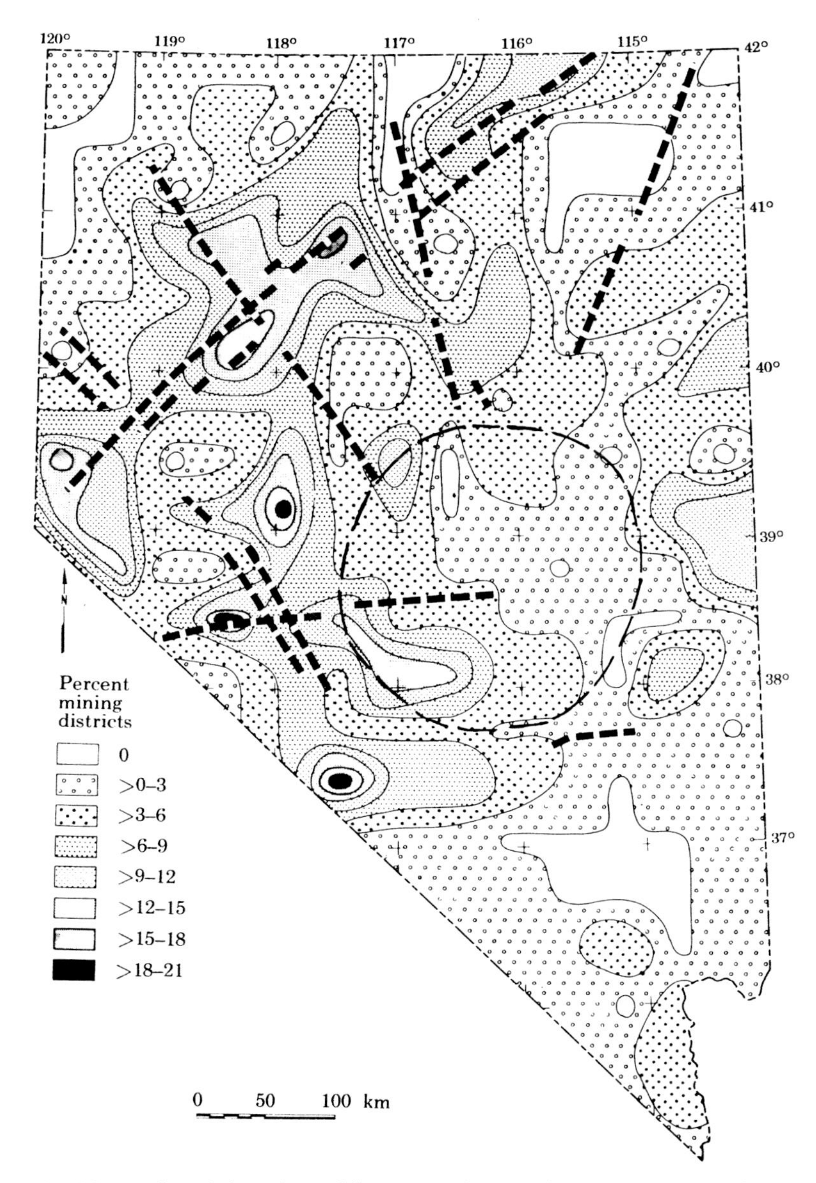

Fig. 4.27 Mineral deposits and lineaments in Nevada. From Rowan and Wetlaufer (1979).

evaporite deposits, as shown in Fig. 4.25. The compositional approach to mineral exploration is thus a large and extremely complex topic. A now-classic investigation by Abrams *et al.* (1977) of the Cuprite area in Nevada and the detailed discussion of the Cuprite study by Vincent (1997) are highly recommended.

A Canadian example from Nova Scotia, by Harris *et al.* (1990)

Fig. 4.28 (See also Plate XVII) IHS transform-modulated radar image of Nova Scotia, with proportions of potassium, uranium, and thorium used for IHS (see text). From Harris *et al.* (1990).

illustrates the use of radar imagery, in this case airborne, in combination with gamma-ray data, to delineate gold-associated structures and rock types (Figs. 4.28, 4.29, 4.30). The "IHS transform" refers to a technique in which different remote sensing data types are used, digitally, to modulate the intensity, hue, and saturation of another type of data, in this case the radar imagery.

4.5 Environmental geology

Environmental geology is a new term for several long-established geologic specialties. An authoritative definition (Bates and Jackson, 1980) describes it as essentially the application of geologic knowledge and principles to problems of the physical environment, in particular those caused by man's occupancy of that environment. The Bates and Jackson definition is a broad one, including engineering geology, hydrogeology, topography, and economic geology. As used here, the term will be somewhat more restricted, concentrating on fields not already covered above and on those usefully approached by orbital remote sensing methods. The order of presentation will roughly reflect the degree to which remote sensing has had an impact on various fields of environmental geology.

4.5.1 Active volcanism

Millions of tourists have seen "active volcanism" in reasonably safe locations, such as Yellowstone National Park or Kileauea. But, like the caged and well-fed tigers in a zoo, volcanos can be deadly perils in other circumstances. Remote sensing from space has now become a well-established means to mitigate these perils in a wide variety of ways.

Volcanic hazards take several forms. Lava flows are the most obvious, but paradoxically are among the less hazardous. Molten silicates, i.e., lava, generally move slowly, slowly enough in occasional sad examples to permit homeowners to remove their furniture and photograph lava creeping across their lawns. Ash flows, on the other hand, erupt suddenly and move rapidly, as the 40,000 fatalities in St. Pierre from the 1902 eruption of Mt. Pelee demonstrate. Volcanic mud flows (lahars)) triggered by volcanic eruptions, recently those of Mt. St. Helens and Mt. Pinatubo, can be equally dangerous (Mouginis-Mark *et al.*, 1993). Several large American cities, such as Seattle and Tacoma, are built on material deposited by lahars. Turning from geology to meteorology, we see that volcanic eruptions have produced phenomena such as the "year without a summer" in

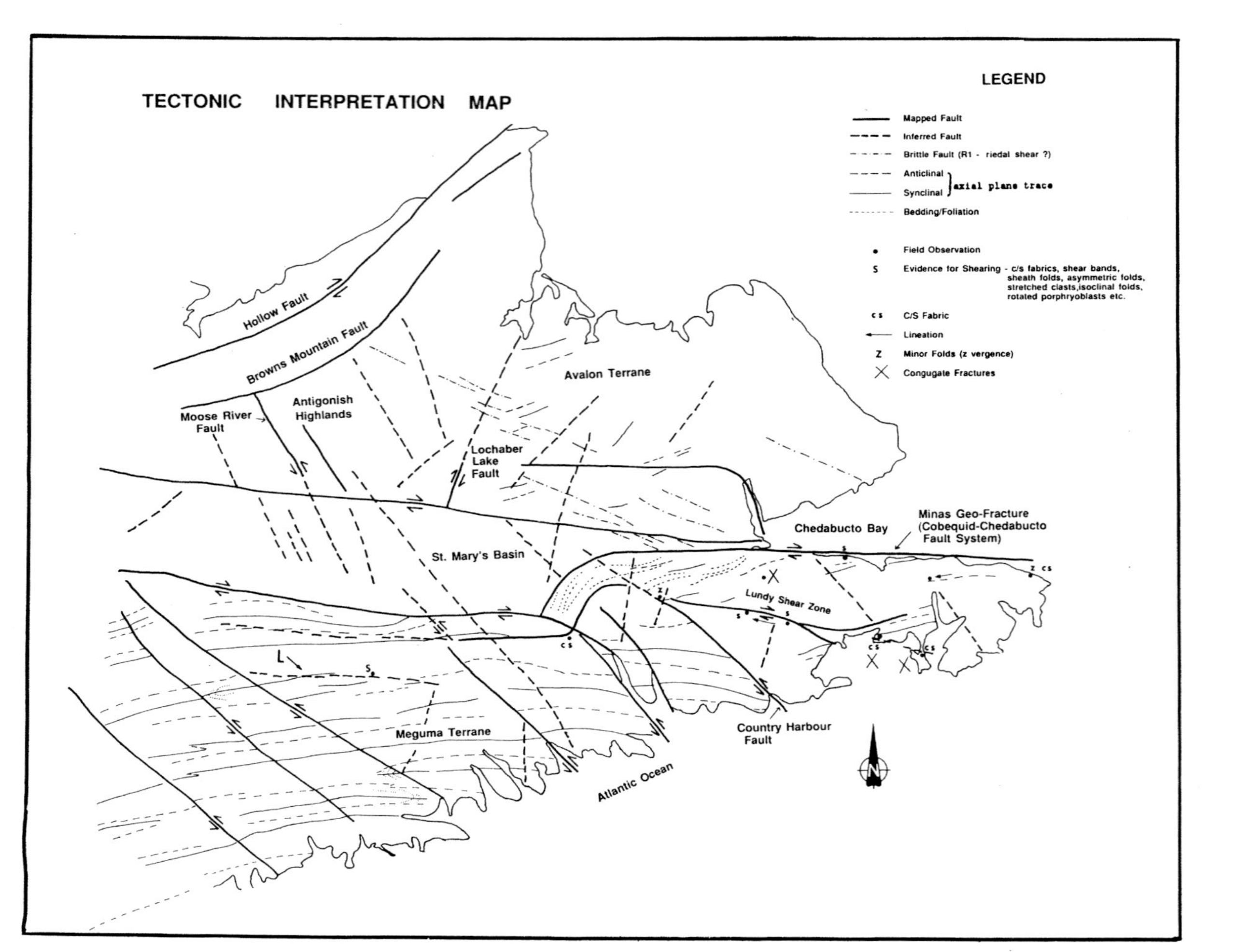

Fig. 4.29 Structure of area shown in Fig. 4.28. From Harris *et al.* (1990).

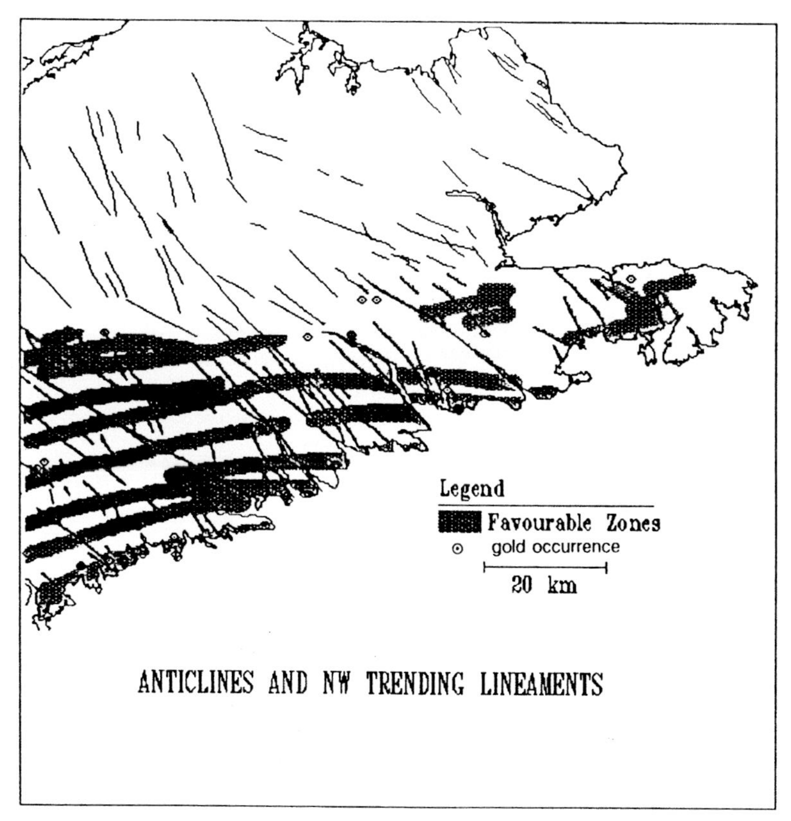

Fig. 4.30 Gold potential map of area shown in Fig. 4.28. From Harris *et al.* (1990).

1815, caused by the eruption of Tambora in Indonesia. Volcanic ash driven into the stratosphere has caused millions of dollars worth of damage to jet airliners, an important problem well suited to monitoring from space, as will be shown.

Remote sensing is now well established as a defense measure against active volcanism and its accompanying phenomena. Only a few of the most important techniques can be illustrated; a more comprehensive summary has been compiled by Mouginis-Mark *et al.* (2000), from which the following examples have been taken. The global tectonic activity map will be referred to again (see Fig. 4.2) for background as showing the long-term distribution of active volcanism around the world.

The most obvious thing to know about volcanic hazards is where the volcanos are, especially the young and potentially active ones. The International Decade for Hazard Reduction listed global mapping of all active or potentially active volcanos as a specific goal

Fig. 4.31 *Landsat* picture of volcanos in Andes of Chile (left) and Argentina at latitude 24 deg. S. Volcanos chiefly stratocones. White area at right is Salar de Arizaro, large salt playa in Argentina. *Landsat* scene 2221-13474, 31 August 1975. From De Silva and Francis (1991).

(Francis, 1989). The value of remote sensing for such a goal was demonstrated by some of the first space photographs acquired for geologic purposes (Lowman *et al.*, 1966), when those taken by astronauts McDivitt and White revealed a large unmapped volcanic field only 100 km west of El Paso. Although not an obvious hazard, judging from the eroded condition of the volcanos (Lowman and Tiedemann, 1971), this finding provided an interesting preview of later discoveries.

The active volcanos of the Andes are well known and demonstrably dangerous, but orbital imagery (Fig. 4.31) has been used by De Silva and Francis (1991) to identify more than 60 major volcanos as potentially active, in comparison to the 16 previously cataloged. The reason for this surprising ignorance of Andean volcanos lies in the sparse population, remoteness, and the fact that the Andes lie in several countries. However, history shows that many volcanos thought to be extinct, such as El Chichon, are only dormant. For example, Mt. Lamington, in Papua New Guinea, was not even recognized as a volcano until it erupted. The tectonic activity map is an attempt to give a more realistic picture of global volcanism by showing not just the last 10,000 years of volcanic activity, but that of the last one million years.

Prediction of volcanic eruptions is another obvious measure to reduce volcanic hazards. Very few volcanos erupt with no warning at all, eruptions being generally presaged by earthquakes, steam venting, and other phenomena. The eruption of Mount St. Helens in 1980 had been approximately predicted several years earlier on the basis of geologic mapping and morphology, and warnings of the actual eruption were issued weeks in advance on the basis of seismicity and steam venting. However, these examples are in a densely-populated and highly-developed country. For the rest of the world, satellite methods are proving valuable in reducing the casualties from volcanic eruptions.

The simplest (in principle) satellite approach is monitoring thermal anomalies over volcanos, to detect the increasing activity before an actual eruption (Rothery *et al.*, 1988; Francis, 1989). An early example of the use of remote sensing, described by Mouginis-Mark *et al.* (1993) was for the volcano Lascar, where a thermal anomaly was discovered accidentally in 1985 on *Landsat* TM infrared imagery. Temperatures were estimated at 800–1000 °C, clearly indicating magmatic activity. The volcano erupted in 1986, but because of its remoteness, the first evidence of the eruption was ash fall 300 km away, in Argentina.

Volcanos frequently swell up before erupting, because of the rise of magma below them. Ground measurements of such swelling, or deformation, have been useful in predicting eruptions. However, it is hardly practical to monitor the 600 or so volcanos active worldwide with ground-based methods. Orbital remote sensing has here found dramatic application, for monitoring pre-eruption swelling, and for tracking the products of eruptions after they happen.

Pre-eruptive deformation has been monitored by means of radar interferometry, a relatively new technique (Zebker and Goldstein,

1986; Massonet *et al.*, 1993b). Useful reviews of this technique as applied to volcano deformation have been presented by Zebker *et al.* (2000) and Massonet and Sigmundsson (2000), which can be described, with extreme simplification, as follows.

Radar is basically a ranging technique, most commonly using pulsed microwaves with wavelengths on the order of 5–30 centimeters. If two pulses are transmitted, at different times, from a single antenna to a stationary target with unchanged back-scatter properties, the return pulses will be in phase. However, if the target has moved, or been deformed, between pulses, the return pulses will interfere, being out of phase. Radar images produced this way will show interference fringes, generally reproduced as colored interferograms. If the antenna moves between pulses, the interferogram can express topography, roughly analogous to stereographic aerial photography. This is however a major topic by itself that will not be covered here.

As applied with orbital radar, interferometry of volcano deformation is carried out from satellites such as *ERS-1* with repeated orbits, supplying the equivalent of a "single antenna." If the target – in this application, the surface of the ground – moves between successive passes, as in an inflating volcano, the ground movement, or deformation, will appear as interference fringes or colors in an image constructed from this repeated coverage. An important advantage of interferometry is that, depending on phase differences, it can detect movements of the same order as the wavelength involved, for radar a few centimeters. This is far better than the spatial resolution, generally tens of meters in range and azimuth, of the same radars. These principles were applied by Massonet *et al.* (1993b) to monitor pre-eruption deformation of Mount Etna. The technique has since been applied to many other volcanos; two examples will be presented here.

For instructional purposes, the study of Fernandina volcano, in the Galápagos, by Zebker *et al.* (2000) is most useful. As shown in Fig. 4.32, radar coverage from *ERS-1* and *2* spanning a five-year period revealed not only the broad deformation of the volcano before the eruption of 1995, but the location of a major dike feeding the eruption. The value of this technique for study of individual volcanos is obvious.

A second example has implications not only for volcanology but for understanding of global tectonic activity. Iceland has been intensively studied for many years, both for its continual volcanic activity and because it is one of the few places (see Fig. 4.2) where a major sea-floor spreading center, the Mid-Atlantic Ridge, is exposed on

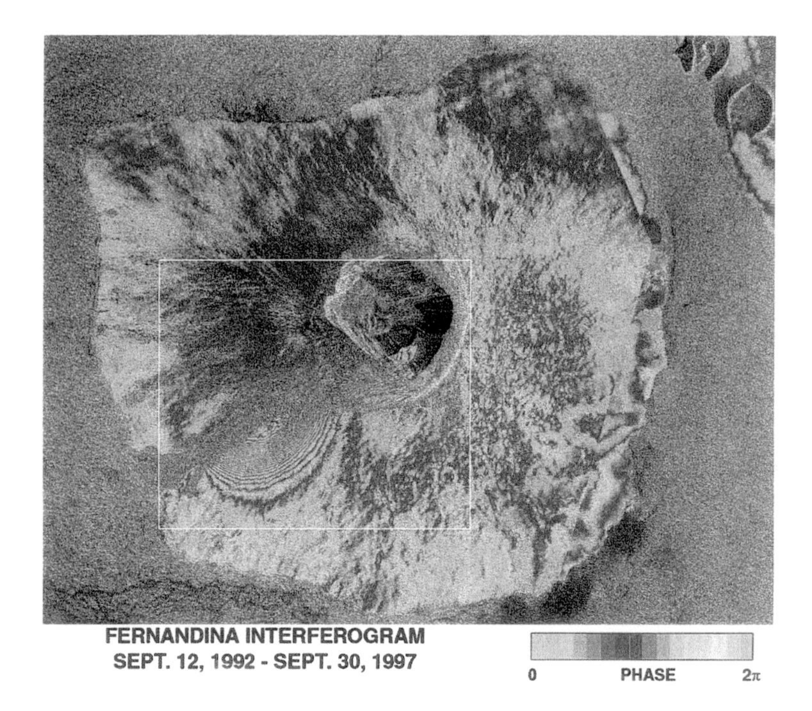

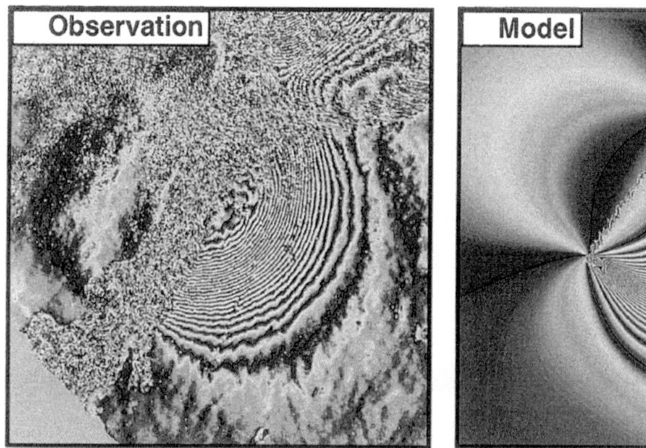

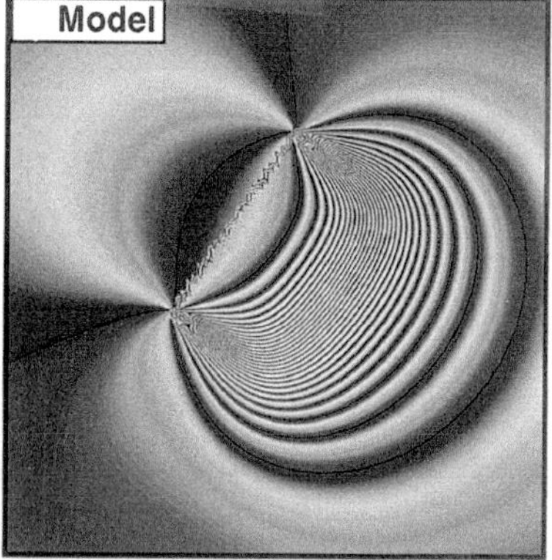

Fig. 4.32 (See also Plate XVIII) *Top:* Differential interferogram of Isla Fernandina, Galápagos Islands, Ecuador. Island is roughly 60 km east to west. Interferogram shows deformation during five-year period, including a flank eruption in 1995. *Bottom left:* Observed interferogram fringes over Fernandina flank. *Bottom right:* Best-fit model from interferogram, indicating a lava source dike striking N 47° E, dipping 33° to SE. From Zebker *et al.* (2000).

land. Satellite radar interferometry of this area thus has fundamental implications. The Reykjanes Peninsula has been studied with radar interferometry by Vadon and Sigmundsson (1997) using data from *ERS-1*. As shown in Fig. 4.33, they were able to detect not only localized subsidence over the Reykjanes volcanic field, but oblique movement along the central rift, i.e, the landward expression of the Mid-Atlantic Ridge. This study thus demonstrates that not only volcanic phenomena but the actual motion of plates can, in principle, be monitored with satellite radar interferometry.

Remote sensing has become an operational method for tracking the products of volcanic eruptions, specifically ash clouds. Volcanic ash is produced in immense clouds by andesitic volcanos, such as those overlying subduction zones around the Pacific Ocean. Ash flows, the volcanic equivalent of turbidity currents, have long been known as potential catastrophes; the 40,000 deaths at St. Pierre in 1902 were caused not by ash fall but by ash flows. However, the volcanic ash injected into the stratosphere has in recent years become a major hazard because of its effect on high-flying jet aircraft (Casadevall, 1994; Schneider *et al.*, 2000). The problem can be summarized, in brief, as follows.

The volcanic ash clouds from andesitic eruptions are conspicuous (Fig. 4.34), and no pilot would intentionally fly through one. However, it has been found from satellite monitoring (Krueger *et al.*, 2000) that such clouds retain their identity for days or weeks, and after traveling many thousands of kilometers. Such clouds are not visually obvious at altitudes of 20–40 thousand feet, where most commercial air traffic flies. Jet aircraft, whose engines function by ingesting large volumes of unfiltered air, are especially susceptible to volcanic ash. As documented by Casadevall (1994), damage from undetected volcanic ash at high altitudes had, even in the early-1990s, caused damage estimated at tens of millions of dollars, though fortunately no fatalities at this writing.

The potential danger of volcanic ash presents a striking example of the synergism among geology, meteorology, and aeronautics. The most specific demonstration of this synergism is in the North Pacific Ocean. All commercial air routes between North America and northeast Asia (including Japan) are great circles. These great circles are downwind from, and parallel to, the volcanically active island arcs of the Aleutians and the Kamchatka Peninsula (see Fig. 4.2), and several nearly-catastrophic flame-outs have resulted from this coincidence of air routes with volcanically active areas. Individual volcanos such as Mt. Etna (Fig. 4.34) can similarly be dangerous in heavily-traveled areas.

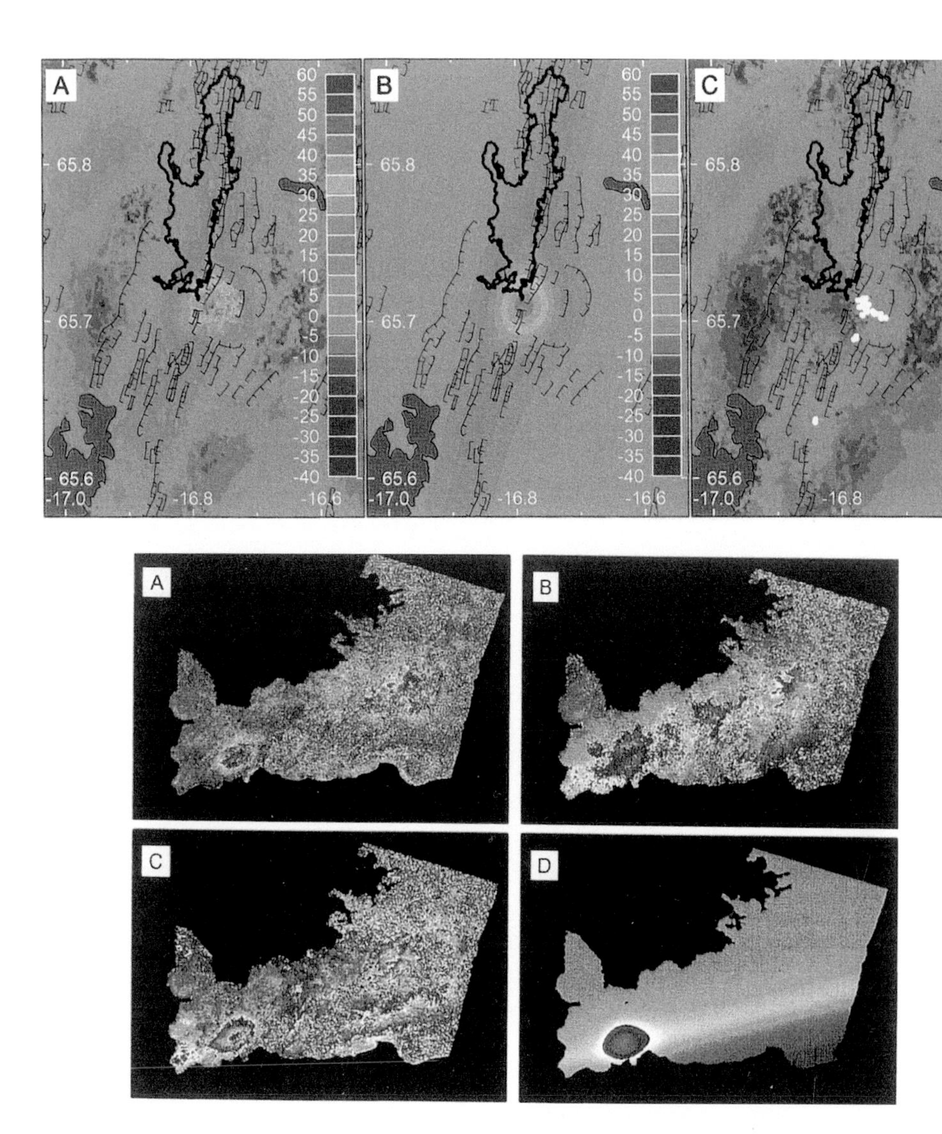

Fig. 4.33 (See also Plate XIX) *Top:* Radar interferogram of Krafla, Iceland, showing (A) range changes in mm over a one-year interval dominated by a 40-mm-deep subsidence bowl over a magma chamber. (B) is best-fit model allowing for pressure decrease in magma chamber. (C) is residuals from the model. Interferogram shows that the subsidence bowl is superimposed on 14 mm subsidence along the Krafla rift axis.
Bottom: Radar interferogram of 50 × 25 km area of Reykjanes Peninsula in southwest Iceland. (A) covers 0.83 years; (B) 2.29 years; (C) 3.12 years; (D) is model interferogram showing best-fit 2.29 year deformation. Figure shows subsidence caused by geothermal energy fluid withdrawal, superimposed on along-axis range increase showing plate separation. From Massonet and Sigmundsson (2000).

Fig. 4.34 (See also Plate XX) *Landsat* picture of volcanic plume coming from Mount Etna, Sicily, during eruption in 1983. Processed by Telespazio, Italy. Acquired 23 April 1983.

The most obvious application of satellite remote sensing to this problem is simply the detection of eruptions in remote areas, a major contribution in itself. The use of satellite thermal instruments for monitoring volcanic hot spots, as described by Harris *et al.* (2000) and Flynn *et al.* (2000) has become an invaluable operational technique. Geosynchronous weather satellites provide routine coverage of nearly entire hemispheres, and can permit Internet posting of eruption alerts within an hour of the event. Non-geosynchronous satellite instruments such as the Advanced Very High Resolution Radiometer (AVHRR) provide information on hot spots due to lava flows, fumaroles, and geothermally-heated lakes.

Satellite investigations originally undertaken to study atmospheric composition and circulation have proven unusually valuable for mitigation of volcanic hazards to air travel, as demonstrated by Krueger *et al.* (2000). The ozone content and distribution are first-order problems themselves, and it was found that orbital observations could effectively monitor ozone. However, it soon developed that ultraviolet observations from space could be applied to the

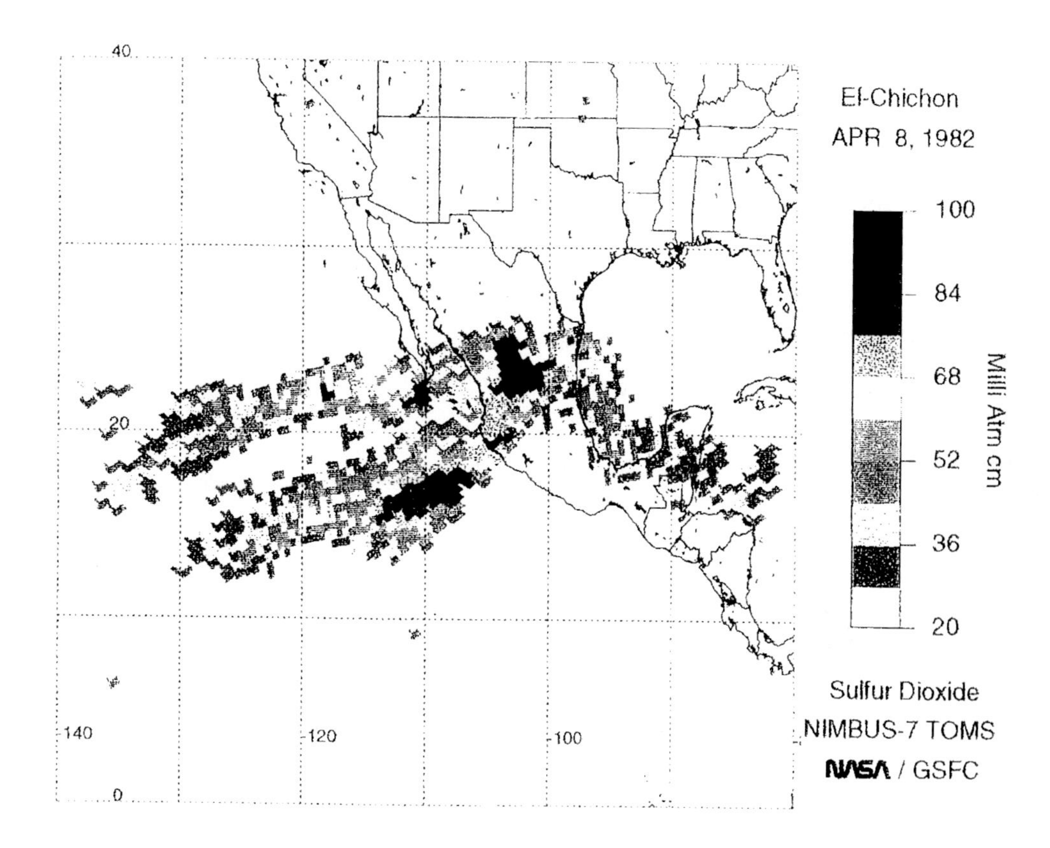

Fig. 4.35 (See also Plate XXI) Sulfur dioxide cloud from the 4–5 April 1982 eruption of El Chichon, Mexico, after drifting southwest for 4 days at 25 km altitude, as tracked by Total Ozone Mapping Spectrometer (TOMS) on *Nimbus 7*. From Krueger *et al.* (2000).

monitoring of ash clouds from volcanos. The Total Ozone Mapping Spectrometer (TOMS) (Krueger *et al.*, 2000) proved effective in mapping the distribution of sulfur dioxide, which in turn showed the location and density of volcanic ash clouds (Fig. 4.35). The result of this and other unexpected discoveries is that satellite observations have become an operational and international method to warn airline pilots of potentially dangerous eruptions. The eruption of Mount Spurr, in Alaska (Fig. 4.36), was tracked by its sulfur dioxide all the way to the Atlantic Ocean, and was still dense enough over Toronto to require diversion of commercial flights from Europe.

The print and electronic news media in the United States almost invariably preface reports of new satellites or other space projects with the cost. For example, the readers were informed in the sixth line of a half-page article that the *Terra* satellite (launched in 1999)

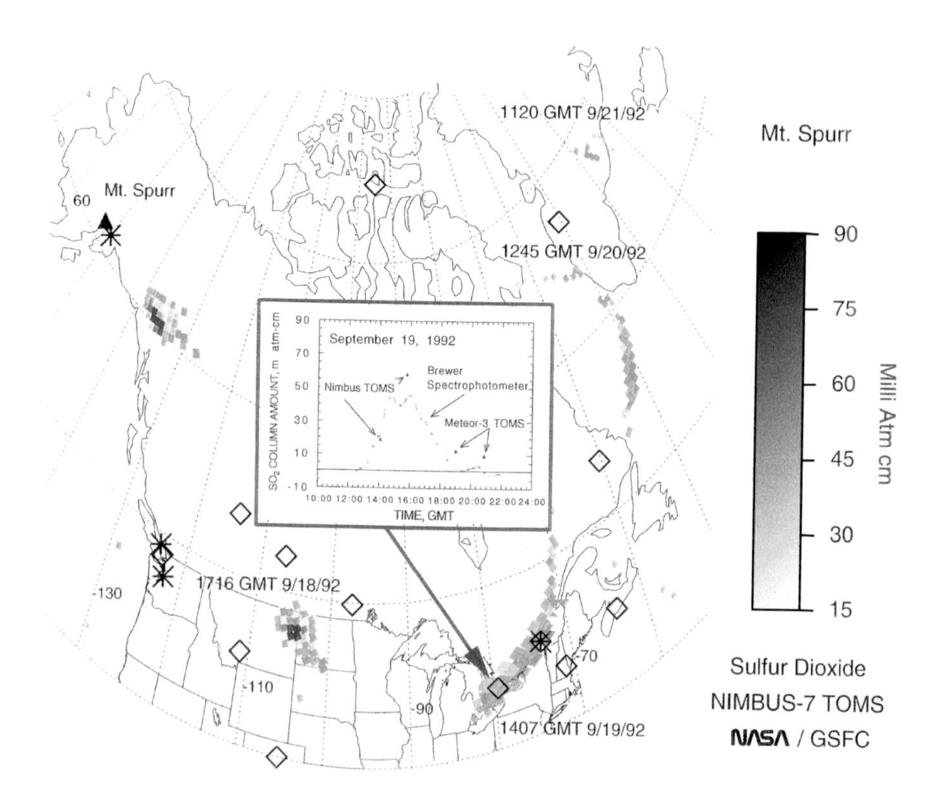

Fig. 4.36 (See also Plate XXII) Composite image of the 17 September 1992 Mount Spurr eruption sulfur dioxide cloud, showing path to southeast across North America. Inset shows sulfur dioxide measured by ground-based instruments as the cloud passed over Toronto. From Krueger *et al.* (2000).

had cost $1.3 billion (Supplee, 2000). Just one of the many *Terra* instruments is already in use for volcano monitoring. A wide-body jet generally carries 200–400 people; roughly 200–400 such jets fly great-circle Pacific routes downwind of andesitic volcanos (Casadevall, 1994). If loss of even one such airplane could be prevented by satellite remote sensing, it would go a long way toward satisfying the taxpayers that their money is being well spent.

4.5.2 Glacial geology

The normal climate of the Earth for the last several hundred million years has been warm from pole to pole; some 60 million years ago, alligators, turtles, and palm trees could be found on Ellesmere

Island, Canada, at 80 degrees north latitude. However, from time to time, ice ages have occurred – "snow-ball earths" in the most extreme case (Hoffman *et al.*, 1998) – driven primarily by a combination of astronomical parameters. The most recent of these ice ages, in the Pleistocene, began about three million years ago, and has not yet ended. A whole continent (Antarctica), the world's largest island (Greenland), and parts of the Canadian Arctic are still covered with remnants of the Pleistocene ice sheets. In addition, there are innumerable valley glaciers in various mountainous regions, even at the equator, but these have been produced primarily by local conditions permitting seasonal accumulation of snow. Finally, the terrain in large areas of North America and western Europe is dominated by glacial or periglacial landforms and deposits related to Pleistocene ice sheets. Glacial geology is thus an extremely important part of environmental geology, even though most inhabitants of the planet have never actually seen a glacier.

Glacial geologists were among the first to make extensive and systematic use of orbital remote sensing data, which provided them with global repetitive coverage utterly impossible with airborne methods. Our coverage of environmental geology thus continues with examples of the application of remote sensing to "glaciers and glacial landforms." This useful phrase is from the chapter by R. S. Williams (1986) in *Geomorphology from Space* (Short and Blair, 1986), recommended to the reader as an excellent introduction to glacial geology as well as an example of remote sensing applications in the field. Following previous practice, this large subject will be selectively sampled, with a few of the best examples.

Mountain (or valley) glaciers are the most familiar to the average reader, and the first orbital picture (Fig. 4.37), from *Landsat*, shows a section of southern Alaska unusually well suited for showing how glaciers form and evolve. The high latitude, coastal location, and high relief collectively produce heavy year-round precipitation, generating the snow fields conspicuous here. As pointed out by Williams (1986), the snow can be considered a sedimentary rock which, with enough pressure, is transformed into glacial ice, the equivalent of a metamorphic rock. The ice flows plastically (as in metamorphic deformation) down the valleys. Variations in flow, in particular glacial surges, can be inferred from the visibly crumpled ends of glaciers, formed by pressure from the upstream active segments. Although not obvious from simple inspection of the *Landsat* photograph, the glaciers in this area have been receding during the 20th century. Hall *et al.* (1995) used *Landsat* imagery of the Glacier Bay area, just south of here, to show that the Muir Glacier receded

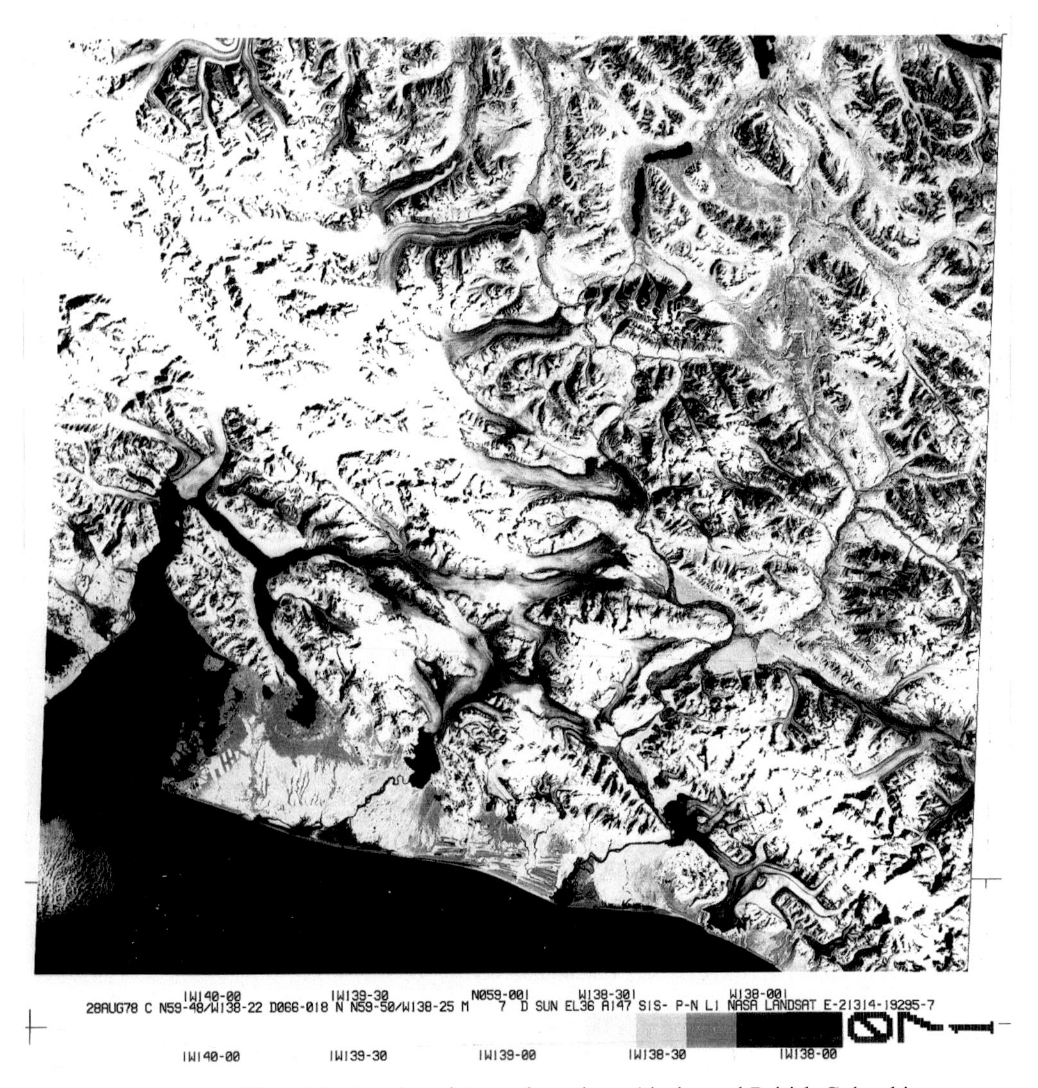

Fig. 4.37 *Landsat* picture of southern Alaska and British Columbia, Canada, showing Yakutat Bay area and glaciers.

7.3 km between 1973 and 1992, an excellent example of the value of remote sensing in glacial geology. Hall *et al.* noted that local influences and positive feedback effects have probably influenced this rate. Similar applications of *Landsat* images to glacial regime studies have been made in Austria by Bayr *et al.* (1994), and several other regions cited by Williams *et al.* (1997). Long-term or global temperature changes can not be inferred from any one area, because of the effects of local conditions. However, collectively, satellite-based glacial studies will be an essential contribution to the question of

Fig. 4.38 (See also Plate XXIII) *Landsat* multispectral composite picture of Vatnajokull, Iceland. From Williams (1986). *Landsat* scene 1372–12080, 30 July 1973.

whether global warming is happening and, if so, what man's contribution to it is.

A different type of glacier is shown in our next example (Fig. 4.38), from Williams (1986). Williams *et al.* (1997) have used *Landsat* images (Fig. 4.39) to measure the Vatnajokull ice cap on Iceland. Vatnajokull is a classic area, the most-studied ice cap on Earth. Its

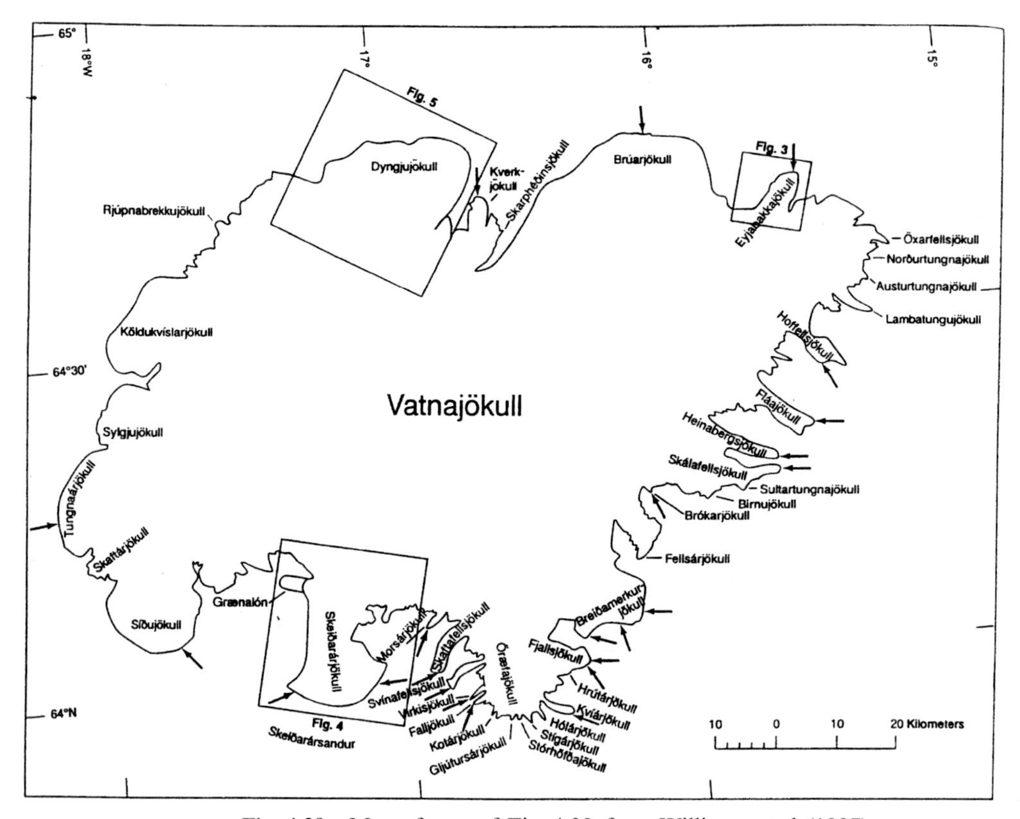

Fig. 4.39 Map of area of Fig. 4.38, from Williams *et al.* (1997).

location is unique, straddling the Mid-Atlantic Ridge (Fig. 4.2), an area of active sea-floor spreading and continous volcanic activity. Like southern Alaska, Iceland has the ideal combination of latitude and precipitation to produce "jokulls" (Icelandic for ice cap). However, these ice caps are unusual because of their tectonic location, with large active volcanos underneath, with consequences to be discussed. Because Vatnajokull has been studied scientifically for almost two centuries by the Icelanders, it provides an invaluable baseline for studies using remote sensing. Williams *et al.* (1997) have used *Landsat* to meaure recession rates for some of the outlet glaciers from Vatnajokull. However, a much more rapid and locally catastrophic glacial phenomenon has also been studied by satellite techniques, in this case orbital imaging radar.

The eruption of volcanos under an ice sheet can produce huge floods, first recognized on Iceland and termed jokulhlaups. Such a flood occurred in 1996, from Vatnajokull, causing $15,000,000 in damage to the Icelandic road system. As shown by Garvin *et al.* (1998), the extent of this jokulhlaup was monitored by repetitive coverage with *Radarsat* imagery (Fig. 4.40). Guided by this imagery, Garvin *et al.* carried out airborne laser altimetry surveys over the "sandur," the alluvial plains formed by the jokulhlaup. These surveys showed that the net result of this flood was substantial deposition, despite the obvious erosion. Apart from the obvious importance for glacial geology studies, the surveys of the 1997 jokulhlaup illustrate the importance of catastrophic events in geomorphology, another demonstration that "uniformitarianism" must include what was once called "catastrophism."

This brief review of the use of remote sensing in glacial geology would be incomplete without a now-classic *Landsat* view (Fig. 4.41) of the Channeled Scablands of Washington state. This deranged topography, visible in its entirety only from space, was produced by the sudden release of glacier-dammed water from the Columbia River in the Pleistocene. This explanation, originally proposed by J. Harlan Bretz in the 1920s, was for decades rejected by geologists of the day, wedded to "uniformitarianism" in the classic sense. It was eventually realized that Bretz was right, in time to award him the Penrose Medal in his 90s. Had synoptic views of the Scablands been available to him, the award might have come much sooner.

4.5.3 Aeolian geology and desertification

The term "aeolian geology" is intended to cover the study of wind-dependent processes and the deposits and landforms produced by wind. It is roughly equivalent to desert geology, for obvious reasons, but it should be pointed out that there are large areas of aeolian features that are not, or are no longer, deserts. Much of Nebraska, for example, is covered by stabilized sand dunes (the Sand Hills), dating from the late-Pleistocene, when winds from the continental ice sheets deposited them.

Many of the world's great deserts lie at low latitudes, and accordingly have been well covered by orbital photography and other types of remote sensing beginning in the early-1960s. Many of the desert photographs taken by *Gemini* and *Apollo* astronauts are unsurpassed even today, possibly because the Earth's atmosphere has become less clear as a result of man-made air pollution. A *Gemini 5* photograph (Fig. 4.42) demonstrates the potential value of such imagery for study of "sand seas." The Namib Sand Sea has in fact been studied with *Landsat* imagery by White *et al.* (1997), who were able to map the iron oxide content of these dune sands. The study of global sand seas has been revolutionized by remote sensing, as shown by the massive compilation of sand-sea investigations edited

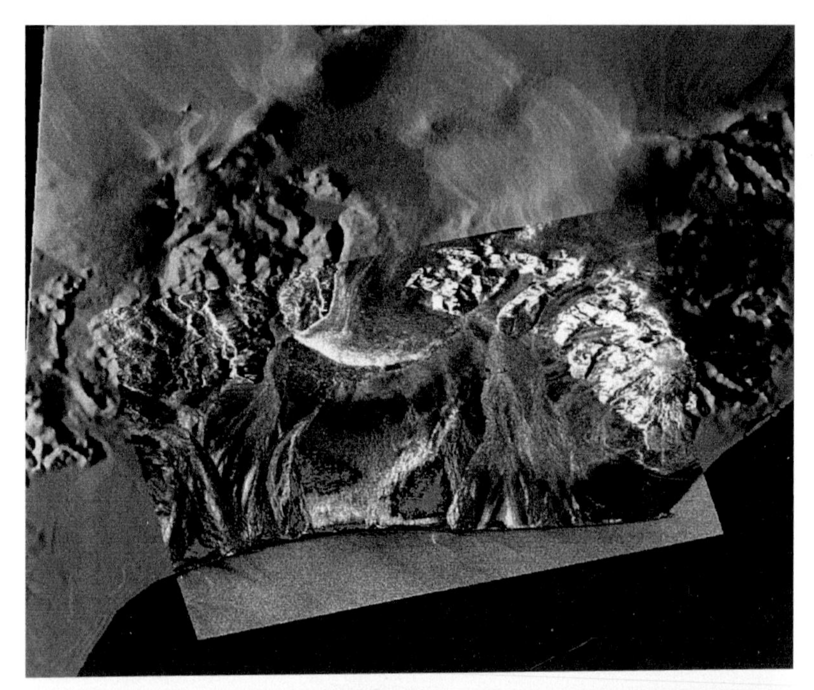

Fig. 4.40 *Radarsat* standard beam images of Skeidararsandur, Iceland, acquired 5 October 1996 (*top*) and 9 November 1996 (*bottom*), pre- and post-jokulhlaup respectively, with 30 deg. incidence angle. Bright areas on 9 November image show flood deposits. *Radarsat* is operated by the Canadian Space Agency. Image courtesy of Dr. James B. Garvin.

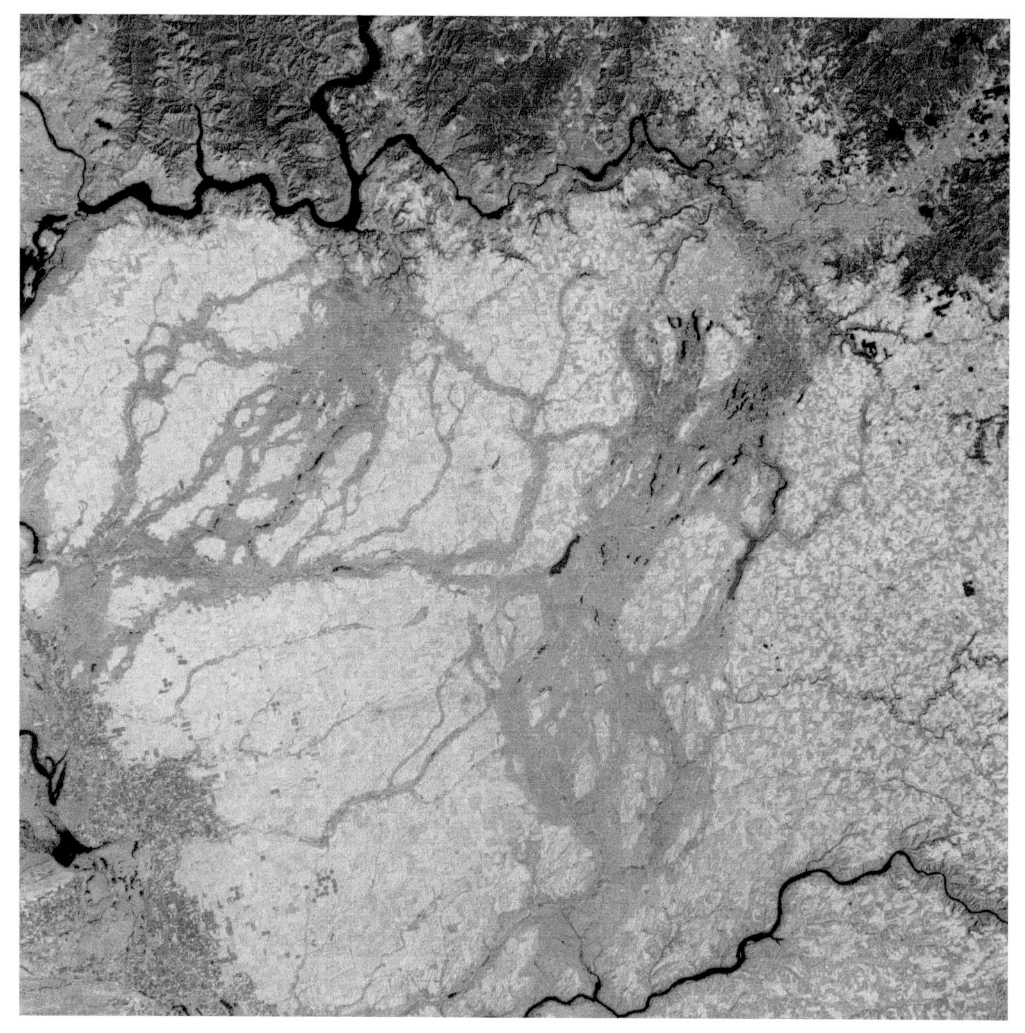

Fig. 4.41 (See also Plate XXIV) *Landsat* picture of Channeled Scablands, Washington; Spokane at upper right, Columbia River at upper left. Light gray and black patterns show valleys carved out by catastrophic glacial floods. *Landsat* scene 1039-18143, 31 August 1972.

by McKee (1979). A comparable compilation of aeolian landforms has been published by Walker (1986), including images of sand seas on Mars.

An environmental geology topic closely related to aeolian geology is desertification, essentially the destruction of arid or semi-arid environments by human activity, aggravated in some areas by natural conditions. It has been estimated by the UN that 35% of the Earth's land area is in danger of desertification (Zhenda and Yimou,

Fig. 4.42 (See also Plate XXV) *Gemini 5* 70 mm photograph of Walvis Bay area, Namibia, showing sand dunes of Namib Desert. Northward transport of dunes is stopped by the Kuiseb River. NASA photograph S-65–45579, 1965. Width of view ~100 km.

1991), and most of this is in developing countries such as India and China, least able to withstand the process. Both countries have developed strong orbital remote sensing capabilities (Kasturirangan, 1985). The Indian Remote Sensing (*IRS*) satellites have proven useful in a wide range of environmental geology applications. China has to date largely used other countries' satellite data, but very effectively. A comprehensive collection of reports on remote sensing approaches to environmental change in Asia was edited by Murai (1991).

Remote sensing was recognized as a valuable tool in monitoring desertification even with the earliest *Landsat* images (Otterman *et al.*, 1976), as demonstrated by Fig. 4.43, showing overgrazed native preserves in Zimbabwe. Even in highly-developed but semi-arid countries, such as those around the Mediterranean, satellite images are being used to monitor desertification and related conditions, such as deforestation and soil erosion (Hill *et al.*, 1995).

Several anthropogenic processes contribute to desertification, including deforestation, overgrazing, sand dune migration, and salinization. The term "wastelands" used by Nagaraja *et al.* (1991) for India encompasses several of these processes. Satellite remote sensing, by *IRS*, has been shown by these authors to be an effective means of monitoring the status of wastelands.

Perhaps the broadest view of desertification, a global one, has been provided for some decades by various meteorological satellites, monitoring radiation in the microwave region. This technique, not previously discussed here, uses passive remote sensing in the centimeter range to map physical characteristics of the Earth's surface, in particular the vegetation distribution (Townshend *et al.*, 1993). Dense vegetative cover is obviously the inverse of desertification. However, passive microwave methods have been directly applied to desertification by Choudhury (1993). The technique, in brief, depends on the difference in detected intensity between horizontally- and vertically-polarized 8-mm wavelength radiation, the greatest difference indicating the most barren terrain. The images produced this way give a good global view of the distribution of vegetation. More localized views, confined to a single continent, show seasonal vegetation changes. Further discussion would be beyond the scope of this review, but it is clear that desertification can be monitored by several different orbital remote sensing techniques.

Our final example of remote sensing as used in environmental geology is from the *Terra* satellite, that is, the \$1.3 billion *Terra* satellite, launched in 2000. Among the instruments carried was the MOderate Resolution Imaging Spectroradiometer (MODIS), which

Fig. 4.43 *Landsat* picture of Zimbabwe. Feature in center is Great Dike. Light and dark patterns represent areas of overgrazing. Bedrock is granite–greenstone terrain; note large elliptical batholith left (west) of Great Dike. *Landsat* scene 1103-07285, 3 November 1972.

produces multispectral images with extremely wide swaths (Fig. 4.44). This scene was acquired March 6, and covers the United States and southern Canada from the Mississippi River to the Atlantic Ocean. The photograph was taken near the time of maximum seasonal change in vegetation, i.e., mid-spring. Apart from giving an outstanding view of regional geology, it also provides an unprecedented picture of deforestation on a subcontinental scale. When the scene was acquired, on a single pass, crops were not yet up

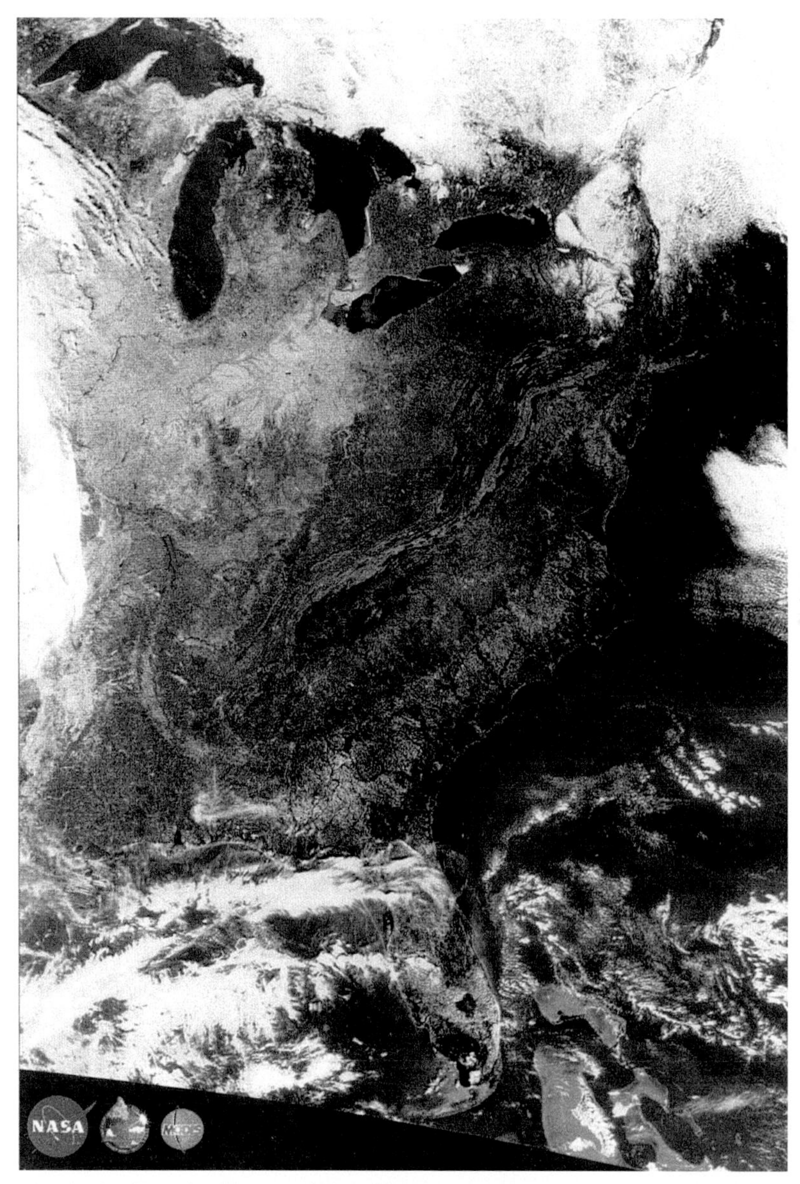

Fig. 4.44 (See also Plate XXVI) MODIS image of eastern United States and southern Canada, acquired 6 March 2000. Natural color.

except in the deep south, and bare fields show up on the picture as light areas in contrast to forests. This is most obvious in the Ridge and Valley Province of Pennsylvania, where the valleys are farms (now fallow) and the ridges are forest. In parts of the Coastal Plain, southern Ontario, Michigan, and Wisconsin, areas of forest cover similarly show up dark. The MODIS scene thus provides, by a lucky combination of season, clear weather, little snow cover, and swath width, an almost frightening view of the enormous amount of forest cover removed since the continent was settled. Environmental problems such as soil erosion and sediment pollution are much more easily grasped with this graphic example.

4.6 Summary

A summary of the impact of orbital remote sensing on geology must begin on a slightly apologetic note: the impact has only begun to be felt by geologists (as distinguished from specialists in remote sensing as such). It should be admitted at once that remote sensing has had nothing like the fundamental effect on geology that, for example, the Deep Sea Drilling program had. In applied geology, the situation is quite different, with orbital data in routine use for petroleum and mineral exploration, and increasingly for environmental geology (especially for monitoring volcanic hazards). However, geology is becoming increasingly an applied science, with basic research being de-emphasized by government funding sources. Geology is even losing its identity to some degree, being merged with "earth sciences," "environmental studies," and other fields to form "earth system science." These trends, whether good or bad, will almost certainly lead to increasing use of remote sensing in "geology."

The future for remote sensing in basic geologic research is less clear. The apparent success of plate tectonic theory has produced a generation of geologists that considers the big problems of geology to have been solved, at least in principle, by this master plan. Continents are now assumed to be formed by accretion of terranes, mountain belts by continental collisions. "Opening of the Atlantic" is referred to as confidently as if the event had happened during the 1969 Geological Society of America meeting in Atlantic City. This unhealthy situation may be remedied by the stimulus of new discoveries from seismic reflection profiling, the *World Stress Map* project, and planetary exploration. Furthermore, remote sensing may come to its own rescue, so to speak, in that the continuing stream of new imagery from satellites can hardly fail to excite interest in geologic problems among students. The *International Space Station* may make it possible for more professional geologists to see the Earth from orbit; at this writing, only two (Drs. Kathryn Sullivan and Harrison Schmitt) have had this privilege.

On balance, it can be said that orbital remote sensing has now taken its place firmly as a useful geologic tool. The new century will show if it will be more than this.

CHAPTER 5

Impact cratering and terrestrial geology

5.1 Introduction

An interstellar explorer examining the solar system would immediately be impressed by one characteristic of the Earth's geology: the apparent scarcity of impact craters, which dominate the terrain of most solid planets and satellites (Fig. 5.1). This scarcity may account for the failure of terrestrial geologists to appreciate the importance of meteoritic or cometary impacts in geologic history. However, stimulated by the results of space exploration, terrestrial studies, and modern astronomy, we now realize that such impacts must be appreciated to understand some of the most fundamental geologic problems. These include the causes of mass extinctions; the anomalous location of some large oil and gas fields; the formation of some major ore deposits; and the origin of the first ocean basins (perhaps also the oceans themselves). There is substantial evidence that the primordial atmosphere may have been brought in by cometary impacts on the early Earth (Chyba, 1990; Owen, 1998), perhaps even pre-biotic organic compounds from which life developed (Chyba and Sagan, 1992).

The role of space research in this development is interesting, not only by itself but because it illustrates the indirect effects that space exploration has had in several fields. Very few terrestrial impact craters have been actually discovered from space, and orbital remote sensing has only recently been applied to their study (Garvin *et al.*, 1992). But the new understanding of impact cratering and terrestrial geology is demonstrably due to space exploration and in particular to *preparation* for it. The purpose of this chapter is to present such a demonstration, by tracing the historical evolution of cratering studies as applied to the Earth. It should be stipulated at once that impact craters are the most common landform in the solar system, and the subject a correspondingly large one. This treatment must therefore be considered only a summary, focussed primarily on

Fig. 5.1 *Viking Orbiter* mosaic of cratered highlands of Mars; center of photograph about 23 deg. S, 348 deg. W. Large sharp crater just right of center is Bakhuysen, about 175 km diameter.

terrestrial craters. An illustrated summary of extraterrestrial impact craters has been published by Lowman (1997).

Several general references are invaluable background sources on impact cratering. Many important older papers were reproduced in facsimile in the collection edited by McCall (1979). The first recognizably modern treatment of the subject is Baldwin's *The Face of the Moon* (1949), a long-out-of-print classic superseded by his *The Measure of the Moon* (1963). The study of lunar craters by Dietz (1946) was a similar treatment by a geologist later noted for development of the sea-floor spreading concept. The most valuable single paper, even after four decades, is Shoemaker's (1962) discussion of Meteor Crater, Arizona, nuclear explosion craters, and the lunar crater Copernicus. Studies of Canadian impact craters, started in the late-1950s, were summarized by Dence (1965). Impact (or shock) metamorphism, now a recognized branch of petrology, was essentially founded with the compilation by French and Short (1966).

More recent summaries of this subject have been published by Koeberl (1997) and by French (1999). The comprehensive text by Melosh (1989) covers all aspects of impact cratering. A review by Grieve (1991) gives a definitive catalog of the roughly 160 terrestrial craters known. The largest impact craters, multi-ring basins, have been treated by Spudis (1993); this is a good reference for impact cratering in general and for extraterrestrial craters.

The first major compilation of papers on the importance of impacts for terrestrial geology was that edited by Silver and Schultz (1982). These papers, given at a conference in Snowbird, Utah, concentrated largely on the Cretaceous–Tertiary boundary event, then recently suggested by Alvarez *et al.* (1980) to have been an impact that finished off the dinosaurs (and many other life-forms). A second Snowbird conference produced a similar collection of papers edited by Sharpton and Ward (1990). A contemporary conference held in Australia concentrated more on the physical aspects of impacts on the early Earth (Dressler *et al.*, 1994). A meeting on large impacts and planetary evolution was held in 1992 on the rim of a 60×30 km crater, the Sudbury Basin (Dressler *et al.*, 1994). The Geological Association of Canada's annual convention, Sudbury 1999, was similarly situated, and saw presentation of many papers on the Sudbury structure. A meeting focussed on economically important impact structures, such as the Ames structure of Oklahoma, produced a comprehensive volume (Johnson and Campbell, 1997), equivalent to a textbook, on impact craters in general.

5.2 Hypervelocity impact

Although this chapter will be organized more or less chronologically, it will begin with a brief discussion of the nature of hypervelocity impact, since most impact craters are of this sort rather than simple low-velocity splash craters.

Man-made explosion craters have been known since the invention of gunpowder. The previously unimaginable artillery bombardments of World War I produced Moon-like landscapes of regional extent (Fig. 5.2), and study of explosion craters was thus a part of Baldwin's (1949, 1963) work. Valuable though the analogy is, meteoritic impact is essentially different from nuclear or chemical explosions (Gault *et al.*, 1966; Cooper, 1977) in being produced by the hypervelocity collison of a projectile with a target rather than the sudden generation of large volumes of gas (or plasma in a nuclear explosion). The essence of hypervelocity impact is generation of

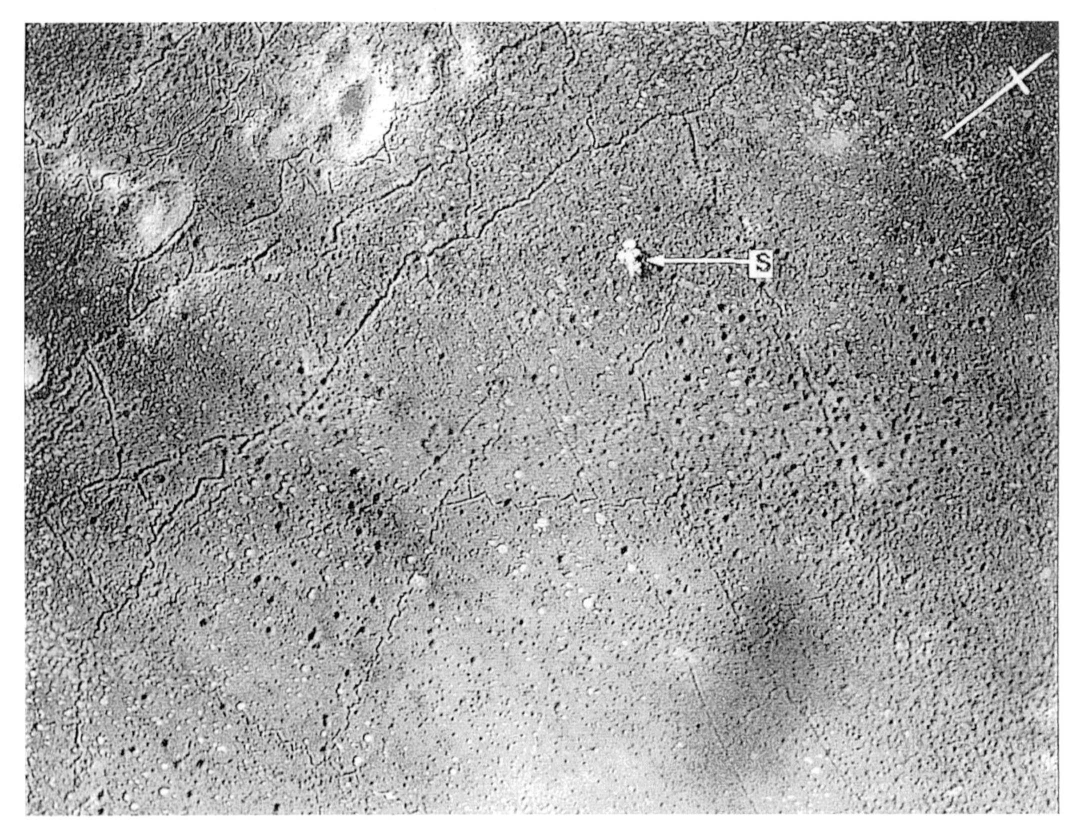

Fig. 5.2 Aerial photograph showing cratered terrain in northern France; bursting shell labeled S. Large craters at upper left were produced by mine explosions. Segmented lines are fire trenches of the German defensive positions. Photograph taken in April 1917, from 3000 feet altitude. Area covered is 0.7 km right to left; arrow shows north. From Watkis (1999).

shock waves orders of magnitude greater than the strength of the target material (Bates and Jackson, 1980). "Shock waves" are supersonic in relation to compressive wave velocities of the target material. Since meteorite impact velocities, neglecting air resistance, will, a priori, be equal to or greater than the escape velocity from the target planet, they will, in principle, be over 2.5 km/s for the Moon and over 11 km/s for the Earth. [Air resistance, for the Earth, is negligible for bodies much over 100 tons (10^5 kg).] The relative velocities of projectile and target may thus be "supersonic" to begin with, and shock-wave generation is an inherent aspect of major meteorite impacts.

The phenemon of cometary impact was observed for the first

time in 1994, when the nine cometary fragments of Shoemaker–Levy-9 hit Jupiter (Hammel *et al.*, 1995). Although the effects of these violent impacts were still visible a month later in Jupiter's atmosphere, these were not "craters" of the sort produced by impacts on solid bodies.

Meteoritic impact is qualitatively different from simple low-velocity impact, as in the common school experiment of pebbles dropped into sand or mud. The kinetic energy of a projectile is dependent on the square of the velocity, and hypervelocity impact thus produces craters much bigger than the projectile (which is generally destroyed and dispersed inside and outside the crater). Craters in the centimeter size range have been formed on windows of the Space Shuttle by orbiting paint flecks from discarded rocket stages, hitting at relative velocities of several kilometers per second.

Shock waves produce a now well-described family of changes in rock-forming minerals, that is, shock metamorphism (Chao, 1967; Stoffler, 1971; Short, 1975; French, 1999). These include instantaneous melting, producing glassy thetomorphic crystals; deformation lamellae ("planar features") in quartz (Fig. 5.3); dissociation of refractory minerals into their oxides; and formation of high-pressure SiO_2 polymorphs such as coesite and stishovite (Koeberl, 1997). The gross products of large impact include shatter cones (Fig. 5.4); complex breccias with glass, rock and mineral fragments; and veins of pseudotachylite (Spray, 1998) that were for many years considered the result of unusually violent volcanism. Under some circumstances, in both terrestrial and lunar craters, large volumes of what appear to be normal igneous rock have been generated, but which have been shown by their geologic context to be impact melts (Dence, 1971; Grieve *et al.*, 1991). The proportion of impact melt produced increases with size of the crater. Most of the apparently igneous rock of the 60×35 km Sudbury Structure, to be discussed in Section 5.8, is now thought to be a high-temperature impact melt rather than a magmatic mantle-derived rock.

The unique nature of shock metamorphism is shown by a pressure–temperature diagram (Fig. 5.5) (Koeberl, 1997), comparing the facies of terrestrial regional metamorphism with the various shock effects just described. To a geologist, an interesting feature of shock metamorphism is that unlike regional metamorphism, which produces the familiar schists and gneisses of Precambrian terrains, shock metamorphism is generally a non-equilibrium process. An example of this is that shock-produced rocks frequently violate the mineralogical phase rule, which says that the number of minerals

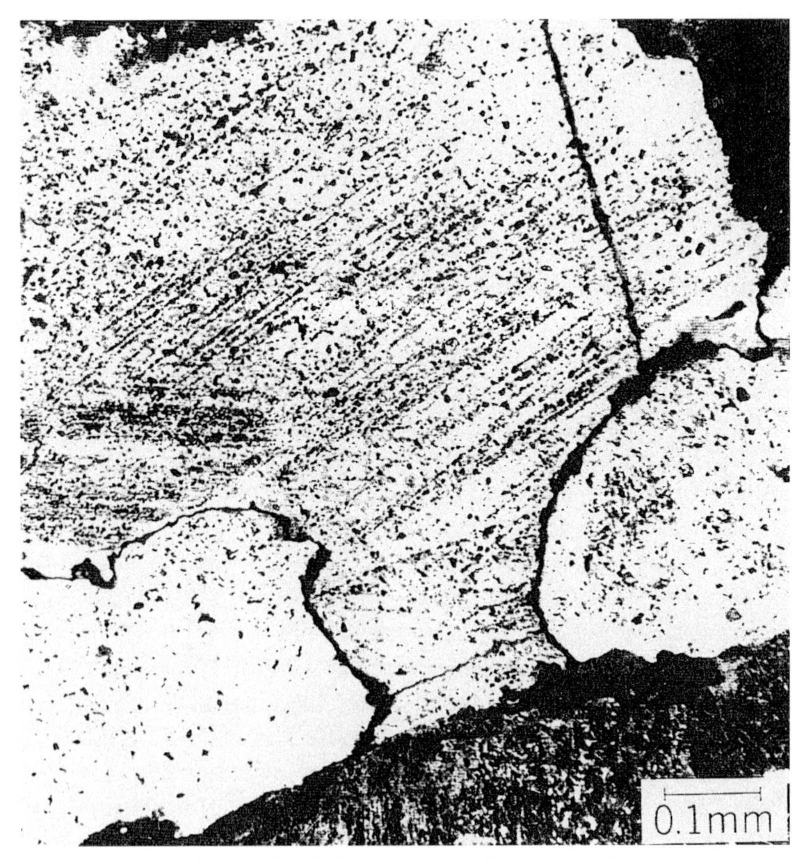

Fig. 5.3 Photomicrograph of planar deformation features in quartz, in inclusion of granitic rock from Onaping Formation, Sudbury Structure; crossed polarizers. From French (1967).

will not exceed the number of chemical components. This means that, for example, a single-component rock composed only of SiO_2, such as a quartzite, will normally have only one form of silicon dioxide, that is, one mineral – in this case quartz. However, shock-metamorphosed rocks, such as the classic Coconino sandstone around Meteor Crater, may have several SiO_2 minerals: quartz, coesite, stishovite, and even silica glass. (Glass is not a mineral.)

The question of whether shock metamorphism can be produced by volcanism was reviewed by French (1990), who showed that the evidence against this is now overwhelming.

5.3 Impact craters

Impact craters range in size from "zap pits" on lunar rocks, visible only with a microscope, to multi-ring lunar basins the size of Texas.

Fig. 5.4 Shatter cones in Mississagi Quartzite south of Kelly Lake, on south side of Sudbury Structure. Pocket knife gives scale. View to south, away from center of Structure. From Lowman (1992).

They have been found on every solid body so far visited in the solar system, with the exception of Io, where tidally-induced volcanism continually resurfaces the topography. About 160 craters or structures of probable impact origin have been identified on Earth (Grieve, 1995), and additional ones continue to be found. Most of these, however, are strongly modified by erosion or covered by later sedimentary rocks, and it is thus hard to see their original morphology. The Moon, in contrast, has been internally inactive for 2.5 billion years, and furnishes a museum of impact craters (Lowman, 1997). The structure of typical craters is illustrated by Fig. 5.6, from Horz *et al.* (1991). For a pristine impact crater, let us examine the

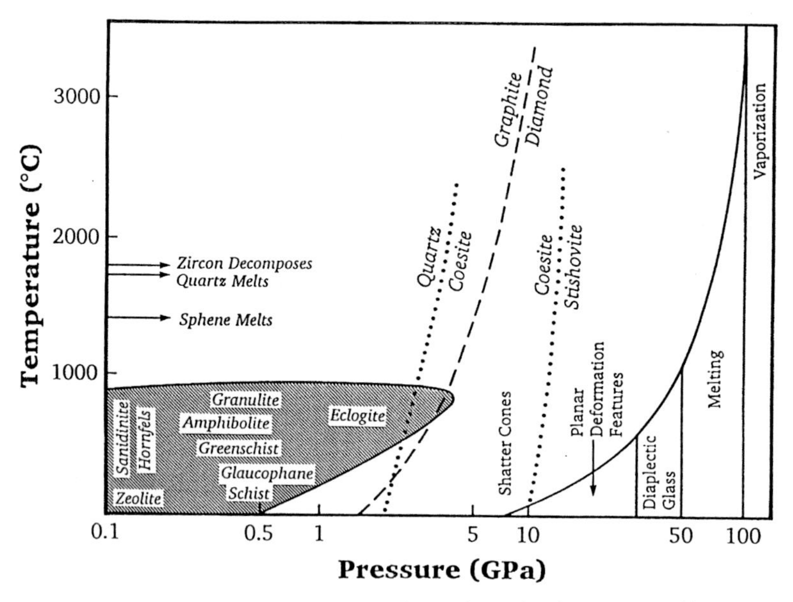

Fig. 5.5 Pressure–temperature ranges for various shock metamorphic features, compared with facies of regional metamorphism at lower left. 1 GPa = 10 kilobar. From Koeberl (1997).

85 km lunar crater Tycho (Figs. 5.7, 5.8). Although several hundred million years old, and technically a "complex" crater (Spudis, 1993), Tycho has few of the complications we will see in other large craters. Detailed descriptions can be found in Lowman (1969), Schultz (1972), and Wilhelms (1987).

Tycho's most conspicuous feature, as seen at full Moon, is not shown here: its enormous ray system, extending most of the way around the visible hemisphere. In Fig. 5.7, the dominant feature is its raised rim, roughly circular in outline. Despite its pristine appearance, the crater we see is not the one formed at the moment of impact (the transient cavity). The rim has visibly slumped inward along a series of normal faults, producing a stepped topography. The floor of Tycho is a complex terrain reminiscent of pahoehoe lava, and which is generally interpreted as impact melt and fallback material, largely molten at one time. The central peak is typical of many lunar craters in this size range; its topography is exaggerated by the low Sun elevation. Outside the raised rim, we see an apparently rugged terrain dominated by radiating ridges, with occasional playa-like level low areas. This material is clearly ejecta from the crater, and the

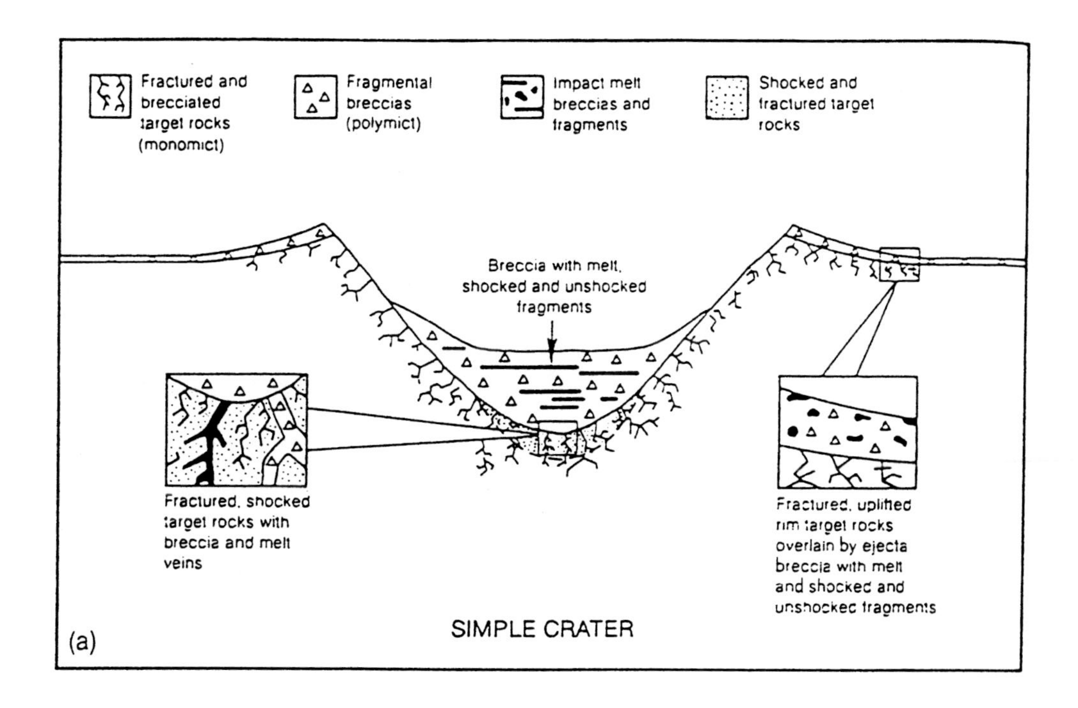

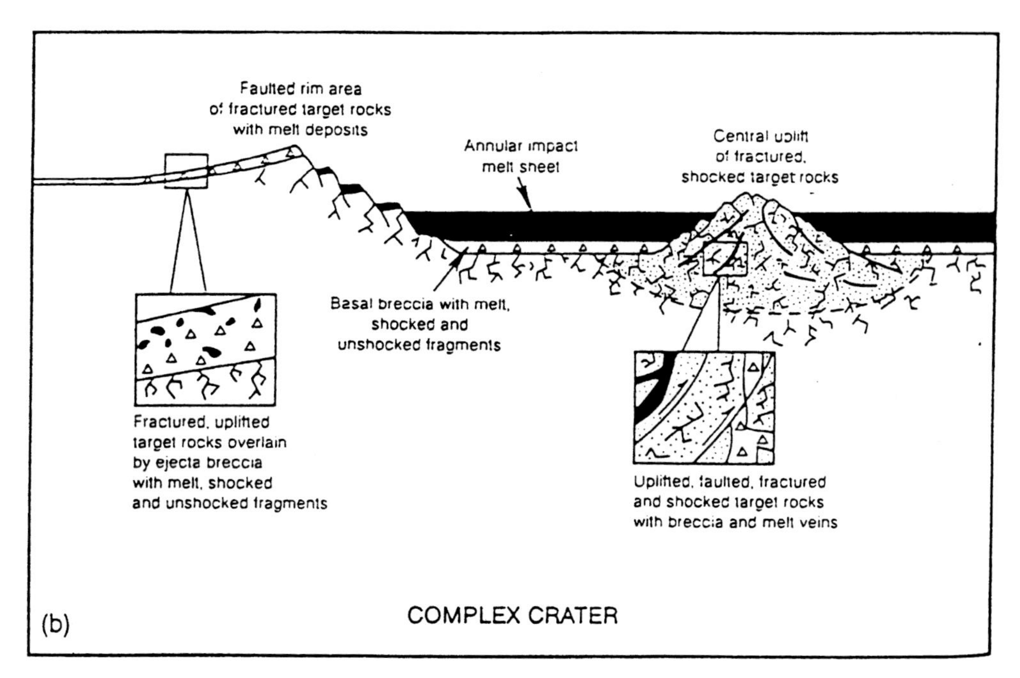

Fig. 5.6 Schematic cross sections of simple and complex craters. From Horz *et al.* (1991).

Fig. 5.7 *Lunar Orbiter V* photograph of crater Tycho, on Moon, 85 km diameter, 4.5 km depth. North at top. From Lowman (1969).

"playas" are considered ponds of fluidized impact melt. Similar features are visible in some of the terraces inside the rim.

It was argued for many years (Green, 1971) that Tycho and its relatives were volcanos, specifically calderas. This once fierce debate has long since stopped. However, O'Keefe (1985) raised a number of critical questions about this topic, and presented arguments for a major role of volcanism in lunar geology. Accordingly, the question must be asked: Why are most authorities so sure craters of the Tycho variety are primarily impact craters?

There are several reasons. Most general is the fact that dozens of large objects, asteroids or comets, are observed to cross the Earth–Moon orbit around the Sun, and the Earth is now known to be *in* an asteroid swarm (Shoemaker *et al.*, 1990). In 1993, a sizable asteroid missed the Earth by only 140,000 km. Some objects do hit the Earth. The 1908 Tunguska, Siberia, object produced effects

Fig. 5.8 *Lunar Orbiter V* close-up of floor of Tycho, showing presumed impact melt and breccia. Width of view about 30 km. From Lowman (1969).

comparable to a hydrogen bomb; fortunately, Czarist Russia was incapable of launching a nuclear counterstrike. The 1947 Siberian iron meteorites formed numerous craters. Given these facts, it is obvious that in 4.5 billion years (the Moon's age), many crater-forming objects *must* have hit the lunar surface. If Tycho and similar craters are not the scars of such impacts, where *are* the scars?

Another general argument in favor of an impact origin is the generally related morphology of craters of all sizes (see Fig. 5.1), from small ones formed by the impact of spent rocket stages (on Earth and Moon), up to multi-ring and partly basalt-filled giants like the Orientale and Imbrium Basins (Spudis, 1993). There is no such progression in known volcanos, which include diverse types such as maars, cinder cones, shield volcanos, and irregular volcano–tectonic depressions. It should be stressed here, however,

Fig. 5.10 Barringer Crater (Meteor Crater), Arizona, view to southeast from about 2500 feet altitude above ground level. Crater diameter is about 1.2 km. Photo by author, 1965.

After October 1957, it was obvious that man or his machines would soon reach the Moon and planets. NASA was organized in 1958, and within 2 years was supporting (and still supports) studies in astrogeology by the US Geological Survey. Among the work carried out was preliminary mapping of the Moon and establishment of a lunar stratigraphic time-scale. A major discovery by the USGS, which can be considered the origin of shock metamorphism as a recognized phenomenon, was the discovery of natural coesite in ejecta from Barringer Crater by Chao, Shoemaker, and Madsen (1960). First prepared artificially by Loring Coes in 1953, coesite had been predicted to occur at Barringer Crater by H. H. Nininger in 1956. Its discovery by Chao and his colleagues was followed shortly by the discovery of stishovite, a still higher-pressure silica polymorph, also in Barringer crater ejecta. Coesite was also

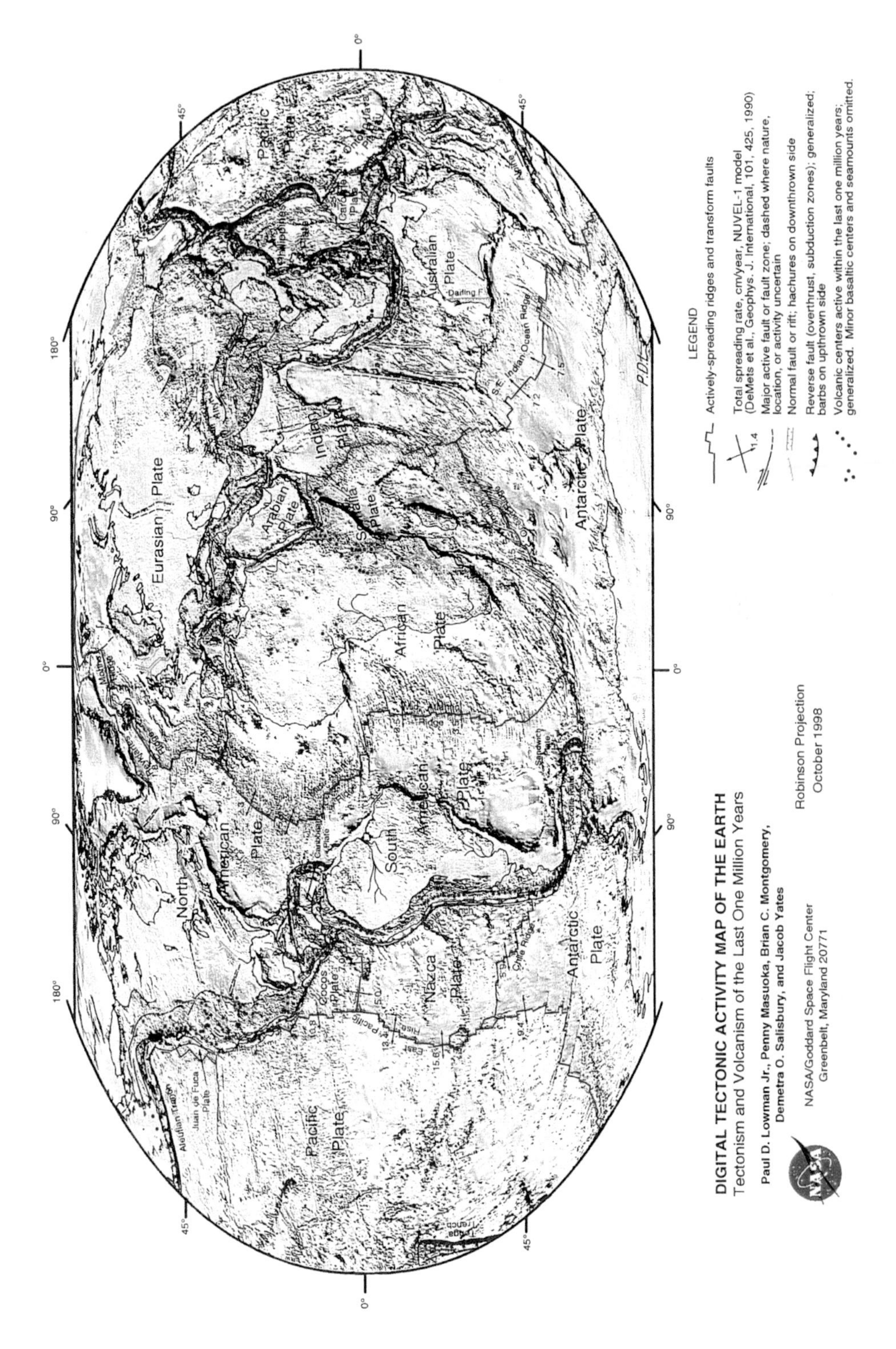

I Digital tectonic activity map (DTAM) of the Earth, based on shaded relief map largely generated from satellite altimetry.

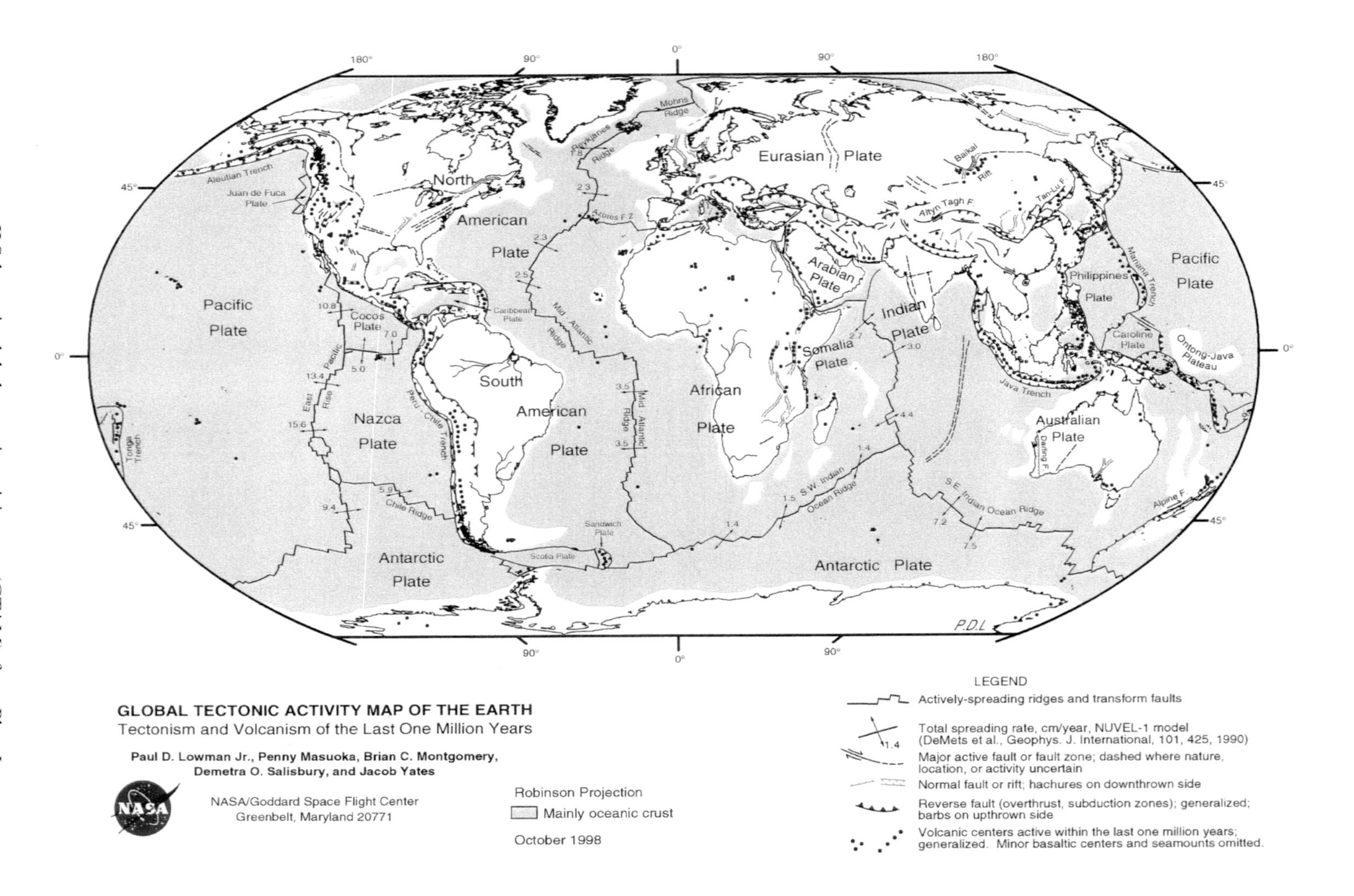

II Schematic global tectonic activity map (GTAM), from Plate I.

III Gemini 12 photograph S66–63082; view to east over the Zagros Mountains (left), Strait of Hormuz, and Makran Range.

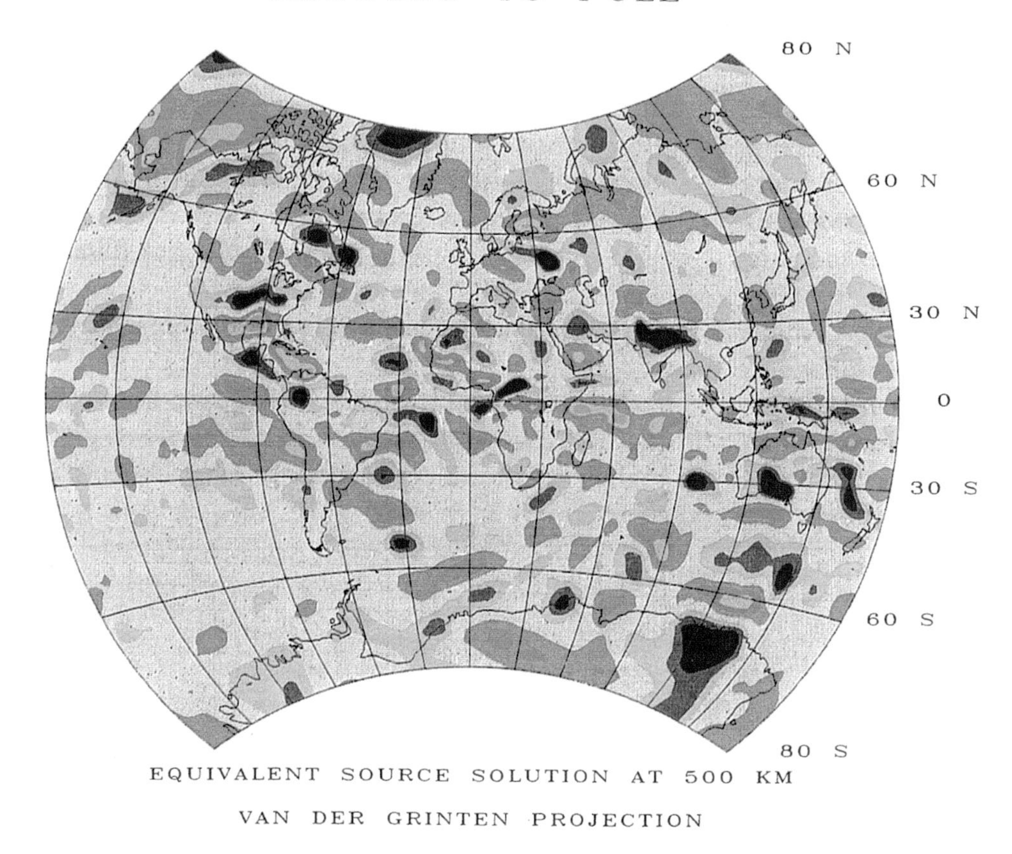

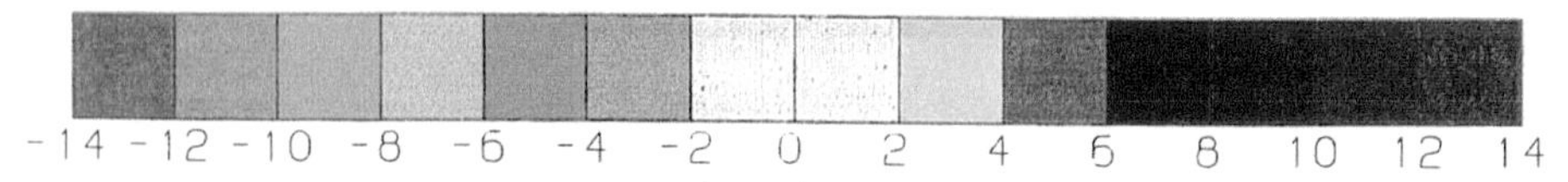

IV Scalar (non-directional) magnetic anomaly map from *POGO* data, equivalent sources at 500 km altitude, reduced to pole. Values in nT.

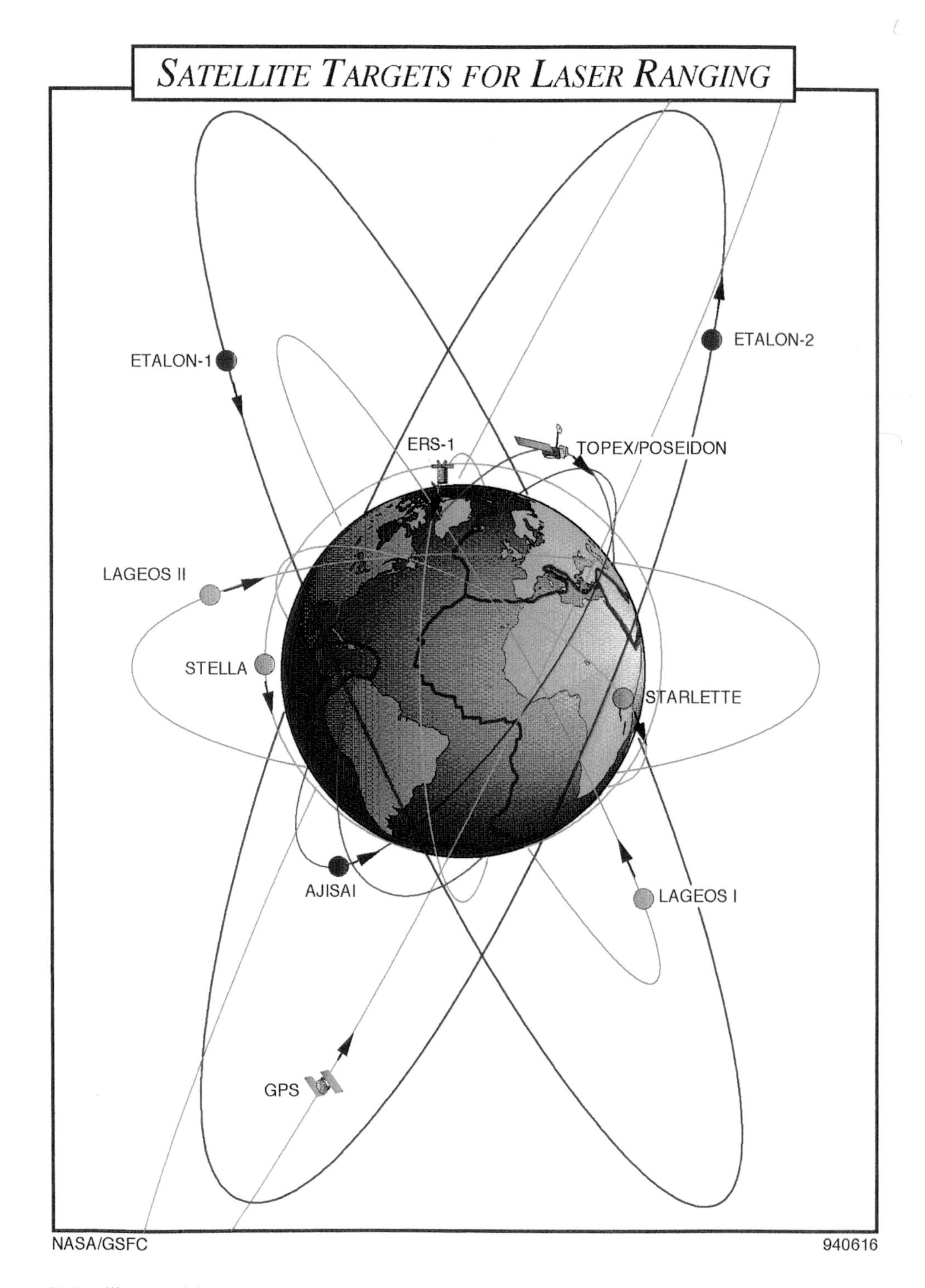

V Satellites used for laser ranging.

Selected VLBI Velocities

VI VLBI station velocities, NUVEL 1A-NNR reference frame. Note similar azimuths of all European stations.

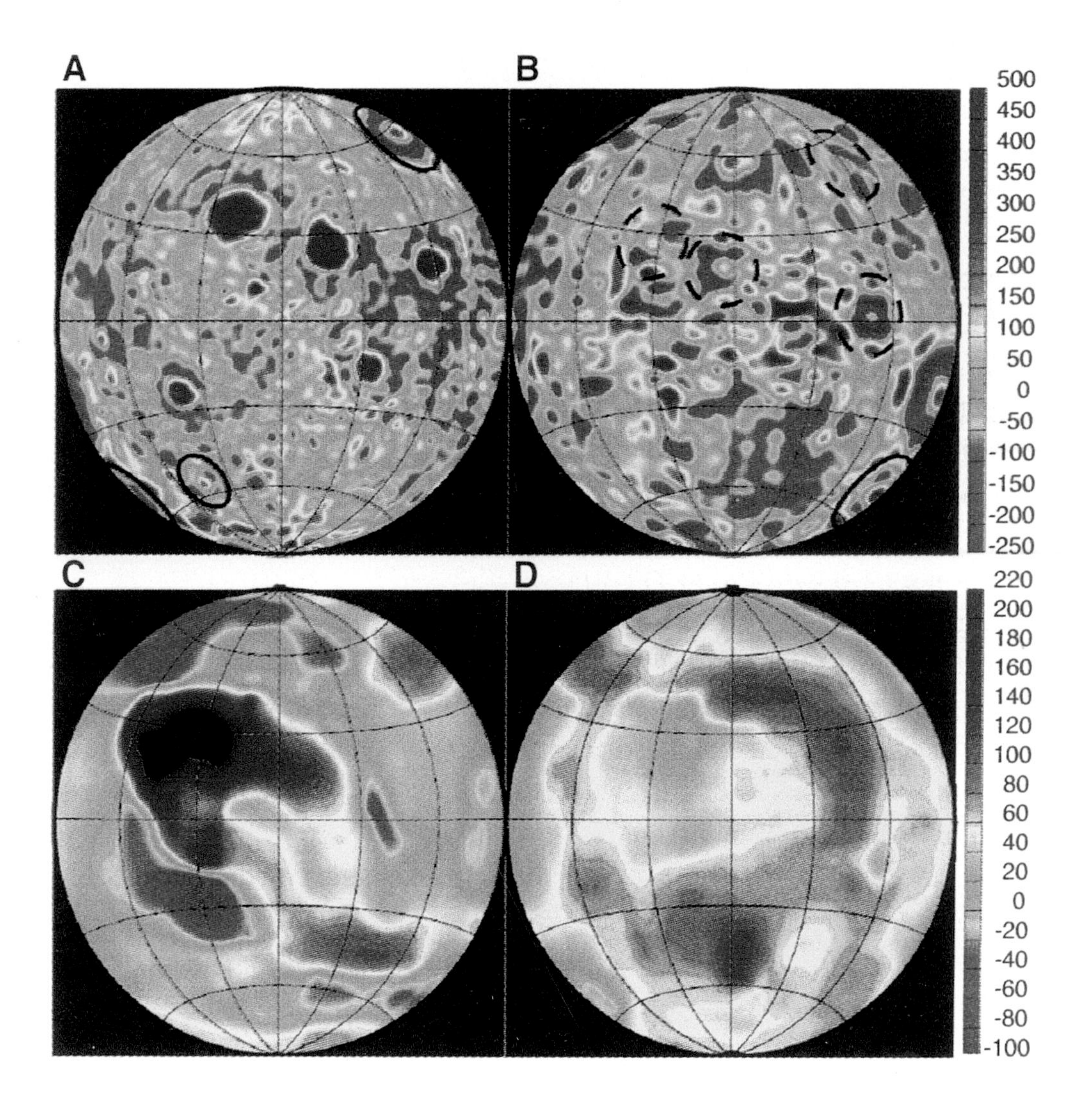

VII *Lunar Prospector* gravity and crustal thickness maps, Lambert equal-area projection. A, C: near side; B, D far side. *Top*: Vertical gravity anomalies, in milligals. Newly-discovered near-side mascons shown with solid circles, far-side ones with dashed circles. *Bottom*: Crustal thickness in kilometers, calculated with an Airy compensation model (constant density) without principal mascons.

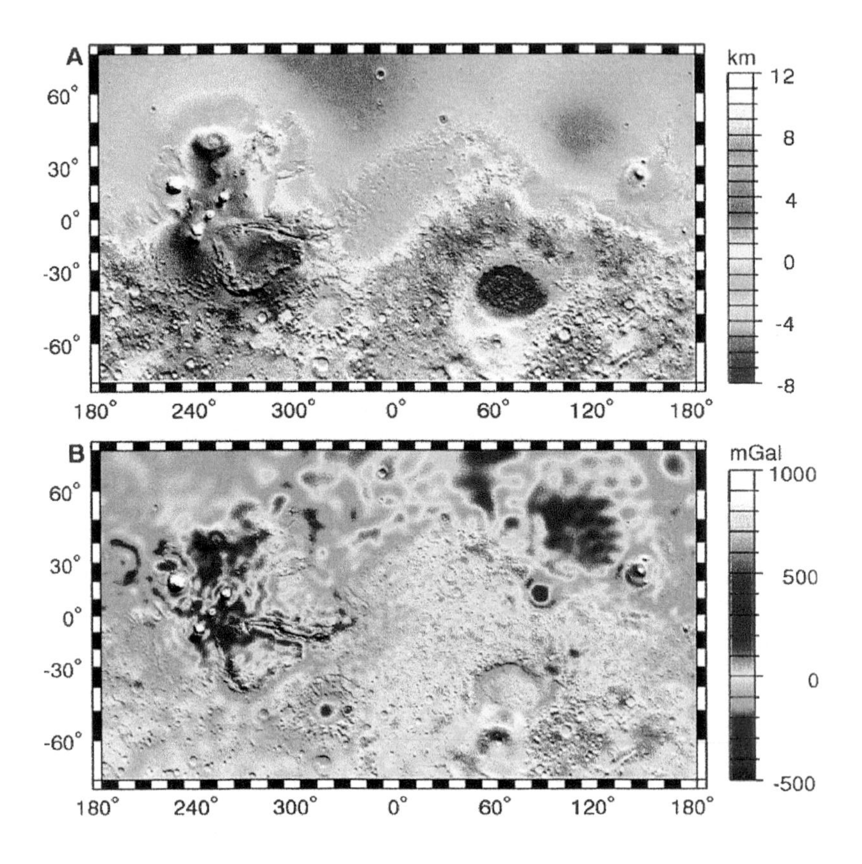

VIII *Mars Global Surveyor* maps of topography (*top*) and free-air gravity values (*bottom*). Tharsis area near the equator between 220 and 300 deg. E. Hellas Basin: 45 deg. S, 70 deg. E; Utopia: 45 deg. N, 110 deg. E.

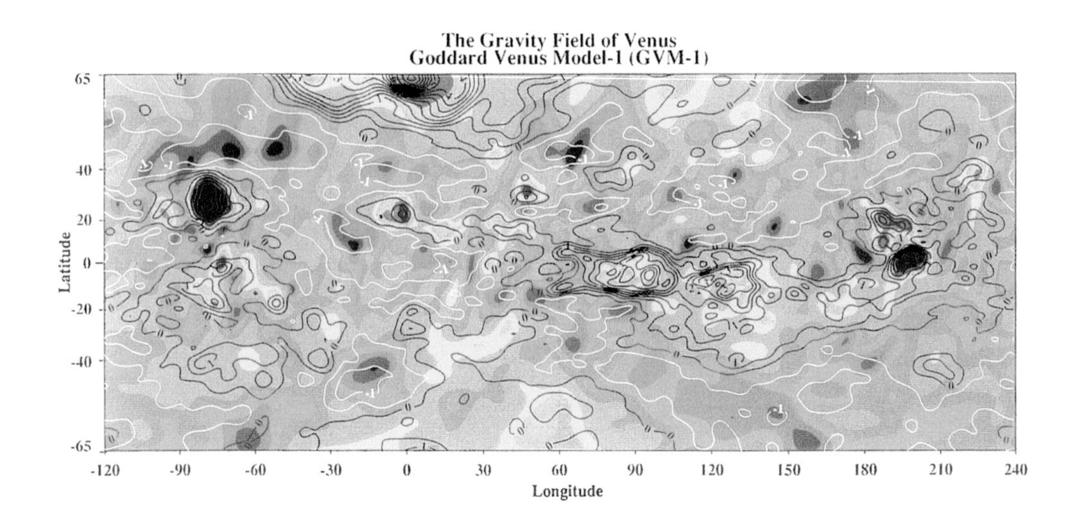

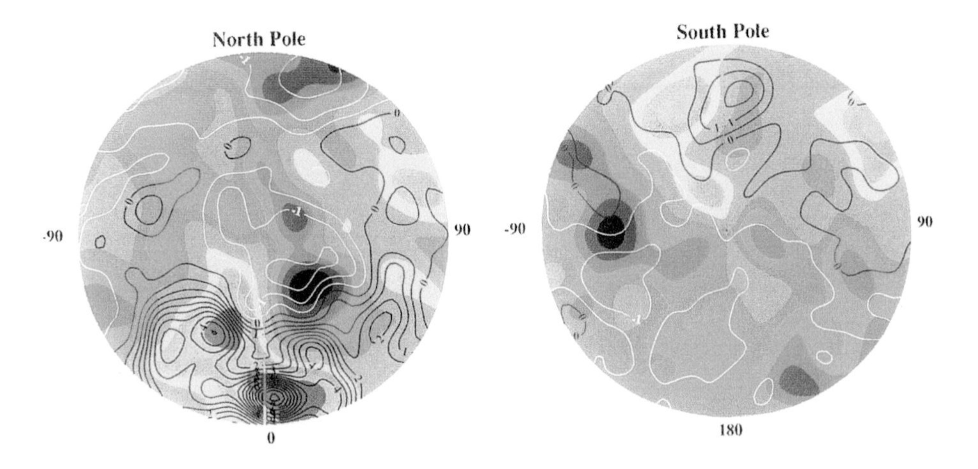

IX (*opposite bottom and above*) Gravity field of Venus, equatorial and polar segments, in milligals.

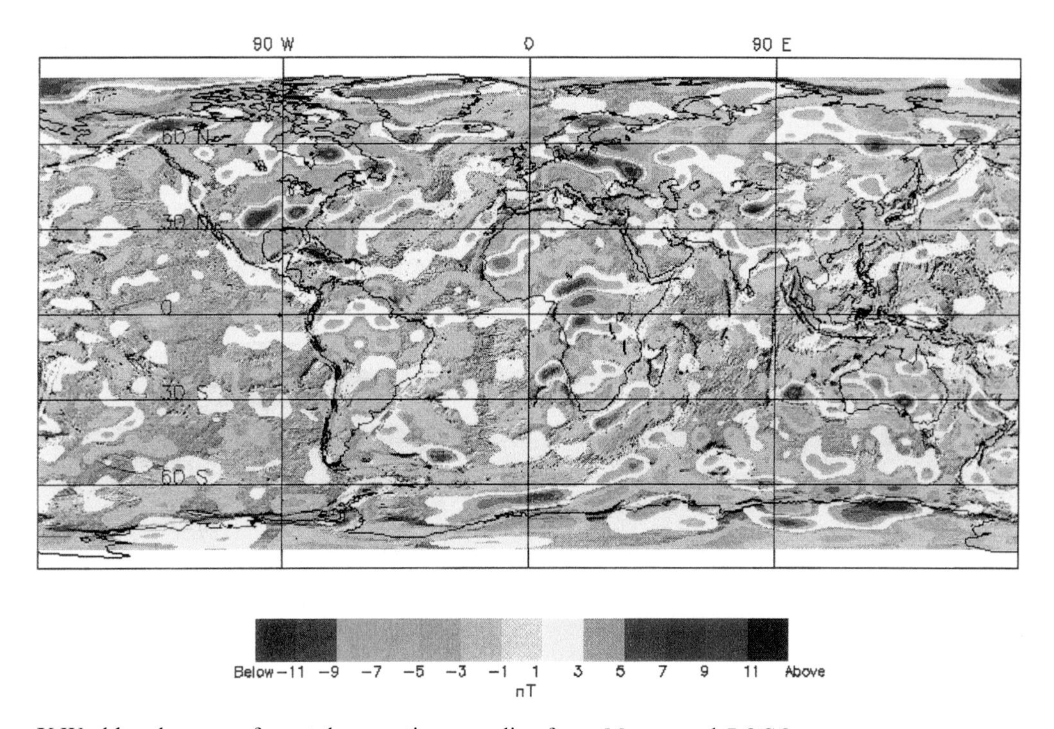

X World scalar map of crustal magnetic anomalies, from *Magsat* and *POGO* data.

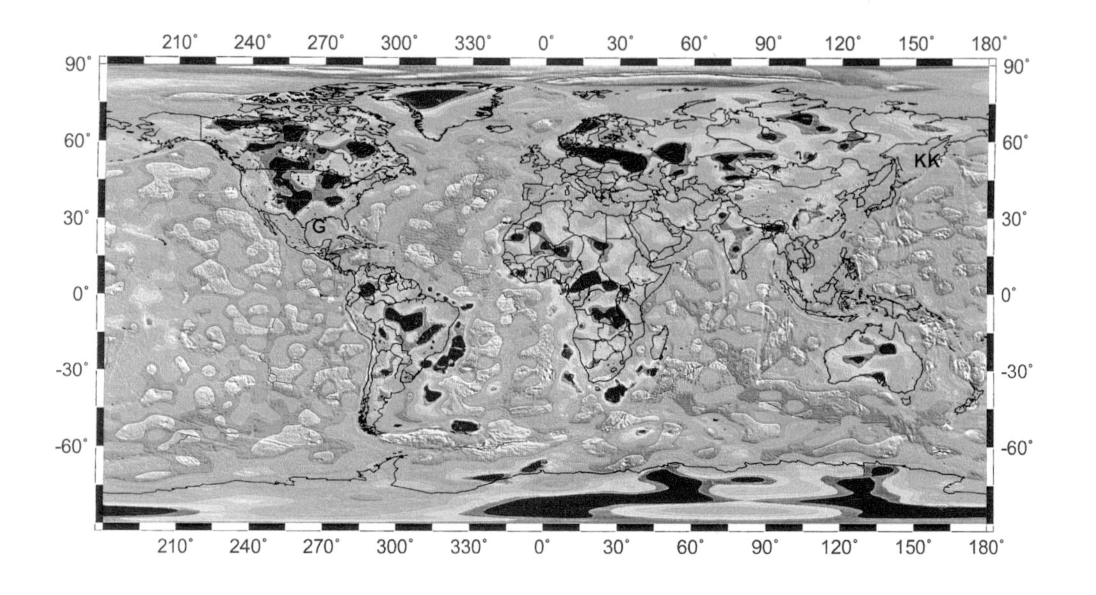
210° 240° 270° 300° 330° 0° 30° 60° 90° 120° 150° 180°
90° 60° 30° 0° -30° -60°
G
KK

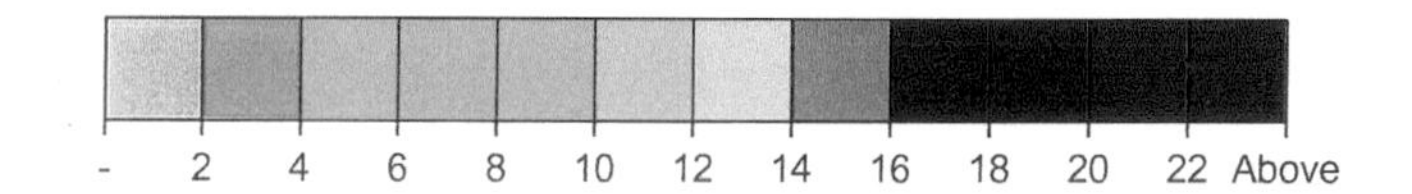
- 2 4 6 8 10 12 14 16 18 20 22 Above

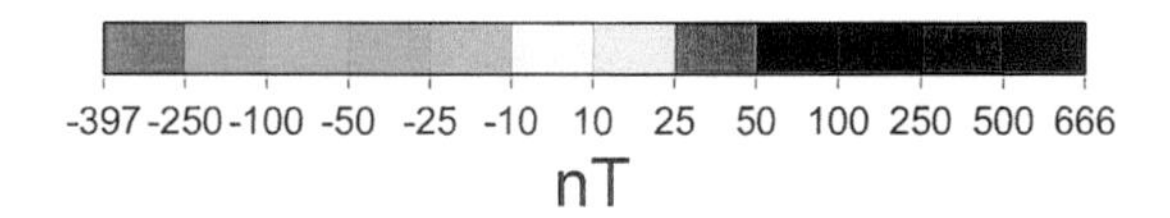
-397 -250 -100 -50 -25 -10 10 25 50 100 250 500 666
nT

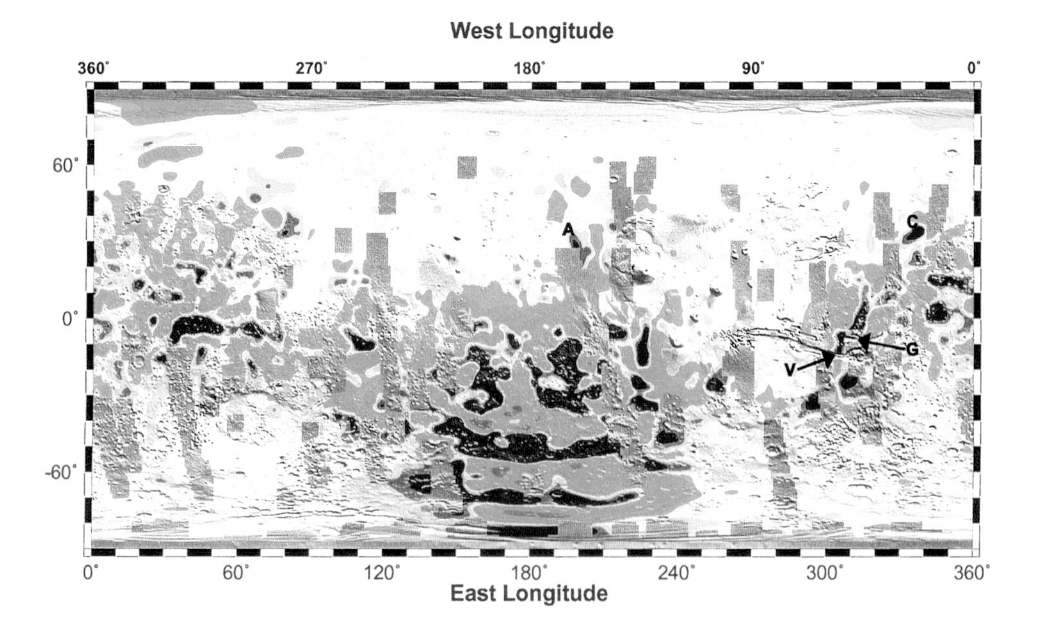
West Longitude
360° 270° 180° 90° 0°
60° 0° -60°
A
C
V
G
0° 60° 120° 180° 240° 300° 360°
East Longitude

XIII *Gemini 5* photograph over Salton Sea, looking northeast. Note linear valleys at lower left, crossing Elsinore Fault without offset. Gyre in Salton Sea is formed by winds through San Gorgonio Pass (*left*). Width of view ~150 km at center of photograph.

XI (*opposite top*) Global map of susceptibility (SI) time thickness times 10 of the SEMM-1 model shaded by surface topography for correlation with major bathymetric and topographic features.Negative values in gray. Units are SI $\times$ km $\times$ 10.

XII (*opposite bottom*) *Mars Global Surveyor* magnetic anomaly map of Mars.

XIV *Landsat* MSS mosaic of Red Sea area.

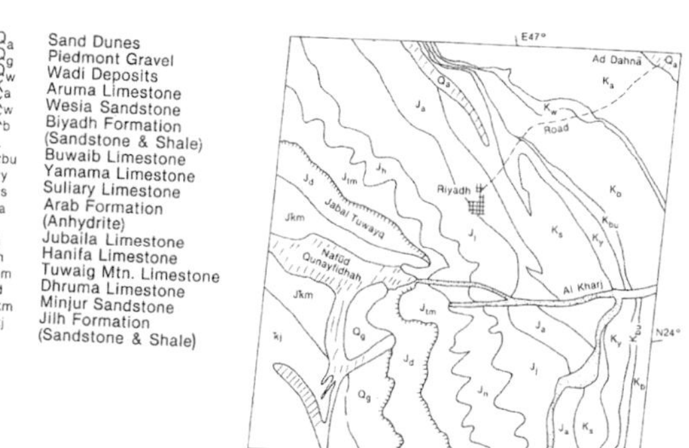

XV *Landsat* picture of Riyadh area, Saudi Arabia.

XVI Aerial view from 37,000 feet, looking northeast in western Nevada near Lake Tahoe, showing typical Basin and Range topography.

XVII IHS transform-modulated radar image of Nova Scotia, with proportions of potassium, uranium, and thorium used for IHS (see text).

N
eU
K
eTh
0
5
10 km

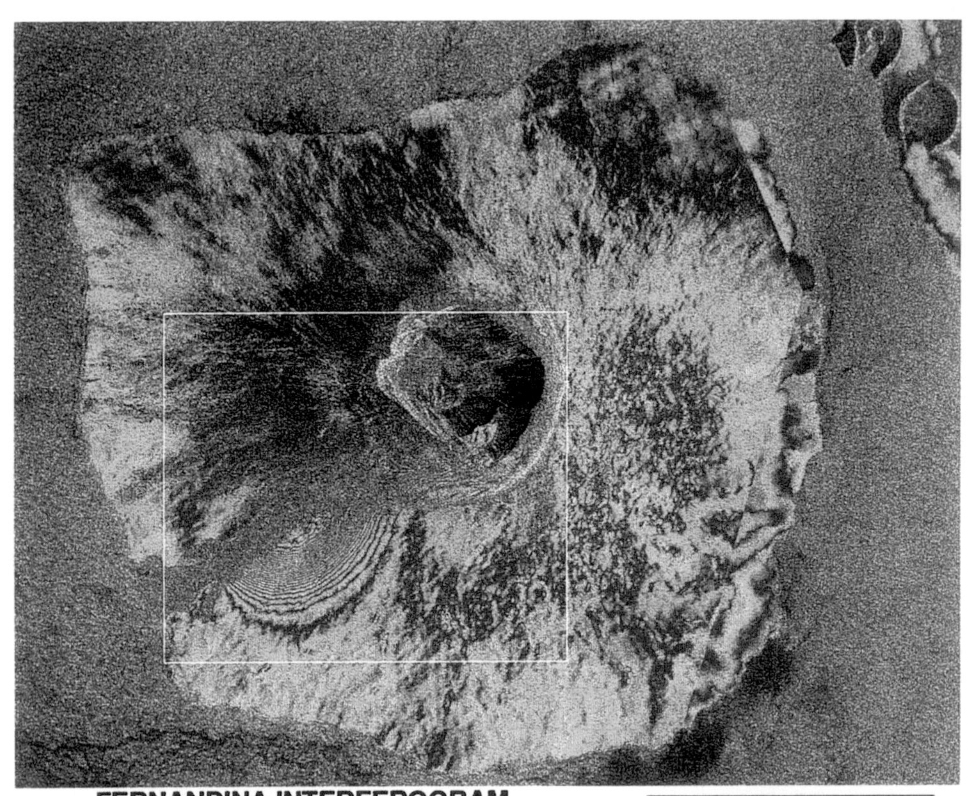

FERNANDINA INTERFEROGRAM
SEPT. 12, 1992 - SEPT. 30, 1997

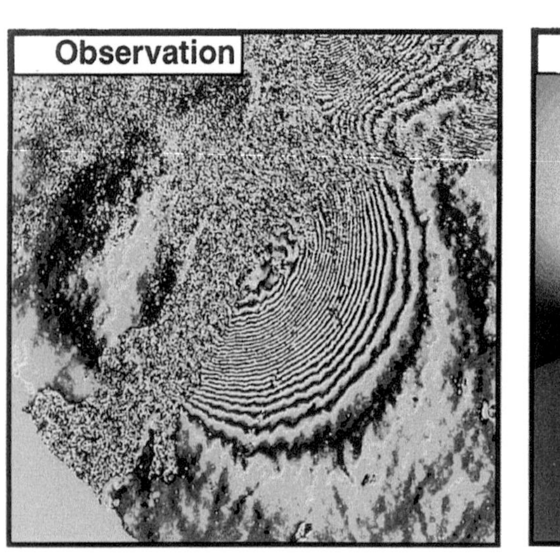

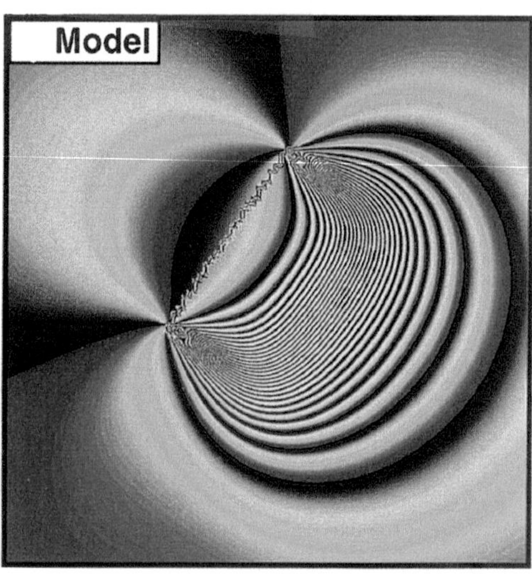

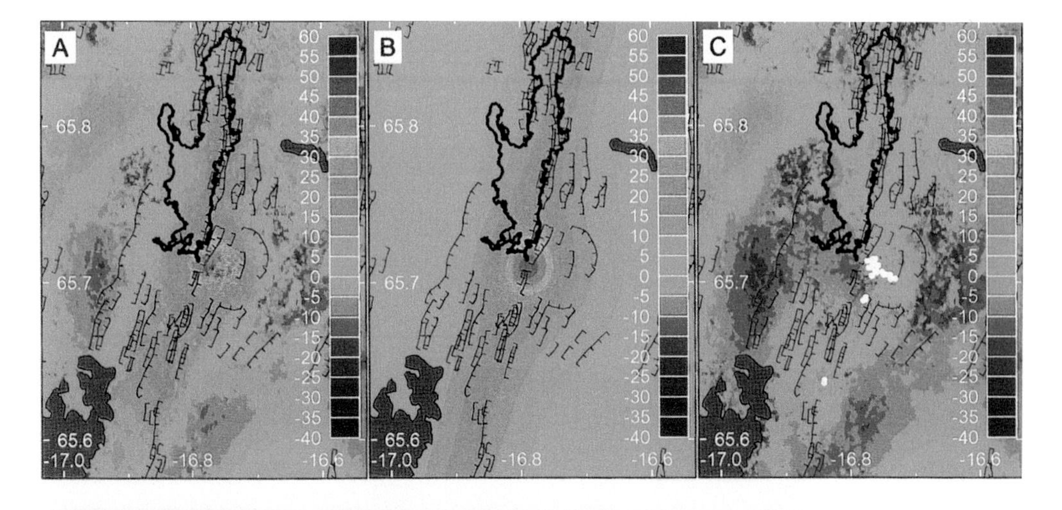

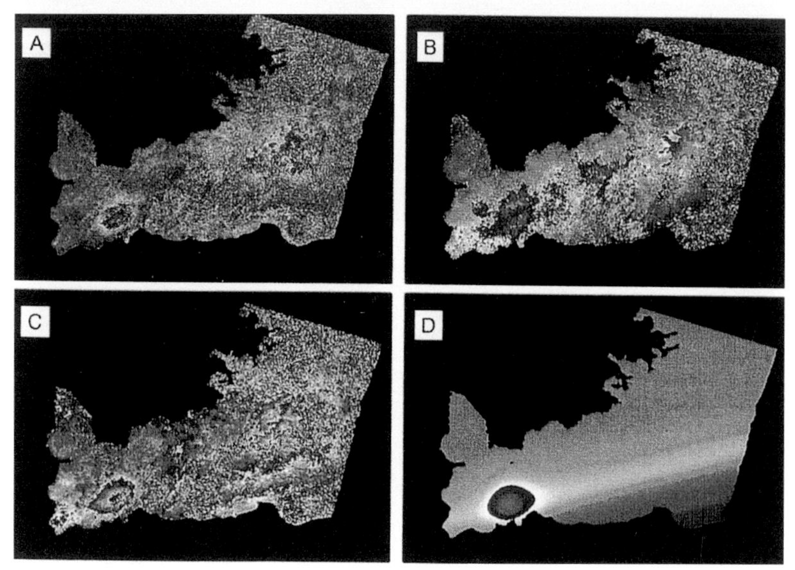

XIX Radar interferogram of Krafla, Iceland (*top*) and of 50 × 25 km area of Reykjanes Peninsula in southwest Iceland (*bottom*). See caption 4.33 for details.

XVIII (*opposite*) *Top*: Differential interferogram of Isla Fernandina, Galápagos Islands, Ecuador. Island is roughly 60 km east to west. Interferogram shows deformation during five-year period, including a flank eruption in 1995. *Bottom left*: Observed interferogram fringes over Fernandina flank. *Bottom right*: Best-fit model from interferogram, indicating a lava source dike striking N 47deg. E, dipping 33 deg. to SE.

XX *Landsat* picture of volcanic plume coming from Mount Etna, Sicily, during eruption in 1983.

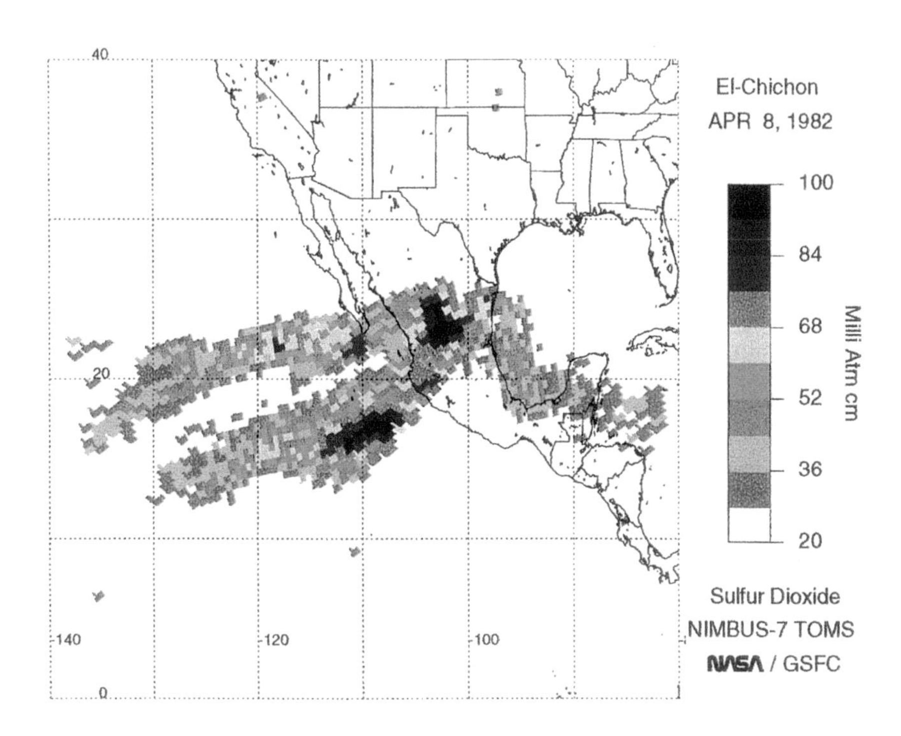

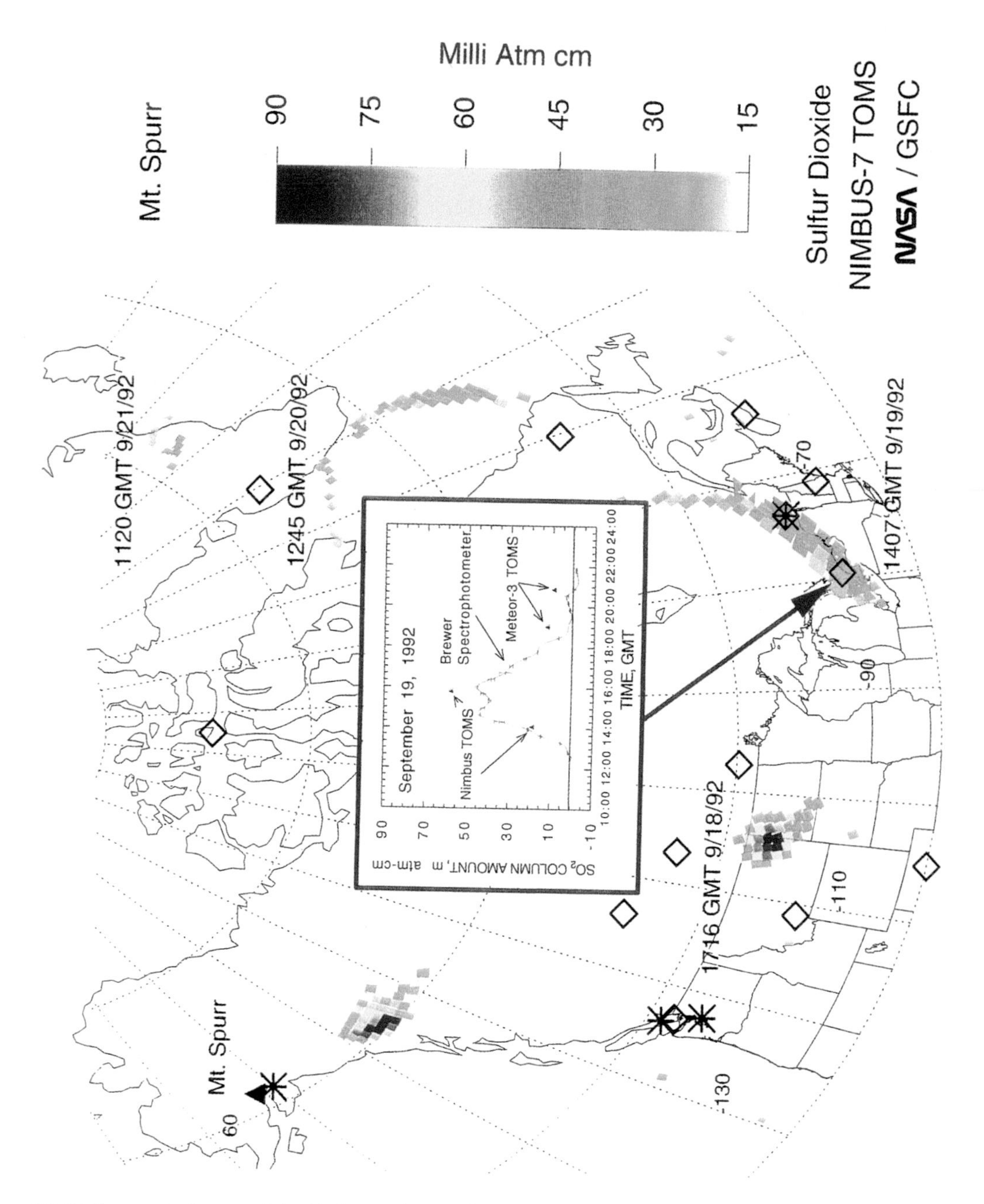

XXII Composite image of the 17 September 1992 Mount Spurr eruption sulfur dioxide cloud, showing path to southeast across North America.

XXI (*opposite*) Sulfur dioxide cloud from the 4–5 April 1982 eruption of El Chichon, Mexico, after drifting southwest for 4 days at 25 km altitude, as tracked by by Total Ozone Spectrometer (TOMS) on *Nimbus 7*.

XXIII *Landsat* multispectral composite picture of Vatnajokull, Iceland. Areas in red are vegetation, green in natural color.

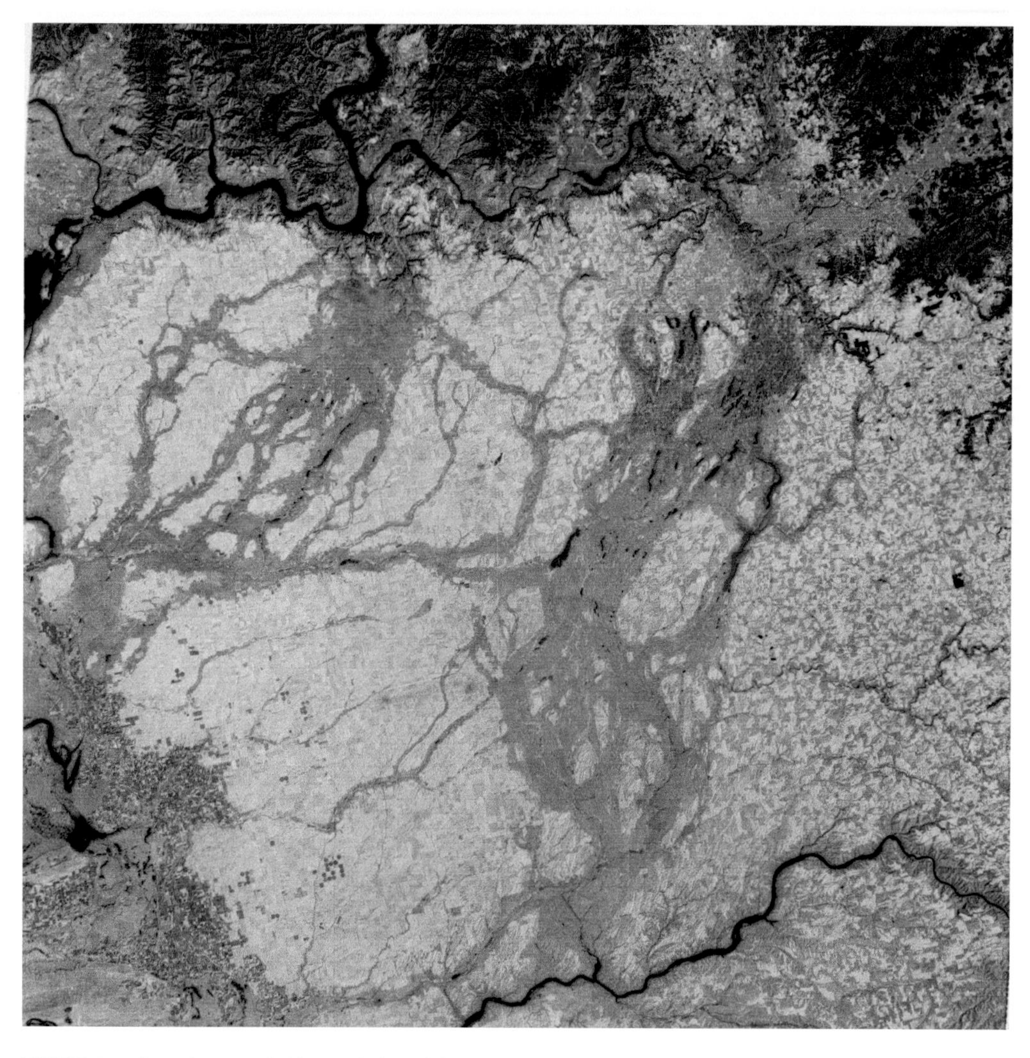

XXIV *Landsat* picture of Channeled Scablands, Washington; Spokane at upper right., Columbia River at upper left. Light gray and black patterns show valleys carved out by catastrophic glacial floods. *Landsat s*cene 1039–18143, 31 August 1973.

XXV *Gemini 5* 70 mm photograph of Walvis Bay area, Nambibia, showing sand dunes of Namib Desert. Northward transport of dunes is stopped by the Kuiseb River. Width of view ~100 km.

XXVI MODIS image of eastern United States and southern Canada, acquired 6 March 2000. Natural color.

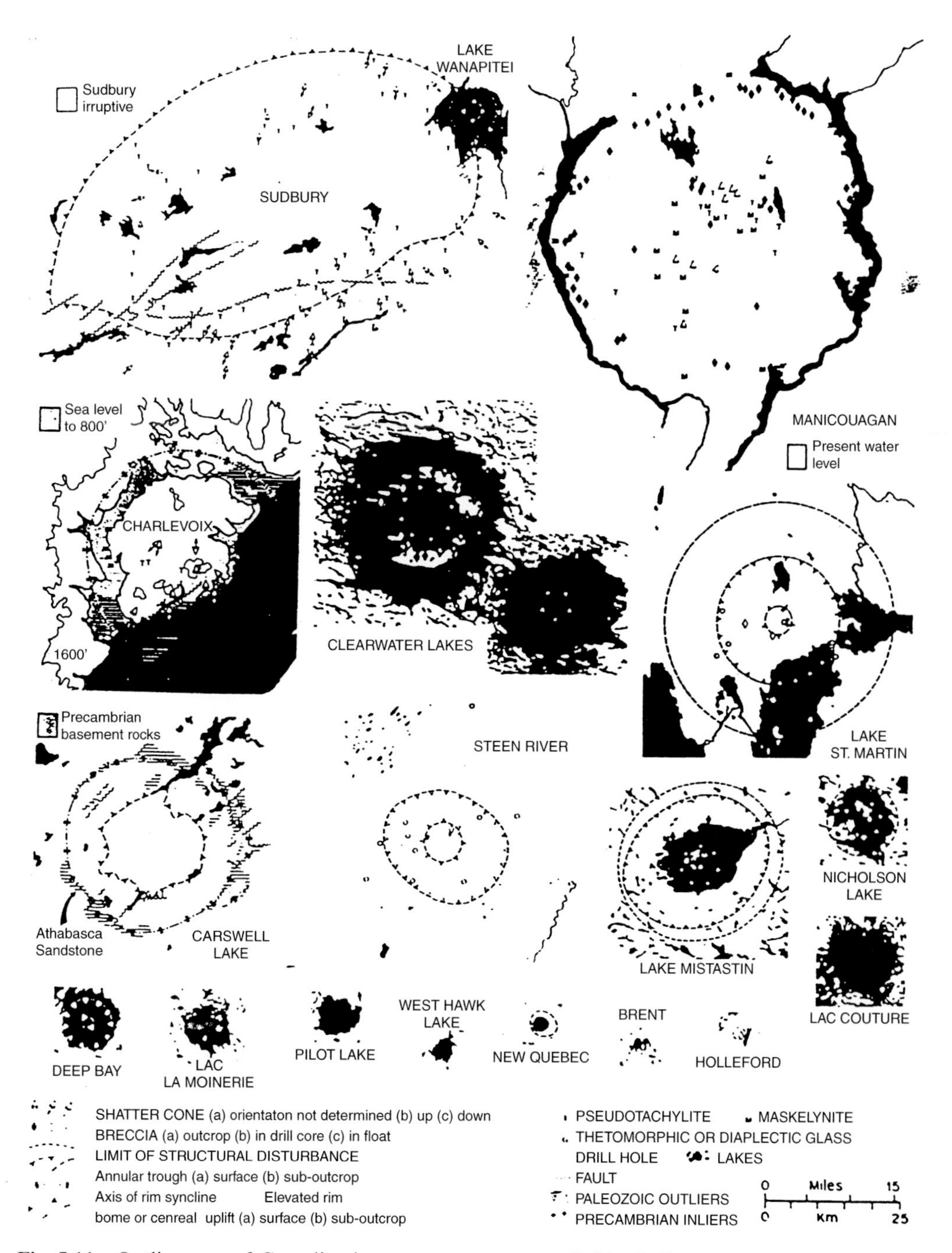

Fig. 5.11 Outline map of Canadian impact structures, compiled by D. P. Gold. Diameter of Sudbury Structure is debatable; line shown is not the outline of the igneous complex. See Grieve *et al.* (1991) and Lowman (1991) for discussion.

found in the Ries Basin, Germany, by Shoemaker and Chao (1961). Investigations of this sort multiplied rapidly, especially after the *Apollo* program began. The great increase in NASA budgets did not all go for rockets, spacecraft, and launch facilities; dozens of university grants were made, many for cratering research. These rapidly bore fruit, leading to the first meeting on shock metamorphism in 1966 at Goddard Space Flight Center (French and Short, 1966).

Developments in cratering studies during the 1960s and 70s were too rapid for even a brief summary here, especially since they overlap the expanding field of lunar and planetary exploration. The situation today has been best described by Melosh (1989): ". . . impact cratering is emerging . . . as one of the most fundamental processes in the solar system . . .". Let us turn now to some specific applications of cratering phenomena to terrestrial geology.

It is generally agreed that impacts represent the latest stages of planet formation (Shoemaker, 1977; Taylor, 1992). The older highland craters of the Moon, for example, may be considered the end of the Moon's accretion (although later impacts may lead to a net loss of mass). The South Pole–Aitken Basin (Fig. 5.12), 2500 km wide, is so old that there is almost no physiographic expression of its structure, apart from the *Clementine* laser altimetry confirming its depth (Spudis *et al.*, 1994). It must have been formed in the first few hundred million years of the Moon's existence, since its site is concealed by heavily-cratered highland crust. An impact this early must have been from one of the planetesimals from which the Moon accreted (or fragments from the "giant impact" thought to have formed the Moon). However, the collective effects of primordial impacts may have been extremely varied and complex. As discussed by Ward and Brownlee (2000), a given body may, instead of increasing in mass, be destroyed, assimilated by another body, or ejected from its orbit. Close range study of asteroids is giving us a new appreciation of these complexities.

The study of impact phenomena thus bears on the most fundamental problems of planetology. A comparably fundamental topic is formation of the Moon, now thought to have been by the impact of a Mars-sized object on the primordial Earth (Stevenson, 1987; Melosh,1989, 1992). These topics, important as they are, as as much cosmological as geological, and accordingly will not be further covered here. There are several specific areas of terrestrial geology, however, in which impact cratering may have played a major role and left behind some trace in the geologic record.

Fig. 5.12 *Galileo* image of Orientale Basin and far side of Moon. North at top. South Pole–Aitken Basin is dark circular area at lower left.

5.5 Origin of continents

Several workers have suggested that the first "continental nuclei" were formed in some way by major impacts early in the Archean (Salisbury and Ronca, 1966). Detailed petrologic proposals have been published by Goodwin (1976), Grieve (1980), and Glikson (1990). The general process began by "giant impacting," in Goodwin's term, about 4.2 to 3.8 Ga ago, roughly synchronous with the impacts that formed the lunar mare basins (1 Ga is one billion years). These impacts stimulated and localized mantle upwelling, producing "accelerated sialic differentiation" that formed protocontinents or nuclei. Goodwin speculated that the rough crescent shape of the hypothetical Pangaea, which is at least roughly similar to the distribution of the lunar maria, outlined the impact sites.

The geologic record becomes much better after 3.8 Ga, and it is generally assumed that the evolution of the impact-formed protocontinents from this time on was governed by more familiar internal processes leading to lateral accretion of island arcs, suspect terranes, and other elements. However, Glikson (1995) suggests that impacts may have played a major role in formation of granite–greenstone terrains through the Archean. The discovery of glass spherules that appear to be impact ejecta in Archean rocks by D. R. Lowe supports Glikson's proposal.

The concept of impact initiation of continent formation is a stimulating one. The strongest argument in its favor is the so-far unproven but near certainty that the Earth did undergo a distinct late heavy bombardment about four billion years ago or earlier. Large multi-ring impact basins are found not only on the Moon, where they have been well dated, but on Mercury, Mars, Venus, and Callisto (Spudis, 1993). The lunar mare basins, in particular, imply that a similar bombardment should have affected the Earth, barring divine intervention.

Beyond this, there are several problems with the impact theory for continental nuclei. The origin of continents, more precisely of continental crust, will be discussed in detail in the next chapter; the following is only a preliminary summary. First, it is by no means clear that there are such "nuclei" in the genetic sense, i.e., extremely old cores around which continents grew. Radiometric ages of exposed continental crust do not give a true picture of its chronological development, because many dates represent metamorphism, not mantle separation times. Furthermore, the concentric arrangement of Phanerozoic orogenic belts in North America is clearly deceptive because these belts appear to be underlain by Precambrian crust. In South America, most of the west coast is Precambrian or Paleozoic rock, yet this is where the youngest rocks should be if the continent is growing laterally. These relationships cast doubt on the very concept of "continental nuclei."

Second, there is the problem that giant impacts on the Moon and on Mars formed nothing like sialic nuclei. The circular lunar maria in particular are now known to be large impact basins filled by many generations of basaltic extrusions, and the gross topography of Mercury and Mars suggests a similar sequence of events. However, the primordial Earth differed from the Moon and probably the smaller planets in several critical aspects: greater heat retention, greater volatile and lithophile content, and probably greater internal activity (i.e., mantle convection and upwelling). Apart from their inherent importance in promoting sialic crust formation, these conditions may have helped differentiation of the impact melt presumably formed by giant impacts on the early Earth. One specific example of such melt differentiation has been proposed (Grieve *et al.*, 1991; Lowman, 1992) to be the Sudbury Igneous Complex, whose bulk composition and internal zoning suggest a process analogous to classic magmatic differentiation (Marsh and Zeig, 1999). It has been shown by Tonks and Melosh (1993) that a "magma ocean" may have resulted from giant impacts on the primordial Earth. Such an event would presumably promote planetary differentiation, although Warren (1993) has argued that a magma ocean would be an impediment to plate tectonics on the Moon. Whether this would be true for the Earth is not clear.

Further speculation about these possibilities would be unproductive, but it is clear that since continental crust began to form at least 4.0 Ga ago, any major processes active then must be taken into account. Giant impacts were almost certainly one of these major processes.

5.6 Origin of ocean basins

It may be that impacts were involved in the formation, not of continents, but of ocean basins. As will be discussed in the following chapter, the fundamental question of terrestrial geology may be not "How did the continents form?" but "Why does the Earth have two kinds of crust?". The problem is then to explain the crustal dichotomy in general. One possible explanation, based on the presumed period of major impacts, has been put forth by Frey (1977, 1980) to account for the Earth's *first* ocean basins. The term "first" is emphasized for obvious reasons. The present ocean basins are apparently not much older than Mesozoic, and their evolution appears to be reasonably well explained by sea-floor spreading. But the nature of pre-Mesozoic oceanic crust and ocean basins is unknown.

First showing that the Moon and Mars had undergone major periods of impact that disrupted roughly half their original crust, Frey investigated the effects that such impacts might have had on the Archean Earth. With plausible assumptions on initial temperatures, and assuming a global andesitic crust, Frey showed that a late heavy bombardment might have triggered mantle upwelling and sea-floor spreading under the impact sites. This would have created the first ocean basins, essentially similar to those of the present except for their mode of initiation. The present ocean basins are, in Frey's view, the descendants many generations removed of the primordial ones. However, a more recent version of this theory proposed by Oberbeck *et al.* (1993) holds the breakup of Gondwanaland (the hypothetical supercontinent believed by many to have existed in the Paleozoic) to have been initiated along crustal fractures resulting from impacts.

Like the impact theory for continental nuclei, an impact origin for the earliest ocean basins is largely unconstrained by geologic evidence. However, the simplest comparison between the Moon and the Earth suggests that the consequences of the late heavy bombardment would certainly have been quite different on the Earth. A possible weak link in the argument is the assumption that the Earth developed a differentiated (i.e., andesitic) global crust in the early-Archean. This is admittedly a minority view. However, it is supported by the prediction of an andesitic crust on Mars, verified in a general sense by the *Pathfinder* and *Mars Global Surveyor* missions, to be discussed in the next chapter.

Regardless of the validity of the impact hypothesis, it demonstrates the value of comparative planetology in refocussing a fundamental question, from origin of the continents to a broader one: origin of the continent–ocean basin dichotomy.

5.7 Economic importance of terrestrial impact structures

The term impact *structures*, rather than craters, is used here to emphasize at once that because of erosion, deposition, and post-impact tectonic or igneous activity, impact craters on Earth are frequently no longer recognizable as such (Rondot, 1994, 1995). The name Sierra Madera, meaning wooded mountain, describes the present appearance (Fig. 5.13) of what has long since been proven to be the center of a 13 km impact crater in west Texas (Wilshire *et al.*, 1972) corresponding to the central peaks of lunar craters such as Tycho. Sierra Madera was evidently formed in the Late Cretaceous, and its present expression is the result of erosion (Lowman, 1965). Its recognition as an impact structure, first argued by Dietz (1960) on the basis of shatter cones, was an early example of the difficulty in confirming impact origins for very old structures. But Sierra Madera turned out to have more than scientific value, for since the initial hydrocarbon discovery in 1977 (Donofrio, 1997), twenty gas wells drilled on it have been brought into production.

Impact structures on Earth have now been recognized to have great economic importance, both as hydrocarbon sources and mineral deposits. A landmark report on this topic has been edited by Johnson and Campbell (1997), containing several valuable reviews including those by Grieve (1997), Donofrio (1997), and Buthman (1997). A useful classification of impact-related economic deposits

Fig. 5.13(a) Photomosaic of Sierra Madera, Texas, taken from US 385 looking east. Field of view at horizon about 4 km left to right. Photograph by author, 1961.

has been proposed by Grieve (1997), based on the time of formation of the deposits relative to the initial impact. Progenetic deposits form before the impact, whose effect is to redistribute the deposits. Syngenetic deposits form during or immediately after the impact. Epigenetic ones form well after the impact, by, for example, the migration of fluids such as oil and gas into the breccias produced by the impact.

By far the most important structures from an economic viewpoint are epigenetic ones localizing oil and gas deposits in which the impact-formed breccia serves as a reservoir rock, or in which impact-formed structures such as the Sierra Madera uplift provide structural traps. The most spectacular example of such economic impact features is the 300 km wide subsurface Chicxulub crater in Mexico, now confirmed as the "dinosaur killer" at the end of the Cretaceous (Sharpton *et al.*, 1992). With 453 producing wells (Donofrio, 1997) as of 1995, the proven oil reserves at Chicxulub, some 25 billion barrels, are more than the total reserves of the

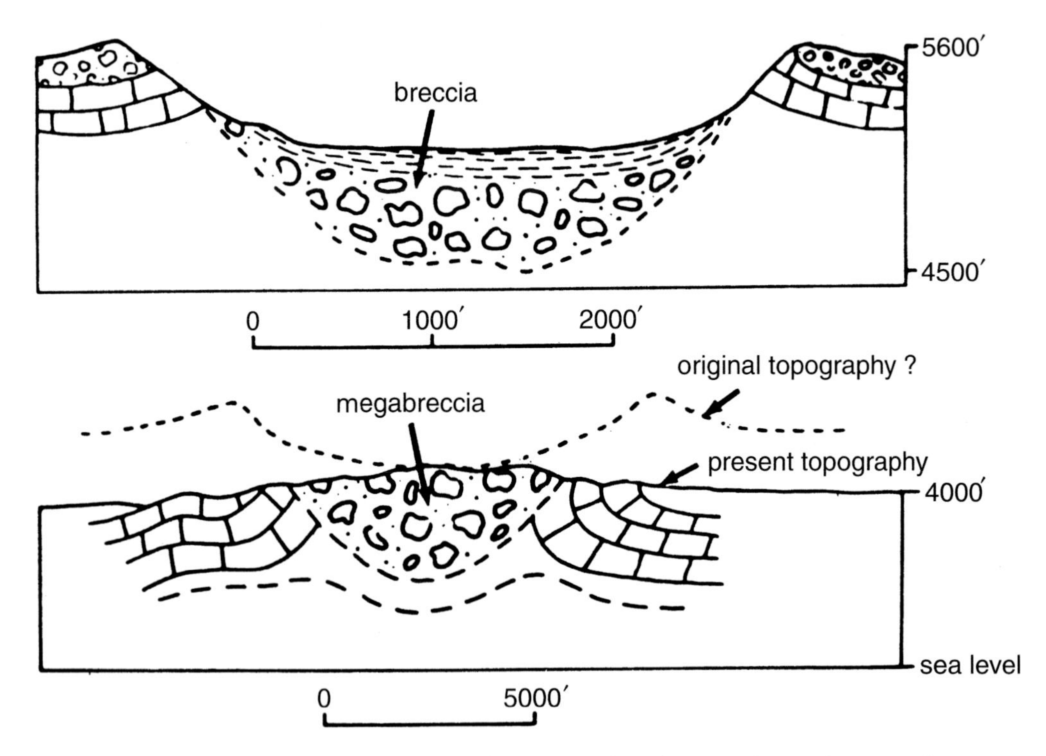

Fig. 5.13(b) Cross sections comparing Sierra Madera and Meteor Crater (1000′ (feet) = 305 m). From Lowman (1965).

conterminous United States including the Alaskan North Slope. It should be pointed out that these wells were not drilled because the crater was identified; the crater was identified on the basis of evidence from already-producing wells. Nevertheless, the enormous value of the Chicxulub structure dramatically illustrates the economic importance of impact structures in hydrocarbon production alone.

The specific ways in which knowledge of impact craters can be applied to petroleum exploration is illustrated by the history of the 13 km wide Ames structure in northwest Oklahoma, described in several papers in the collection edited by Johnson and Campbell (1997). Occurring in the Ordovician, the impact produced a crater now some 3 km down, with no present surface expression at all. Its general shape is obvious only in subsurface, on maps showing, for example, the thickness of sedimentary formations overlying the crater (Fig. 5.14). The first hydrocarbon discovery was in 1991 (Donofrio, 1997). Further drilling and other investigations began to

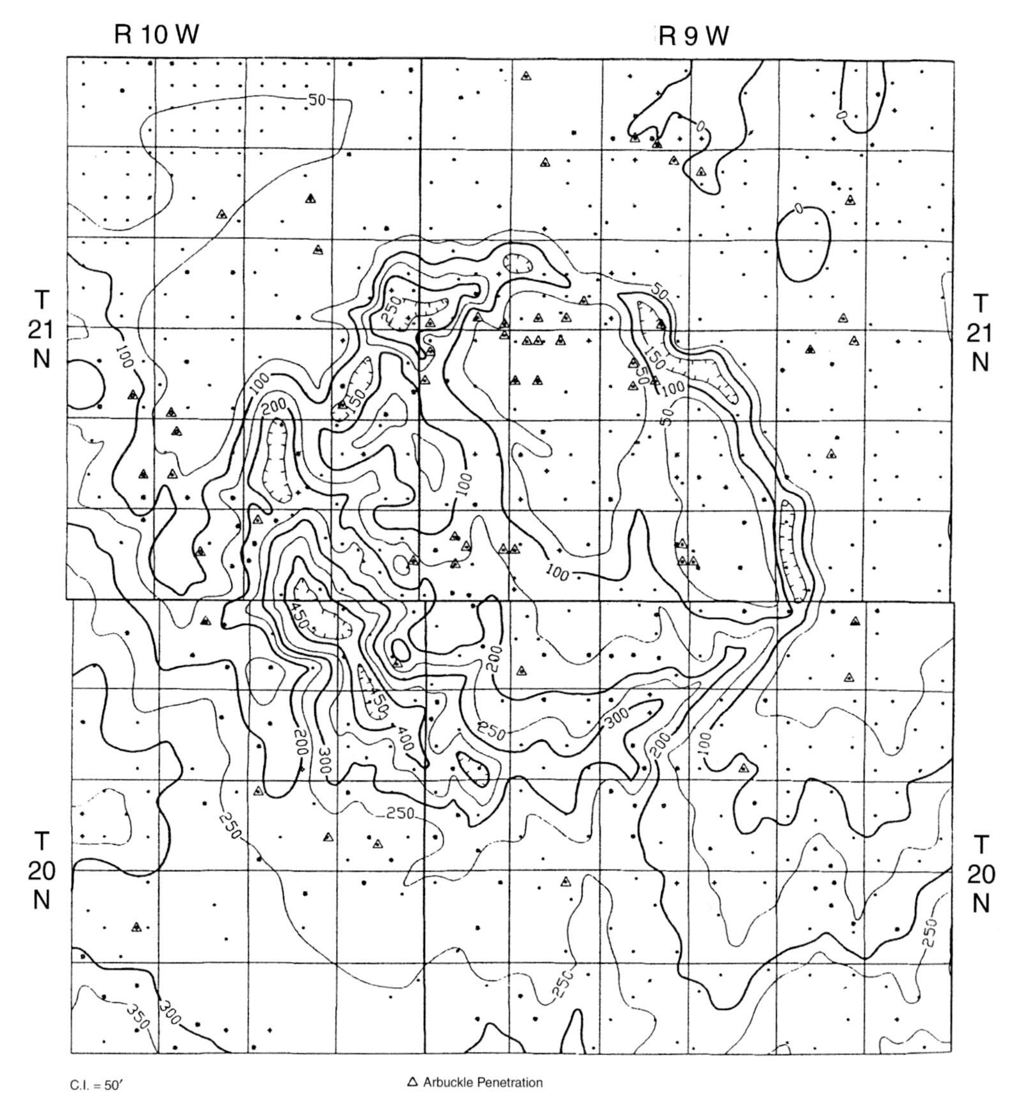

Fig. 5.14 Isopach map showing thickness in feet of Hunton Group, Silurian–Devonian rocks overlying Ames Structure (Ordovician age), interpreted as showing continuing post-impact subsidence over crater. One mile (1.6 km) squares (sections). From Coughlon and Denney (1997).

reveal the circular shape of the structure, raising the possibility of an impact origin. Drilling brought up granite from the center of the structure, and petrographic studies showed evidence of shock metamorphism. Seismic studies were much more effective after the outlines of the structure and its probable impact origin became clear, as brought out by Sandridge and Ainsworth (1997). By 1995, there were 40 producing oil and gas wells on the structure.

The 35 km wide crater underlying the southern Chesapeake Bay, first identified by Poag (1997), is similarly invisible on the surface, despite its relatively young age of 35 million years.

Donofrio (1997) lists nine confirmed impact structures in North America producing oil or gas. Although at this writing there has not been a discovery based solely on knowledge of an impact origin, it is clear that petroleum-exploration geologists must be aware of the nature and occurrence of impact structures for maximum effectiveness.

The classic Meteor (or Barringer) Crater of Arizona was first drilled by the Barringer Crater Company early in the 20th century in hopes of finding a large iron meteorite. This never materialized, but several impact-localized metallic mineral deposits are now known. The Vredefort dome of South Africa, for example, now generally agreed to be the deep root of a Precambrian impact structure, appears to have controlled the present distribution of the pre-impact gold and uranium deposits, progenetic in the classification of Grieve (1997). However, the most famous impact-related ore deposit occurs in the Sudbury Structure of Canada, for many years the most productive single source of nickel in the world and still a major producer of nickel, copper, and platinum group elements. The story of how this feature was found to be of impact origin deserves a dedicated section.

5.8 Origin of the Sudbury Structure

In 1964, Robert S. Dietz proposed that the Sudbury Igneous Complex, which hosts one of the world's largest nickel–copper deposits, was an extrusive lopolith occupying an impact crater. An impact origin is now accepted by the great majority of geologists familiar with the structure, but when proposed the idea was rejected as "nonsense" by a journal reviewer. The evolution of this theory is an unusually valuable one, typifying the effect of space exploration on geology, and it will therefore be recounted briefly. The following summary is based on a longer review (Lowman, 1992).

We begin with what may seem an utterly unrelated subject: tektites (Fig. 5.15). Tektites are glassy bodies found on the Earth. That is the only point on which all authorities agree; from there on the subject has traditionally been a jungle of controversy, unanswered questions, and misleading clues. It could fill books by itself, and in

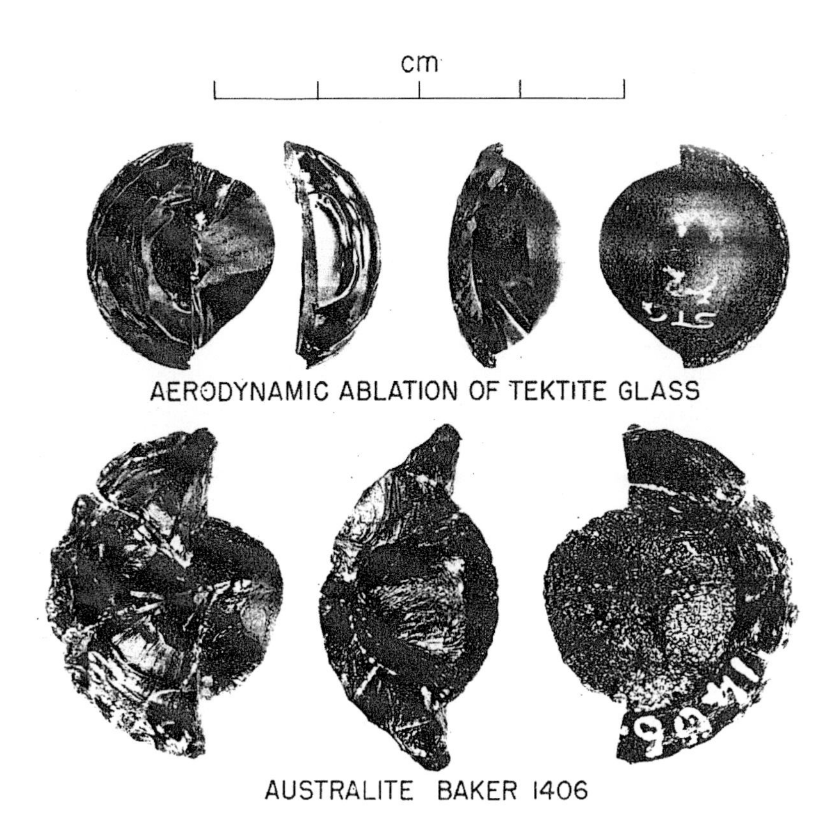

Fig. 5.15 Natural Australian tektites (*bottom*), and artificial tektites (*top*) produced by hypersonic ablation in an arc jet at the NASA Ames Research Center by Chapman (1964). Similarity demonstrates that australites entered the atmosphere as solid bodies at several kilometers per second.

fact has (Barnes, 1940; Barnes and Barnes, 1973; O'Keefe, 1963, 1976). Good summaries have been presented by King (1977) and Glass (1982, 1990). The main points of the tektite controversy are the following.

Glasses in general are supercooled liquids, i.e., quenched melts. Tektite glass has chemical compositions ranging from, roughly, basalt to granite. Unfortunately, liquids retain little visible memory, so to speak, of the material from which they were formed. Relict minerals do occur in some tektite glass from Viet Nam, chiefly high-temperature ones such as zircon, quartz, and rutile (Glass, 1990). The bulk composition of tektite glass can be matched by a wide range of parental rocks and soils; the granitic compositions of tektites as originally defined can also be matched by various sedimentary and igneous rocks (Barnes, 1940; Lowman, 1962). For a number of reasons, it is essentially certain that tektites are not terrestrial volcanic glass, i.e., obsidian. The problem is then to find out how and where the parent material was melted to produce the liquids which, when quenched, formed tektites at various times from 35 million to about 700 thousand years ago.

Australian tektites have distinctive shapes that were shown experimentally by Chapman (1964), a noted authority on hypersonic aerodynamics, to have been produced by ablation in the Earth's atmosphere at close to 11 km/s. This is escape velocity for the Earth and also re-entry velocity, and remains a major argument for an extraterrestrial origin for at least the australites. A lunar origin was championed by O'Keefe (1963, 1976, 1985), most recently from hydrogen-driven lunar volcanos. However, no tektites have been found on the Moon, and there is little evidence to date of any large amount of parent material for the granitic tektites.

Majority opinion holds tektites to be droplets of melt from major impacts on the Earth (King, 1977; Koeberl, 1994). Microtektites have in fact been found in association with several impact craters, and classic tektites have been linked with the Ries crater of Germany and the Bosumtwi crater of Ghana. Tektites have been found associated with the Chixculub structure, whose impact origin is now undoubted as reported by Izett (1991). The long-standing problem, stressed by O'Keefe, of apparent Holocene ages for Australian tektites, whose radiometric age is about 770,000 years, has now been apparently settled by field work in the Port Campbell area by Shoemaker and Uhlherr (1999) showing that their stratigraphic age is similar to the radiometric one.

The subject is considered closed by almost all authorities, except as an interesting aspect of impact cratering. However, the original work by Chapman, demonstrating ablation at re-entry speeds, has never been convincingly refuted. In addition, it has never been shown how material that is almost all glass can be formed within seconds from the heterogeneous crystalline mixtures making up most possible parental rocks. As O'Keefe repeatedly pointed out, industrial glass makers would be very happy to find out how this is done.

The tektite problem was, in the 1960s, studied by many of the great scientists of the day, generating an enormous literature. Arguments on both sides have been compiled by Varrichio (1999). However, it can be shown that tektite research helped revolutionize thought on the origin of the Sudbury Structure which, to remind the reader, is the heading of this section.

Fig. 5.16 Side-looking airborne radar image of Sudbury area, Ontario, acquired by Canada Centre for Remote Sensing Convair 580 at 20,000 feet altitude. North at top; illumination to south. Sudbury Basin at center is about 60×30 km. From Lowman (1992).

The Sudbury Structure had been mined for nickel, copper, and platinum group elements for about a century, after its discovery in the 1880s (Pye *et al.*, 1984). Many aspects of Sudbury geology were puzzling, in particular the origin of the ore deposits, but the Structure (Fig. 5.16) was universally interpreted as an igneous intrusive feature (e.g., a lopolith). A new interpretation for the Sudbury Structure (as well as the Bushveld Complex) was proposed by Hamilton (1960), who suggested that the "lopolith" was actually an extrusive, roofed by its own differentiates. Shortly after, the author (Lowman, 1962) applied Hamilton's concept to the Moon, as part of an effort to find a petrologic explanation for a lunar origin of tektites. He suggested that the lunar maria, even then recognized as occupying large impact basins, were extrusive lopoliths surfaced by ash flows that had been the source of tektites (O'Keefe and Cameron, 1962).

Following his earlier work on lunar geology, Dietz had been looking for terrestrial counterparts of the maria. As a marine geologist, and early supporter of sea-floor spreading, Dietz realized that the ocean basins are quite different from the maria in structure and composition. The proposal of Lowman supplied a possible answer to his quest. Having recognized the probable impact origin of the mare basins much earlier (Dietz, 1946), he realized that the terrestrial "maria" might also be impact craters. A proponent of shatter cones (Dietz, 1960) as index fossils, so to speak, of impact craters, he visited Sudbury in 1962 to see if there were shatter cones around the "irruptive." He found them at once, a striking example of scientific prediction, especially striking since the area had been mapped for many decades without shatter cones having been noticed.

Fig. 5.17 Outcrop of Onaping Formation, Gray Member, north of Sudbury, Ontario. Now interpreted as breccia formed by impact that produced the structure (French, 1967). White inclusions are chiefly quartzites of the Huronian Supergroup. Geologist in foreground (the late Herb Blodget) was 6 feet (1.8 m) tall.

Dietz (1964) therefore proposed that the Sudbury Structure was an extrusive lopolith occupying a large impact crater. The proposal was greeted skeptically. However, French (1967) shortly after made a detailed petrographic study of the Onaping tuff, as it was then termed. He found abundant evidence of shock metamorphism in the Onaping Formation (Fig. 5.17), in particular planar deformation features (see Fig. 5.3). He showed that the general texture, composition, and field relations of the formation were far better explained as an impact breccia than by the then-current volcanic origin. Similar evidence of shock metamorphism was at the time being found for many craters and crater-like structures, such as the Ries Basin of Germany (Shoemaker and Chao, 1961), and several of the Canadian craters discovered by the aerial photography search previously mentioned (Dence, 1965). Consequently, French's conclusion, that the Sudbury Structure was an impact feature, was rapidly accepted. By 1972, a sizable majority of geologists familiar with the Structure agreed that impact had played a key role in its origin. Continued mapping, with petrographic studies, continued to strengthen the impact theory (Dressler, 1984; Dressler *et al.*, 1994). Isotopic and chemical studies of the igneous complex indicate a crustal derivation, implying that this supposed igneous intrusion is actually largely impact melt (Naldrett, 1999), an extremely hot one as shown by Marsh and Zieg (1999). The focus of Sudbury studies has shifted somewhat from its origin to its original size and shape (Grieve *et al.*, 1991; Lowman, 1991, 1992; Rousell and Long, 1998). The Sudbury area was chosen as a test site for a Canadian–American radar investigation, and for a LITHOPROBE traverse (Milkereit *et al.*, 1993). However, further discussion of these efforts would take us too far afield.

The Sudbury story, which we can now view with hindsight, is a striking example of how space research completely overturned conventional thinking about the origin of one of the world's largest nickel–copper deposits. As a thought experiment, imagine how a research project on the origin of the Sudbury Structure might be organized in 1960. We would expect recommendations for a systematic program of diamond drilling, geophysical surveys, petrographic studies of the ore deposits, isotopic analyses of the igneous complex and, for good measure, additional mapping of the regional geology. No research manager in 1960 would have suggested that this new program start by investigating the origin of strangely-shaped glassy buttons found in the Australian outback and other exotic locations. Yet that is essentially what triggered a completely new approach to the problem of Sudbury geology, resulting in a genuine geological breakthrough. It seems clear that we would still be searching for a convincing theory for the Sudbury Structure's origin had it not been for the completely new perspective provided by study of tektites, the Moon, and impact craters.

The Sudbury case-history is a classic example of the value of basic research in general, and space research in particular. However, it has implications for how research should be planned. In particular, to the extent that one example can be used, the Sudbury story illustrates the erratic and unpredictable way in which science actually progresses. Federally-funded research is often criticized for lack of coordination and planning, with occasional suggestions for a cabinet-level Secretary of Science or equivalent. However, the Sudbury impact discovery was the unplanned result of separate efforts by a handful of scientists who were not even Canadian, and who in 1960 had never even been to Sudbury. Dietz's repeated applications for NASA funding were rejected, an ironic note for the 1988

Penrose Medal winner, throwing some doubt on the reliability of peer review.

The history of impact research at Sudbury, in summary, is worth serious study not only by science historians but by anyone involved in high-level research planning and funding.

5.9 Impacts and basaltic magmatism

A recurrent theme since the 1960s has been the role of impact in triggering or at least localizing basaltic magmatism, which has clearly happened repeatedly on the Moon and probably on Mars.

A proposal comparable to those made for the Sudbury Structure was published by Green (1972), influenced directly by the results of the early *Apollo* missions that demonstrated the age and extent of basin formation and subsequent mare basaltic volcanism. Green suggested that major impacts on the Archean crust, a priori hotter and more active than that of later times, might trigger mafic volcanism. After deformation and metamorphism, these accumulations of basaltic rocks – terrestrial maria – became the greenstone belts found everywhere in Precambrian terrains. Although not generally accepted, Green's hypothesis illustrates the post-*Apollo* applications of new knowledge from the Moon to terrestrial geology.

An imaginative proposal by Oberbeck *et al.* (1993) links major impacts to supposed tillites, flood basalts, and the breakup of Gondwanaland. There is a very general association of these features in the late-Paleozoic section of the southern hemisphere, and Oberbeck *et al.* suggest that major impacts may have been responsible for many of them. The link to basaltic volcanism and breakup of the supposed "Gondwanaland" is essentially similar to the proposal of Frey (1980). The possibility that the Pacific Basin is essentially a primordial crater, modified by plate tectonic processes, will be discussed in the next chapter. Glikson (1995) suggested that many Precambrian events, such as rifting, tectonic/thermal episodes, and rifting or incipient rifting may have been triggered by "mega-impacts."

Another suggestion of possible connections between impact and basaltic magmatism has been discussed in Chapter 3. As described there, Girdler *et al.* (1992) have shown that the Bangui magnetic anomaly, generally agreed to result from deep crustal basaltic rocks, may actually mark the site of a Precambrian impact that triggered the intrusion of these basalts.

The most recent and by far the most comprehensive effort to link basaltic volcanism with impacts has been that of Shaw (1994), as part of a monumental synthesis involving celestial mechanics, chaos theory, geophysics, and petrology. Shaw's work is far beyond the scope of this chapter. However, his correlations – geographic and chronological – between impacts and flood basalts are at the very least thought-provoking. An event post-dating the publication of Shaw's book, the 9-impact Shoemaker–Levy comet collision with Jupiter, demonstrates the reality of multiple and geologically nearly-simultaneous impacts, one element of Shaw's theory. New craters continue to be found on the Earth, even in nominally well-mapped areas such as the Chesapeake Bay. The increasing number of known and dated craters should permit within a few years an evaluation of the proposed connections between impacts and various geological/geophysical phenomena.

5.10 Impacts and mass extinctions

One of the perennial unsolved problems of terrestrial geology is the cause or causes of mass extinctions in the geologic record. Over geologic time, species become extinct frequently, genera less frequently, families even less frequently, and so on. However, the geologic record is marked by what appear to be global mass extinctions of many life forms in a wide range of environments. The possible catastrophic effect of major impacts, of meteorites or comets, was recognized (Russell, 1979) at least as early as 1970 by Digby McLaren, a Canadian paleontologist, who outlined the possible results of an impact on marine organisms. A similar proposal had been put forth by Harold Urey who had, after winning a Nobel Prize in Chemistry, gone on to become a leader in space research. Urey pointed out that a major cometary impact might produce widespread heating of the atmosphere and oceans, leading to widespread extinctions.

One of the greatest mass extinctions in the geologic record took place at the end of the Cretaceous. Best known is the final disappearance of the dinosaurs (Russell, 1979), but a wide range of other organisms down to foraminfera – 90% of all living species – also disappeared at this time (Spudis, 1993). A landmark study by Alvarez *et al.* (1980) found that a thin sedimentary unit at the Cretaceous–Tertiary (K–T) unconformity in Italy had an unusually high concentration of iridium, characteristically enriched in chondritic meteorites. Alvarez and his colleagues suggested that a major meteoritic impact 65 million years ago generated the equivalent of a nuclear winter, leading to the extinction of the dinosaurs and many other organisms.

The Alvarez *et al.* paper was primarily responsible for an enormous surge of interest, research, and fundamental rethinking on the nature of mass extinctions. Its most immediate result was a search for the crater or craters that were presumably formed at the end of the Cretaceous. Within a few years, evidence for a buried structure 300 km wide, the Chicxulub Crater, was found as a result of oil drilling under the Yucatan Peninsula of Mexico (Hildenbrand *et al.*, 1991; Florentin *et al.*, 1991). This feature is widely considered the "smoking gun" for the K–T extinctions, since later discoveries (Smit *et al.*, 1991; Sharpton *et al.*, 1992) have found microtektites, anomalously high iridium content, and planar deformation features in quartz from an unusual clastic unit at the K–T boundary. This unit was apparently deposited in water 400 m deep, far below wave base, and testifying to some sort of extremely violent submarine event. The magnitude of the impact is frightening to imagine; Florentin *et al.* (1991) proposed that tsunamis initially *4 kilometers high* may have been generated, felt in oceans around the world. Other effects of this impact have been suggested, including dust loading, fires, earthquakes, global acid rain from nitrates and sulfates vaporized by the impact, and greenhouse warming decades after the impact. An imaginative but well-reasoned description of such impacts has been published by Lewis (1996), who demonstrated that contrary to popular scientific opinion, there have actually been a number of damaging meteorite falls in recorded history.

The mass extinction at the Permian–Triassic boundary, 250 million years ago, was as great or greater than that at the end of the Cretaceous. This event was evidently short in terms of absolute time, probably $<500{,}000$ years duration. Several causes for this extinction, which included ferocious creatures such as the productid brachiopods, have been suggested, for example the Siberian plateau basalt eruptions at about the same time. An impact had been suggested, but it was only with the work of Becker *et al.* (2001) that definite evidence for a major impact at the Permian–Triassic boundary was found. This group detected fullerenes, C_{60}, at three places on the boundary, confirming other reports. However, Becker *et al.* also analyzed trapped helium and argon in the fullerenes with isotopic ratios similar to those in carbonaceous chondrites. They concluded that a body about 9 km in diameter had delivered these materials, with an impact comparable to that at Chicxulub.

The pace of discovery in this field has become too rapid for proper treatment in this book; a useful review was published by Galvin (1998). However, some of the most general developments should be cited as a second or even third generation result of space exploration.

Most important is the sudden revival of what can only be called catastrophism, supposedly replaced by uniformitarianism as the dominant paradigm of geology early in the 19th century (Marvin, 1990). It was becoming realized in the late-1970s that a number of phenomena, some extraterrestrial (e.g., nearby supernovae) might have caused mass extinctions (Russell, 1979). However, the revival of catastrophism was greatly accelerated by the Alvarez *et al.* (1980) discovery of the iridium anomaly, and other supporting geologic evidence, for a major impact(s) at the end of the Cretaceous.

Acceptance of major impacts as the cause of the Cretaceous–Tertiary extinctions is by no means universal. Officer and Page (1996), for example, have shown that there appears to have been an increase in iridium content of sediments well before the end of the Cretaceous, which they attribute to massive volcanism. Basaltic volcanism did increase at this time, as shown by the earliest of the Deccan Traps of India (Rampino, 1992). There have been other well-documented arguments against the impact theory (see proceedings of both Snowbird conferences), and the argument is unsettled at this writing. However, it is clear that the impact theory for mass extinctions has led to wide rethinking of the foundations of paleontology, such as the quality of the stratigraphic record, the absolute duration of unconformities, and the actual nature of the extinctions themselves: stepwise, gradual, or catastrophic (Flessa, 1990).

These problems are of more than academic interest. They bear on the origin of our species, the latest development in the rise of mammals, which had to await extinction of the giant reptiles. Better understanding of biological evolution is absolutely essential at a time when the biosphere itself may be undergoing major changes that could result in *our* extinction. In the long run, such a better understanding may be among the most important effects of space exploration on the earth sciences in the 20th century.

5.11 Summary

Impact craters were recognized on the Earth several decades ago, from unmistakable evidence such as meteorite fragments and melted rocks. But these craters, if mentioned at all, were generally treated in geology texts as curious anomalies. In contrast, at the end of the 20th century, impact cratering was generally agreed to be, in Shoemaker's (1977) words, "the most fundamental process that has taken place on the terrestrial planets." For the Earth, impact is now considered of major importance in mass extinctions, and possibly essential to formation of "continental nuclei," the earliest ocean basins, some major ore deposits, and even the familiar flood basalts such as those of the Deccan Plateau. Impact structures have localized several large producing oil and gas fields, one of them (in Mexico) with reserves greater than the total reserves of the conterminous United States. The term "revolution" has been been used, by Alvarez (1991), for the sudden revival of catastrophism resulting from evidence for the role of impact in biological evolution.

The role of space exploration in opening this new field of geology was primarily as a stimulus – an enormous one – to developments that might have happened anyway, but much more slowly. The *Apollo* program in particular, by injecting massive amounts of support into cratering studies, produced a broad infrastructure of factual data and analytical techniques in the 1960s. The *Apollo* missions themselves, immediately returning overwhelming evidence for the dominance of impact cratering on the Moon, at once converted almost all skeptics. An interesting example is the testimony of Grieve (1991), who tells how he was immediately convinced that the Onaping "tuff" at Sudbury was an impact breccia when he saw nearly identical material returned from the lunar Fra Mauro Formation by the *Apollo 14* astronauts. The many unmanned missions to Mercury, Mars, and other bodies reinforced the importance of impact cratering. Even Venus, dominated by volcanic and tectonic features, has many large impact craters essentially similar to those elsewhere.

These developments typify the serendipitous nature of space exploration. Exploring previously-unknown planets, we expected the unexpected, so to speak. But there was no real anticipation of the new insights to cratering and terrestrial geology produced by space exploration. Furthermore, the study of terrestrial craters has only begun to benefit directly from orbital observations and measurements. It seems safe to say that geologists of the present century will have to be as familiar with impact cratering as those of the 20th century were with folding, faulting, and metamorphism.

A final way to summarize the importance of the impact crater studies described here is to step back, in effect, and consider the ominous picture emerging from these and related studies. We have learned, first, that there are many more large impact craters on the Earth than had been realized at the start of the Space Age. Second, we have discovered the high probability that some of these large impacts, such as the one responsible for the 300 km Chicxulub structure, have been devastating events with world-wide effects, of which the final disappearance of the dinosaurs is only the most spectacular. Third, it has been found, by several years of telescopic search

(Shoemaker *et al.*, 1990), that crater-forming objects – asteroids and comets – are not just sporadic accidental arrivals from the Mars–Jupiter asteroid belt. The Earth is *in* an asteroid belt (Rabinowitz *et al.*, 1993), and the danger of catastrophic impacts correspondingly greater than formerly realized.

Perhaps the most important line of evidence on this problem is the simple fact that we have not (at this writing) detected any evidence of intelligent life elsewhere in the universe. There have been as of 1996 seventy searches (Tarter, 1996) for extraterrestrial intelligence (SETI), which have so far yielded nothing. Given, in addition, the enormously increased capability of astronomy in general, with space- and ground-based telescopes, it is becoming rather unlikely that the universe is teeming with life and intelligence. The title of the recent book by Ward and Brownlee (2000), *Rare Earth*, expresses this view concisely. They summarize several lines of evidence that complex life itself is extremely rare, much less intelligent life. It has been argued by Tipler (1979) that since robotic interstellar colonization would be rapid in geologic terms, if there were intelligence elsewhere in the universe we would know it by now, and since we do *not* know it, we are the only intelligent beings in the universe. There are, however, other possibilities obvious to a geologist.

First, the probability that *communicative* life will arise in the first place is extremely low. For fundamental physical reasons (Oliver, 1979), interstellar communication will probably use microwaves, although optical frequencies are now being scanned for laser-borne communication. Life itself may arise on a wide range of water-bearing planets, but a microwave-developing civilization can only exist on land. This narrows the range of potential planets with communicative ETI greatly, perhaps to planets with an ocean–continent dichotomy like that of the Earth. As discussed in the next chapter, this dichotomy is unique in the solar system, and may have resulted from a combination of very unlikely results.

Second a reason for the failure to date to detect ETI, obvious to any geologist in the late-20th century, may be that communicative intelligent life simply does not survive very long because of uncontrollable natural events, such as glaciation, volcanism, nearby supernovae, magnetic field reversals, and especially catastrophic impacts. (This ignores the obvious possibilities for self-destruction, now a more than familiar concept.)

The "ominous picture" emerging from crater studies and related investigations (McLaren and Goodfellow, 1990) is thus one of a violent and dangerous universe, and a geologic history racked by "global catastrophes" in the terms used for the Snowbird II conference (Sharpton and Ward, 1990). Civilization began something like 10,000 years ago, an instant geologically; its future, if not that of the species, may be similarly brief. Like the residents of St. Pierre before the 1902 eruption of Mt. Pelee, we may be living in the shadow of disaster. Countermeasures are being taken, such as the Spaceguard Survey (Morrison, 1992), started in 1998 as a joint NASA/Air Force program designed to find 90% of the near-earth objects over 1 km in diameter. Drastic defense measures have been well publicized in the popular media and the movies, and need no discussion here. In the long run, the only real countermeasure may be to spread out. Dispersal of mankind beyond the Earth was suggested in 1939 by J. D. Bernal, whose words can hardly be improved upon: "*If human society . . . is to escape complete destruction by inevitable geological or cosmological cataclysms, some means of escape from the earth must be found. The development of space navigation, however fanciful it may seem at present, is a necessary one for human survival.*" Proposed programs for species dispersal focussed on the Moon have been published by Lowman (1996), and on Mars by Margulis and West (1993) and Zubrin (1996). Pending such ambitious endeavours, an obvious first step, now being taken in the Spaceguard Survey, is to assess the danger from major impacts (Gehrels, 1997). The new science of cratering phenomena and shock metamorphism, largely a result of space exploration, will be a major aid to such assessment and thus to the survival of our species.

CHAPTER 6

Comparative planetology and the origin of continental crust

6.1 Introduction

The term "comparative planetology" was whimsically coined by George Gamow (1948). As usual, Gamow's whimsy concealed an important concept: that to fully understand our own planet, we must study others. Gamow was not the only one to appreciate this, for geologists such as Barrell (1927), Wright (1927), and Poldervaart (1955) tried to apply what was known of the Moon's geology to terrestrial petrologic problems such as the origin of the original continental crust. Poldervaart's words are worth quoting: "*An adequate picture of this original planet and its development to the present earth is of great significance, is in fact the ultimate goal of geology as the science leading to knowledge and understanding of earth's history.*" However, many geologists today feel that terrestrial geology is unique, continental crust being "generated in a totally different way" (Weaver and Tarney, 1984) from the crusts of other planets.

Comparative planetology now encompasses eight of the nine solar-system planets, dozens of natural satellites, and even a few asteroids that have been visited by spacecraft (Fig. 6.1). The subject has expanded from an occasional paper to a huge and flourishing field filling many books and journals. The relevance of bodies composed largely of ice, such as Dione, or hydrogen, such as Jupiter, to terrestrial geology appears marginal although it has been argued by Hunt *et al.* (1992) that silane emanations from a hydrogen-rich core may influence many terrestrial tectonic processes . However, silicate bodies, in particular the Moon and terrestrial planets, have much to tell us, by analogy, inference, and speculation, about crustal evolution in the Earth (Solomon, 1980). Two general problems can be studied with this approach: the growth of continental crust; and the role of plate tectonics in this growth. These can be treated under the collective title: origin of the continental crust.

Fig. 6.1 Planetary size comparisons. Diameter of Earth, 12,760 km. From Meszaros (1985).

6.2 Origin of the continental crust

This phrase has been carefully worded. As pointed out by Wasserburg (1961) in a penetrating paper, it is extremely important to express problems in the right way, as well as to pick the right ones in the first place. The problem treated here has traditionally been termed "the origin of continents" or "growth of continents." Both phrases tend to shape one's approach to the problem, and even to prejudge the issue. The very word "continents" emphasizes the size and shape of sialic blocks, and favors a tectonic approach. The word "growth" assumes that the continents have in fact grown over geologic time. The phrase "origin of continental crust" has been used instead to avoid these implicit constraints, and to put more stress on the petrologic question of how the continental crust was extracted from the mantle. But even "origin of continental crust" is polarized, so to speak, in that the fundamental question of terrestrial geology is more general: the origin of the oceanic (*basaltic*)/continental (*granitic*) dichotomy. The general evolution of terrestrial crust could, in principle, involve formation of ocean basins at the expense of continents, rather than growth of continental crust (Hamilton, 1993).

In this chapter, comparative planetology will be combined with terrestrial geologic data to produce a theory for *the origin of continents and ocean basins*, stressing the continental crust. The author has followed this combined approach for three decades (Lowman, 1969, 1976, 1989), and the views presented here are largely personal ones unless labeled otherwise.

Credit (or blame) having been assigned, let us examine the problems of continental crust and contemporary theories of its formation, starting with the customary caveat that the subject is a large one that can only be summarized here. Petrologic terms will be used freely, and the reader referred to the glossary for definitions. For non-geologists, introductory geology texts by Wyllie (1976), Condie (1989a), and Ernst (1990) are recommended. A well-illustrated non-technical account of new developments in global geology, based on a BBC radio series, was published by Redfern (1991). For geologists, the review by Bickford (1988) gives a good contemporary summary. Extensive technical treatments have been published by Taylor and McLennan (1985) and Meissner (1986), stressing geochemical and geophysical approaches respectively. The comprehensive text by Howell (1995) is valuable in that it combines several lines of evidence on continental origin, advocating the terrane accretion mechanism, as will be discussed in the next section.

Good reviews of comparative planetology were published by

Head (1976) and Consolmagno and Schaefer (1994). The authoritative treatments of the subject remain those of Taylor (1982, 2001). The origin of the Moon, a topic touching many aspects of terrestrial geology, is covered by collections edited by Hartmann *et al.* (1986) and by Canup and Righter (2000). The geologic importance of the Moon is also discussed in a book focussed on the occurrence of complex life in the universe by Ward and Brownlee (2000). The "magma ocean" concept, first applied to the Moon, was covered by papers in a special issue of the *Journal of Geophysical Research* (**98**, 1993).

Our knowledge of Venus has increased greatly, thanks to a series of Soviet and American radar missions. The most recent of these, the *Magellan* radar survey, was covered in two special issues of the *Journal of Geophysical Research* (**97**, E8 and E10, 1992). The geology of Venus in light of the *Magellan* mission's results has been discussed by Basilevsky and Head (1998).

The first detailed exploration of an asteroid, 433 Eros, was accomplished by the *NEAR* (*Near-Earth Asteroid Rendezvous*) mission, in 2000–2001, producing chemical data (Trombka *et al.*, 2000) showing that this 34-km-long body has a chondritic composition. As such, it represents an undifferentiated end-member among silicate planets or fragments thereof.

6.3 Previous studies

The study of continental origin dates in recognizable form back to the mid-19th century, specifically to the proposal of J. D. Dana (1856). Dana proposed that North America had grown outward from an "Azoic nucleus" by the addition of geosynclines and mountain belts. This concept has dominated tectonic thinking since then, and is, in somewhat different terms, the theory favored by most geologists at the present. The chief weakness in continental growth as visualized by Dana was the short time scale, "tens of millions" of years since the Cambrian. The discovery of radioactivity, and its application to radiometric dating in the early-20th century, showed that the Earth was *thousands* of millions of years old. However, tectonic theorists soon accommodated their thinking to the expanded time-scale (e.g., Wilson, 1954). The sudden development of the plate tectonic theory in the mid-1960s was followed almost immediately by its application to the lateral accretion concept (Dewey, 1969; Dewey and Burke, 1973).

Skimming apologetically over several decades of post-World War II work, we can summarize contemporary thought as follows. Continents are today generally believed to have formed over geologic time by terrane accretion: the transport, collision, and suturing of crustal fragments or island arcs to continental nuclei by sea-floor spreading. First developed on the west coast of North America (Jones *et al.*, 1977; Nur and Ben-Avraham, 1982), the concept of "suspect" terrane accretion (Kerr, 1983; Hoffman, 1988) was soon applied to the Appalachians (Williams and Hatcher, 1982; Williams *et al.*, 1991) and to other continents (Howell, 1995). The *Journal of Geophysical Research*, **87**, B5, 1982, was devoted to "accretion tectonics." An entertaining popular account of this theory, and its weaknesses, was published by McPhee (1982).

The terrane accretion concept has been extended to individual crustal provinces of the Canadian Shield, such as the Superior (Card, 1990) and Slave (Kusky, 1989) Provinces. As outlined in detail by Card, the mechanism is "subduction-driven accretion of Archean crustal elements," now the greenstone belts of the granite–greenstone area. The belts are considered analogous to present-day island arcs such as those of the Pacific rim. Comprehensive treatments of greenstone belts have been published by Condie (1981) and De Wit and Ashwal (1997).

Terrane accretion to some extent is an inescapable consequence of plate tectonic theory, in particular sea-floor spreading. Like plate tectonics, and Newtonian physics in 1890, terrane accretion has reached a stage of impressive coherence and nearly universal acceptance. The 1998 Toronto meeting of the Geological Society of America took as its theme "Assembly of a Continent," a *double entendre* reflecting this acceptance. The terrane accretion mechanism has in effect been given official status in the 1992 maps of the Geological Survey of Ontario, with titles such as "Tectonic Assemblages of Ontario," explicitly based on plate tectonic models.

Terrane accretion has been convincingly demonstrated in several areas. The initial stage, detachment of a crustal fragment, is happening now with the northward movement of Baja California, driven by sea-floor spreading in the Gulf of California. The eventual docking of such fragments is demonstrable in several places. The geology of southern Alaska, for example, can be best understood as resulting from the accretion of crustal fragments carried northwest by sea-floor spreading and wedged against the continent. The Olympic Range of Washington, a massive accumulation of basalts (many erupted underwater) in reverse fault contact with the crust to the east (McKee, 1972), is a classic example of accretion of a Hawaii-like volcanic pile to the continent. Similarly, the geology of western Colombia is explainable as the accretion of a basaltic oceanic plateau (Kerr *et al.*, 1997).

An uncritical observer might conclude that the long-standing problem of how continents form has finally been solved. However, it is important to distinguish "true continental growth" from "re-arrangement of pieces of already extant sialic crust" (Ernst, 1988), and from the "juggling and stacking of thrust sheets (Cook *et al.*, 1980). The primary question that must be answered is: **How and when was continental crust extracted from the mantle?** Viewed in this context, terrane accretion has a number of serious weaknesses requiring a critical review. Hamilton (1993) has presented such a review, using different lines of argument such as the absence of across-strike systematic radiometric age relationships in granite–greenstone terrains.

6.3.1 Crustal province boundaries: are they sutures?

Taking the maps by Hoffman (Fig. 6.2) and by Howell (Fig. 6.3) as representing the terrane accretion concept, let us examine briefly the crustal province boundaries that are, in this concept, sutures, i.e., zones along which former oceans closed (Dewey, 1977; Dewey and Burke, 1973). To focus the discussion, the area around Sudbury, Ontario, is chosen as an excellent example, where three crustal provinces (or terranes) meet: Superior, Southern, and Grenville (Figs. 6.4, 6.5). The Sudbury area is unusually well understood not only because of the nickel deposits and the 60-km-long impact structure (Lowman, 1992), but because LITHOPROBE seismic profiles have been run across it to study crustal structure including the Grenville Front (Clowes, 1993; White *et al.*, 1994; Ludden and Hynes, 2000a, b).

Geologic evidence shows clearly that these interprovince boundaries are not sutures, although explicitly labeled such by Condie (1989b) and Howell (1995) (see Fig. 6.3). None of them, in the Sudbury area, is marked by the remnants of oceanic crust, specifically ophiolites, although as pointed out by Windley (1984) their absence is not conclusive evidence against suturing. More definite is the fact that the most prominent crustal boundary, the one billion year old Grenville Front – the longest and most distinct one in North America – is demonstrably not a suture. Pre-Grenville rocks and structures, such as the Labrador Fold Belt and remnants of the Sudbury dike swarm (Dudas *et al.*, 1994), have been traced across the Front, which is essentially a zone of ductile faulting along which pre-existing rocks have been reworked. Isotopic studies (Dence *et al.*, 1971; Dickin and McNutt, 1990) show that rocks of Archean age, over 2.5 Ga old, occur in altered form up to 130 km south of the

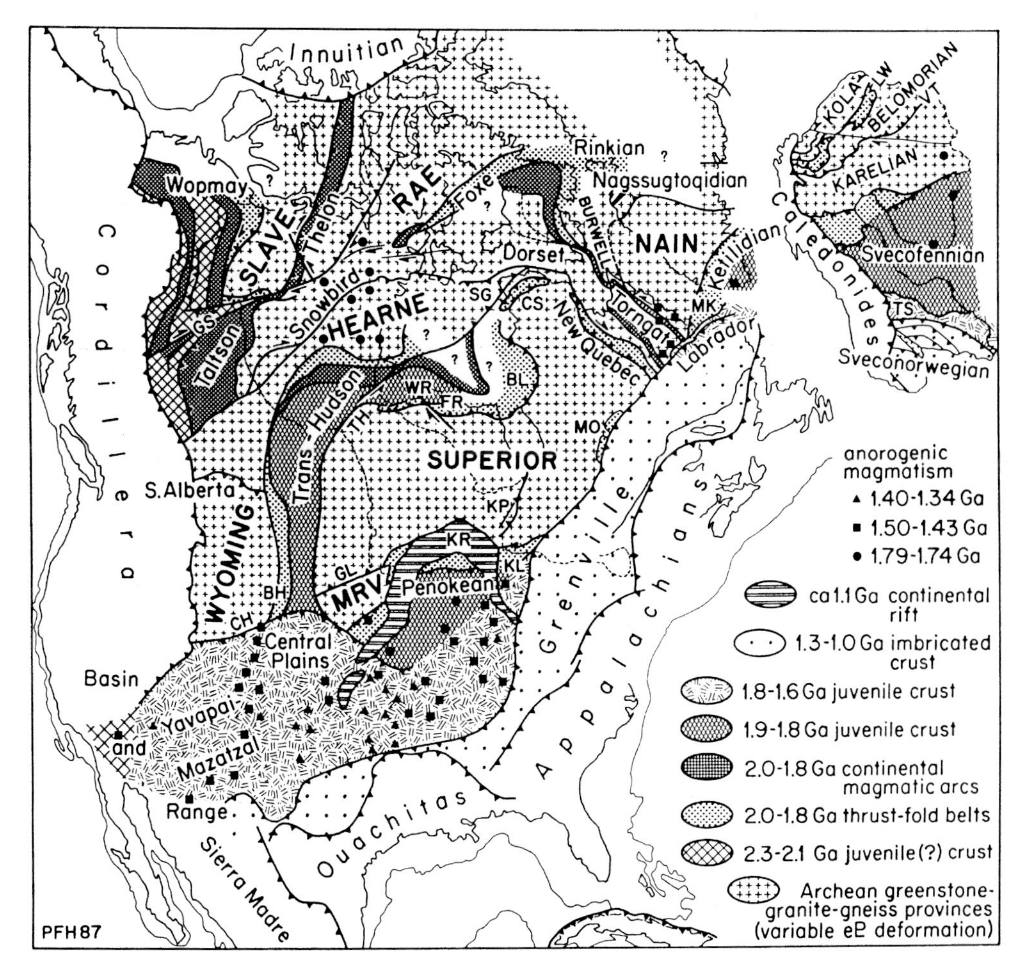

Fig. 6.2 Terranes of North America. From Hoffman (1988).

Front. (Dickin 2000 labels the boundary of Archean samples a "suture," but a Paleoproterozoic one, not Grenvillian.)

In a detailed review, Davidson (1998) termed the location of the supposed suture one of the major unanswered questions of Grenville Province development. A comprehensive discussion of the problem, based on geophysical evidence, was published by Thomas (1985), who suggested that almost the entire Grenville Province in Ontario was a "suture zone." Hanmer *et al.* (2000) presented geologic and isotopic evidence specifically contradicting "accretionary tectonics" after 1.4 Ga. Seismic-based cross sections made as part of the LITHOPROBE program show the Grenville Province all the way to the Adirondacks to consist of a complex series of sheets

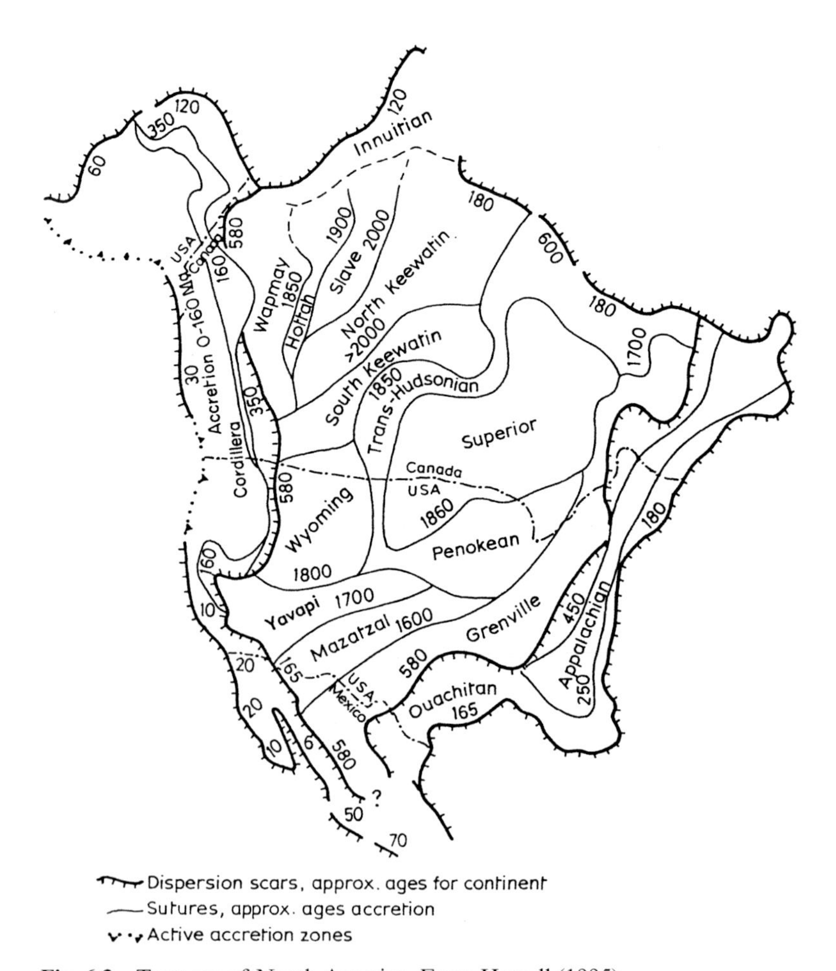

Fig. 6.3 Terranes of North America. From Howell (1995).

overthrust to the northwest. Pre-Grenville continental crust underlies the Central Gneiss Belt and the Central Metasedimentary Belt, almost the entire width of the Grenville in Ontario. No suture has been identified by LITHOPROBE geophysical techniques.

Dewey and Burke (1973) and Burke *et al.* (1977) suggested that the Grenvillian suture was far to the southeast, under the Appalachians. However, no suture has subsequently been convincingly identified anywhere in the Grenville Province for eight hundred kilometers to the southeast, in Canada or the United States (Bartholomew *et al.*, 1984). Rocks of the Blue Ridge, for example, show radiometric ages around 1.1 Ga (Bartholomew and Lewis, 1984), agreed to be metamorphic ages produced in the Grenville Orogeny by reworking of older crust. If the Grenville Orogeny was

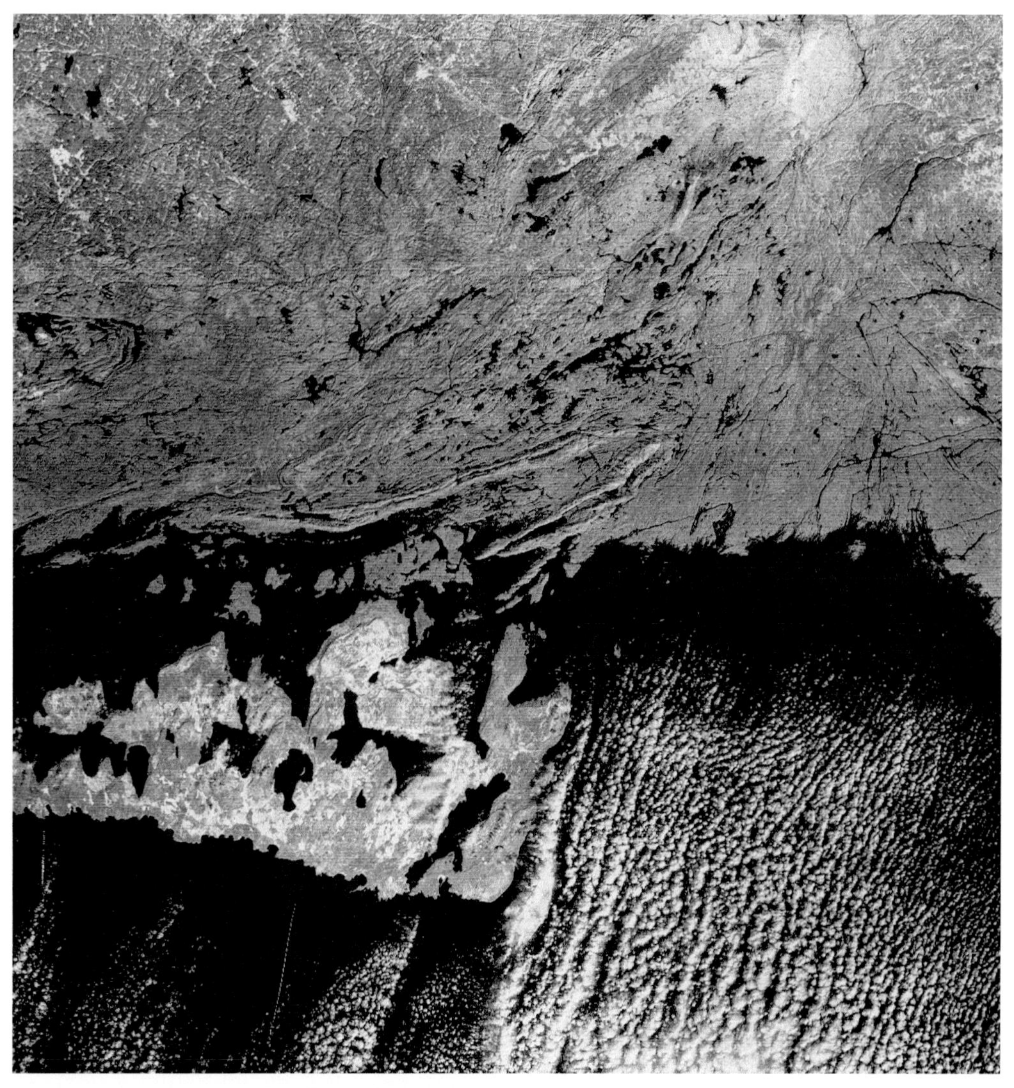

Fig. 6.4 *Landsat* view of Sudbury, Ontario. From Lowman (1992); area 160 km on a side.

a continental collision, the impacting terrane must have been still farther to the southeast.

The former existence of an Andean-type subduction zone has been plausibly argued for the Grenville Province in Canada and the adjacent United States (Hanmer *et al.*, 2000). However, the Andean margin itself is notable for the apparent absence of accreted terranes, with the exception of the northwest Colombian margin as previously discussed. The weight of geological, geophysical, and isotopic evidence has led most workers researching the Grenville

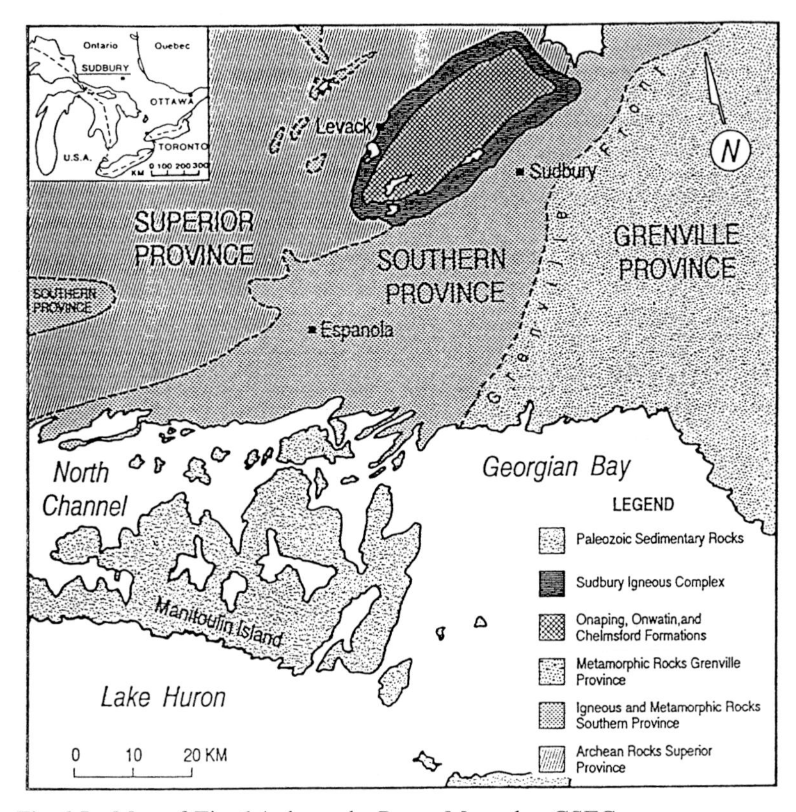

Fig. 6.5 Map of Fig. 6.4, drawn by Penny Masuoka, GSFC.

Province to agree that terrane accretion has not enlarged the continent in the Sudbury area or, more importantly, anywhere in the Grenville Province to the southeast. Let us now examine the evidence for terrane accretion in another part of the Sudbury area, the Huronian rocks representing the Southern Province.

The Southern Province as a whole has been interpreted as an accreted terrane (see Fig. 6.3). However, in the Sudbury area, its Penokean fold belt has been shown by Card (1978) to represent a folded and faulted geosynclinal sequence deposited on pre-existing Archean crust. There is abundant gabbroic rock represented by the Proterozoic Nipissing diabase, but nothing corresponding to an ophiolite sequence in Ontario, although the Niagara fault zone in Wisconsin has been interpreted as ophiolites (Sims *et al.*, 1989). The relationship between the Southern and Grenville Provinces, a long-standing problem ("disappearance of the Huronian"), has been resolved to some extent by the demonstration (Dickin and McNutt, 1990; Dickin, 2000) that the Archean rocks of the Superior Province can be identified radiometrically southeast of Sudbury, i.e., southeast of the Southern Province. The Southern Province rocks, locally the Huronian Supergroup, can not represent an accreted terrane, but supracrustal rocks deposited on pre-existing Archean continental crust.

In summary, the actual terrane boundaries or supposed sutures in a limited but unusually well-studied area have been shown conclusively not to be sutures. A subducting continental margin can still be reasonably argued, and there has evidently been some vertical growth of the continental crust by basaltic magmatism. However, the continental crust in this area has not grown as a whole by terrane accretion, a conclusion also reached for the western US by Ernst (1988).

6.3.2 Ensialic greenstone belts

The terrane accretion theory treats orogenic belts as exotic terranes sutured to continents during collisions, or as foreland belts adjacent to such terranes. However, there is abundant evidence that, viewed objectively, most intracontinental orogenic belts have always been intracontinental, or ensialic – formed on pre-existing continental crust. This problem obviously can not be adequately covered here, but if we focus the discussion on Precambrian greenstone belts, another weakness in terrane accretion as a mantle extraction process emerges. As mentioned previously, these belts are now interpreted as accreted terranes, probably formed by island arc–continent collisions. If they represent new continental crust, recently extracted from the mantle, they should, in principle, be ensimatic, formed on or very near to oceanic crust.

Before discussing their original tectonic setting, it must be pointed out that the Archean crust, collectively, is not simply a greenstone collage. Greenstones are in fact only a minor part, by area, of Precambrian terrains, as shown by Table 6.1, compiled from the collection edited by De Wit and Ashwal (1997). Only the Yilgarn Craton of Australia has a granite : greenstone ratio as high as 3:1. In the latest and best-exposed Precambrian terrains, such as the Superior Province, West African Shield, and Sao Franciso Craton, the granite : greenstone ratio ranges from 8:1 to 20:1.

Volume ratios are similar. Greenstone belts have been known for some decades to be relatively shallow structures, not over 20 km thick at present exposure levels and generally much less, as shown in Fig. 6.15. Obviously, these relationships must be interpreted with caution, since original geometry has generally been disturbed by thrusting and imbrication. Nevertheless, greenstone belts are only a minor component, by area and volume, of the Archean continental

Table 6.1 *Granite : greenstone ratios, Precambrian Shield areas*[a]

Region	Total area (km^2)	Areal ratio
Superior Province	1,572,000	11:1
Slave Province	190,000	20:1
Wind River Range	14,000	5:1
West African Shield	4,500,000	Archean 20:1
Sao Francisco Craton	660,000	8:1
Amazonian Craton	4,260,000	7:1
Zimbabwe Craton	180,000	5:1
Tanzanian Craton	500,000	Northern 4:1
		Southern 20:1
Indian Shield	6,500,000	7:1
Yilgarn Craton	10,000,000	3:1
Pilbara Craton	60,000	6:4
Greenland (exposed shield)	20,000	10:1
Baltic Shield, Finland	320,000	8:2
Karelian Terrain, Russia	350,000	7:3
Inari–Kola Craton	9,000	20:1 to 10:1
Aldan–Stanovik Shield	95,000	9:1
Ukrainian Shield	40,000	10:1

Note:
[a] From compilations in respective chapters of *Greenstone Belts*, edited by De Wit and Ashwal (1997). See references cited therein for sources.

crust, as suggested by the eloquent description by Turner and Verhoogen (1960): ". . . vast formless seas of gneissic granite in which swim islands and rafts of metamorphic rocks . . .".

Greenstone belts were named because they are commonly dominated by greenish metamorphic rocks, chiefly metavolcanics (Wood and Wallace, 1986). The greenstone belt literature is enormous because most Precambrian shield mineral deposits occur in these belts, rather than in the surrounding "granite" (more precisely, granitoid) rocks. They can be briefly described as thick, intensely folded and faulted, sequences of metavolcanic rocks – dominantly basaltic, although andesites are locally abundant (Shirey and Hanson, 1984) – and subordinate sedimentary or metasedimentary rocks. They have been deformed, in part, by the diapiric intrusion of granitoid rocks making up most of the area of granite–greenstone terrains (Drury, 1977; Hamilton, 1998). The now-discarded term "eugeosyncline" (Engel, 1963) is still convenient shorthand to describe the original setting of greenstone belts.

Several lines of geophysical and geological evidence point firmly to an ensialic origin for greenstone belts in two well-studied Precambrian shields, in Canada and southern Africa. On the Canadian Shield, geophysical studies, in particular reflection profiling done as part of the LITHOPROBE program (Clowes, 1993; Bursnall *et al.*, 1994), have shown that greenstone belts are essentially shallow features, up to about 15 km thick, underlain by about 25 km of continental crust. They are thus demonstrably ensialic as they exist today. This objection can obviously be met by arguing that when formed the greenstones were at the edges of continents or in ocean basins. However, there is equally strong evidence that the greenstone belts were originally intracontinental (Donn *et al.*, 1965; Hargraves, 1976; Hamilton, 1993).

The sedimentary rocks of greenstone belts are complex assemblages, well described for the African occurrences by Cooper (1990). They include a wide range of marine and continental clastics, carbonates, banded iron formations, and glacial tillites, interspersed with volcanics. As shown by Cooper, these assemblages were originally deposited on pre-existing continental crust, chiefly in fault-bounded basins ("taphrogeosynclines" in older literature). They may represent a degree of vertical continental growth by addition of mafic volcanics, to be discussed, but not simply accreted terranes. Ernst (1988) reached a similar conclusion for rock associations in the southwest US, although he suggested continental growth over long-lived subduction zones.

The Pacific margin of South America, overlying an unusually well-studied subduction zone, has demonstrably not grown laterally by addition of volcanic rocks. For example, as shown most recently by Cobbing (1999), the magmatic arc now occupied by the coastal batholith of Peru was formed on pre-existing continental crust, well inboard of the continental margin.

To summarize this large topic, greenstone belts were originally formed on pre-existing continental crust, thinned by faulting, not by terrane accretion. They may represent a degree of true continental growth, in that they include mantle-derived metavolcanics. But the early continental crust, as represented by Archean provinces, was not produced by simple accretion of greenstone belts.

6.3.3 Terrane accretion vs. reworking

Another fundamental question, conveniently approached with Precambrian shield examples, is whether such shields have evolved through terrane accretion or by reworking (including remelting) of pre-existing continental crust. A traditional line of evidence, isotope geology, has been frequently cited to support continental growth by accretion of mountain belts (Hurley and Rand, 1969), with obvious extrapolation to terrane accretion. However, the consensus (Armstrong, 1968; 1981; Faure, 1986; Sylvester, 2000) is that isotopic evidence does not point to growth of sialic crust by direct extraction from the mantle. Petrologic evidence (Wyllie, 1988; Luais and Hawkesworth, 1994) similarly indicates that most granites and granitoids formed by remelting of older sialic or mafic crust. Island-arc accretion, for many decades a favored mechanism of continental growth, has been questioned on chemical grounds by Rudnick (1995), pointing out that modern island arcs produce basalts, not andesites, and thus can not directly form the andesitic bulk composition continental crust.

The view that most continental crust has evolved by reworking, i.e., metamorphism, deformation, and remelting, is supported strongly by a wide range of field evidence that can be only briefly summarized here. The most convenient approach is to return to the Sudbury area previously discussed (see Figs. 6.4, 6.5). Taking a more general view, we see that this relatively small area of continental crust – covered by one *Landsat* scene (160 km on a side) – has, over geologic time, undergone repeated reworking as shown by Rousell *et al.* (1997) and, for the Abitibi–Grenville area, by Ludden and Hynes (2000b). There is abundant direct evidence, from isotopic studies, petrography, and geologic mapping, for repeated tectonism and magmatism covering nearly 2000 million years (Fig. 6.6). The oldest event for which there is evidence in the Sudbury area was the early stages of the Kenoran Orogeny, responsible for the bedrock features of the Superior Province in general, starting about 2700 million years ago. These rocks – which are underlain by some 35 km of possibly older sialic crust – were subsequently reworked, remelted, and intruded in many geologic episodes including the Penokean Orogeny, some 300–400 million years duration. The final Precambrian event in this area was the Grenville Orogeny, culminating about one billion years ago.

This example of reworking in a relatively small area can be matched elsewhere in the Canadian Shield, as in the classic Grenville Province Mont Laurier paper by Wynne-Edwards (1969). An intensively studied LITHOPROBE area of the Superior Province, the Kapuskasing uplift, reveals a "protracted polyphase deformation history" some 1300 million years in duration (Bursnall *et al.*, 1994). The Isua Supracrustal Sequence of southwest Greenland (Dymek, 1984) underwent four major metamorphic events covering roughly 2000 million years. Cooper (1990) presents similar examples from

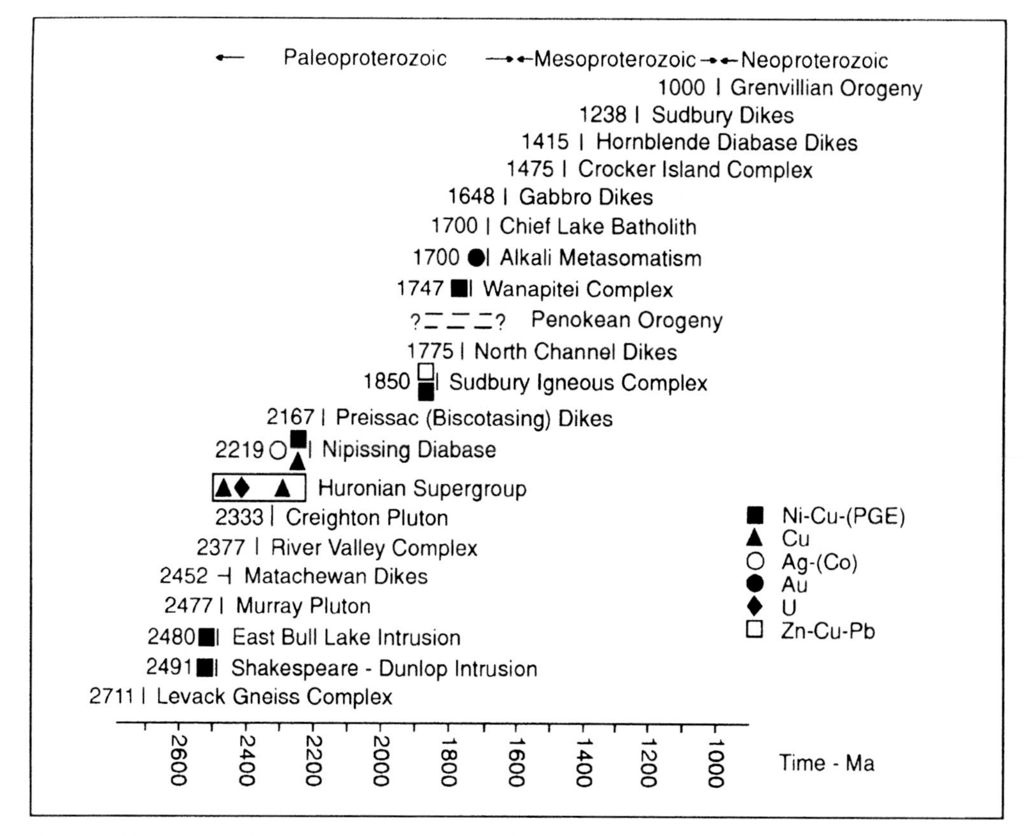

Fig. 6.6 Major tectonic and magmatic events of Sudbury area. From Rousell *et al.* (1997).

the Precambrian of southern Africa. Almost all Precambrian shield areas, and many younger ones, reveal a similar pattern of overlapping structures and igneous or metamorphic events. There has probably been some continental growth as defined here – extraction of new crust from the mantle – but the dominance of reworking argues against simple terrane accretion.

The occurrence of prolonged and recurrent reworking of continental crust is understandable in terms of the Earth's thermal behaviour. Oceanic lithosphere is relatively ephemeral, lasting not more than about 200 million years before it is subducted and returned to the mantle. After creation at spreading centers, it moves outward, cooling as it passes over the mantle. Continental crust, in contrast, is demonstrably long-lived. Whether it is fixed over the mantle or not, it acts as an insulator (Anderson, 1981, 1984), and promotes the buildup of heat in the underlying mantle, heat released in cyclic episodes of magmatic and orogenic activity. Cooper (1990) has shown the existence of 11 such "megacycles," each 320 million years duration, in southern Africa.

To sum up this critical review, the theory of terrane accretion has major weaknesses, amounting to failure to answer the central question: How and when was continental crust extracted from the mantle? There is a real need for a new approach to the question, which can be provided by comparative planetology. This now-enormous subject will be applied selectively, concentrating on those aspects of planetary evolution relevant to the origin of the Earth's continental crust.

6.4 Thermal histories of planets

The primary factor governing crustal evolution in a silicate body, after its mass, is its thermal history: its original temperature, the rate at which its internal temperature changes over geologic time, and the mechanisms by which heat is lost (Solomon and Head, 1991). Before the *Apollo* landings on the Moon, the concept most generally favored for the Earth, the Moon, and the other terrestrial planets was cool accretion, championed by Harold Urey (1952). Urey argued convincingly that these bodies had accumulated from the solar nebula under relatively low temperature conditions (a few hundred degrees absolute), warming up gradually later in their history from radioactive decay and other causes. (MacDonald's (1963) review gives a quantitative analysis of these processes, still valuable in hindsight.) Urey expected the Moon to prove a primitive, undifferentiated body, probably of chondritic meteorite composition, and popularized the term "Rosetta stone," meaning that the Moon would reveal conditions existing in the early solar system. As mentioned above, such a body has now been found by the *NEAR* mission, the asteroid 433 Eros. As applied to the Earth, cool accretion implied that global differentiation, and formation of continental crust, was delayed for some considerable time after the Earth's accretion.

The cool accretion theory was overthrown within months of the *Apollo 11* landing, by the finding from returned samples that the Moon was a highly differentiated body that had undergone extensive melting very early in its history. This evidence came not only from the mare basalts, but from the evidence for anorthosites (Wood, 1972) and related rocks in the older lunar highlands (Fig. 6.7), rocks that apparently form in no way other than magmatic differentiation. Later missions confirmed a high temperature for the early Moon (Hanks and Anderson, 1979), though showing that the highland

Fig. 6.7 *Apollo 16* metric camera view of Moon (diameter 3476 km), earthward side to left. Shows global nature of differentiation (highland crust).

crust had a much more complex composition than anorthosite (Adler *et al.*, 1972). A composite model by Spudis and Davis (1986) includes norites, basalts, and anorthositic gabbros; the term "anorthositic gabbro" is often used as a simple composite description. The compositional mapping of the Moon by post-*Apollo* missions, notably *Clementine* and *Lunar Prospector*, has refined but in general confirmed the earlier crustal models (Pieters *et al.*, 1994; Feldman *et al.*, 1998). Whether the Moon was melted enough to form a magma ocean, or only locally and intermittently ("serial magmatism" (Walker, 1983)), is still not clear. The distribution of lunar KREEP, a potassium–rare-earth-element–phosphorous enriched rock, apparently occurred at depth after formation of the feldspathic global crust, probably by deep-seated magmatic differentiation (Binder, 1998). Collectively, the now-extensive data on lunar crustal compositions point to very early and continuing global melting, even before the eruption of the mare basalt (Cattermole, 1996). The early Moon and, by implication, all comparable bodies were intensely hot.

A later and indirect development reinforced belief in high-temperature origins for all the planets: the studies of Supernova 1987A, the first that could be observed from the ground and from space. The discovery of iron 56, clearly formed in the explosion, raised the theory of explosive nucleosynthesis of heavy elements to an experimental fact (Clayton, 1968, 1989). This in turn supported views that the heavy elements of the solar system had been formed not long before the system itself (Cameron and Truran, 1977), providing still another heat source for the primordial planets in the form of short-lived isotopes, long since decayed (Wasserburg and Papanastassiou, 1982). Such isotopes would raise the temperature of the primordial Moon by hundreds or thousands of degrees by themselves (MacDonald, 1963; Jacobs, 1980). Radioactive decay of aluminum 26 has been shown by McSween (1999) to be a plausible heat source for melting of the achondrite parent bodies, presumably in asteroids. To radiogenic heat sources must be added the heat of accretionary impacts and core formation (Kaula, 1980), themselves highly effective.

It was recognized rapidly that a high-temperature origin for the Moon implied an even higher temperature origin and early evolution for the Earth (Hamilton, 1993). As discussed by Wetherill (1972), Smith (1976), and several others, the Earth's greater mass, much smaller surface to volume area, and other factors absolutely dictated that if the Moon, whose mass is only 1/81 that of the Earth, had been extensively melted, the Earth had been much more so. This, in turn, has major implications for the crustal evolution of the Earth. Let us now consider the general topic of planetary crustal evolution.

6.5 Crustal evolution in silicate planets

The period from 1970 to 2001 has been one of explosive progress in exploration of the solar system, and in particular the terrestrial or "silicate" planets, including the Moon as a planet despite its possibly unique origin. Astronomical studies of the planets of course continued, benefiting from new developments in technology and from the ground truth for at least one "planet," the Moon. Understanding of comets, asteroids, and meteorites advanced enormously, as shown by Taylor (1992) and McSween (1999). Collectively, these lines of progress have produced a surprisingly coherent picture of crustal evolution in the Moon, Mercury, Mars, and Venus. The revival of lunar and planetary exploration in the 1990s refined this picture but did not change its main features. Several discrete stages of crustal evolution are now generally recognized.

6.5.1 First differentiation

The first stage for which visible evidence survives has been called the "first differentiation" (Lowman, 1976). This was an event or events that produced the global feldspar-rich crust, the highlands on the Moon (see Fig. 6.7), as previously discussed. The physiography of Mercury is strikingly like that of the Moon, although its high density implies a large iron core. Reflectance spectra of Mercury suggest an average composition similar to anorthositic gabbro, comparable to the lunar highland crust (McCord and Clark, 1979). Given the partial coverage by *Mariner 10*, and the complete absence of *in-situ* or returned-sample analyses, these results are at best only consistent with a Moon-like first differentiation.

Such crusts have been termed "primary" by Solomon and Head (1991), formed early by accretional heating. The age of the lunar highlands has been estimated from radiometric dating of returned samples at 4400–4500 million years (Schmitt, 1974) although there was probably no sharp cutoff of highland activity. The lunar highlands were of course recognized as the oldest part of the Moon long before *Apollo*, from the density of craters and superposition relationships, but the determination of absolute ages showed that the highlands are close to truly primordial crust.

The surface of Mars is a fascinating composite of Moon- and Earth-like topography, suggestive of a Moon that has been modified by terrestrial processes of erosion and deposition. Much of the planet has cratered highlands similar to those of the Moon, although not generally saturated with overlapping craters. The northern plains are much less cratered, closely resembling the lunar maria (Fig. 6.8). Two *Viking* missions landed on these plains, and returned X-ray fluorescence data indicating soils derived from basaltic bedrock (Clark, 1982). However, the composition of the early crust, a critical aspect of the planet's evolution, requires detailed treatment because of its implications for the Earth, as will be discussed later.

The first compositional measurements bearing on the nature of the crust of Mars came from *Mariner 9* (Hanel *et al.*, 1972), which carried an infrared interferometer spectrometer (IRIS). This instrument produced spectra of the suspended dust (Fig. 6.9) which

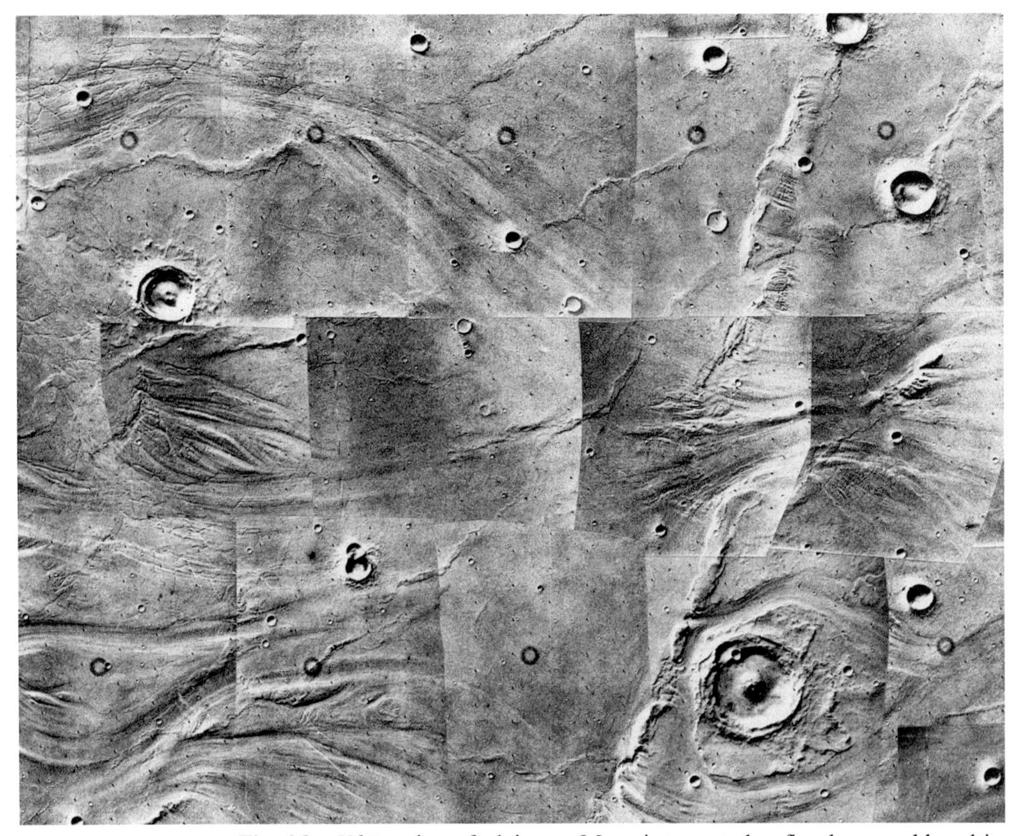

Fig. 6.8 *Viking* view of plains on Mars, interpreted as flood-scoured basaltic lava flows. Area covers approximately 200 × 250 km.

corresponded to the composition of an intermediate igneous rock, estimated at 60 ± 10% SiO_2, and interpreted as indicating planetary differentiation. In 1997, the first rock analyses were produced by the alpha proton X-ray spectrometer (APXS) carried by the *Pathfinder* rover, *Sojourner* (Rieder *et al.*, 1997). A total of five rocks were analyzed, all with compositions averaging 62% SiO_2 after correction for soil, close to andesite and to "the mean composition of Earth's crust." Later study (McSween, 1998) suggested that these might be icelandites, which on Earth form by differentiation of basaltic magmas. Whether the rocks at the *Pathfinder* site were derived from the highlands, and are representative of the martian surface, remains to be seen. Later results from *Mars Global Surveyor* (Bandfield *et al.*, 2000) indicate that martian andesites are widespread although their occurrence in the presumably-basaltic northern plains is not understood. However, in conjunction with the earlier *Mariner 9* data, the *Pathfinder* and *Mars Global Surveyor* results suggest a siliceous crust

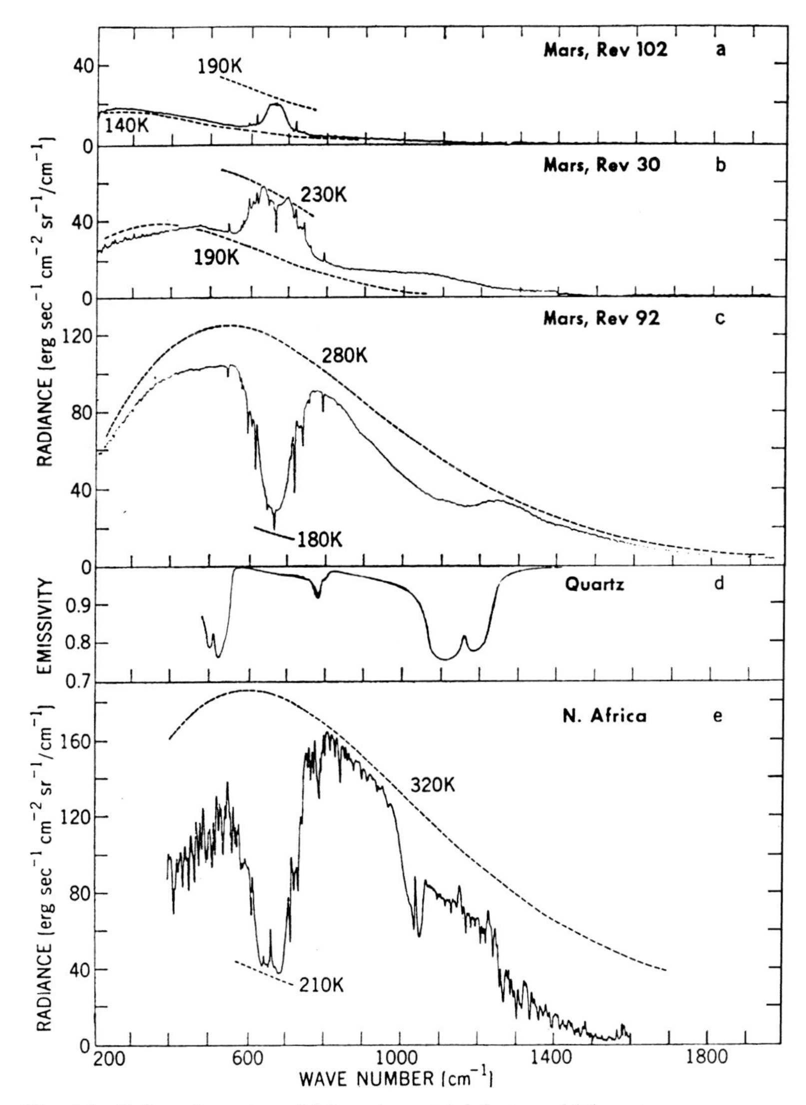

Fig. 6.9 Infrared spectra of Mars, terrestrial dust, and laboratory measurements of quartz. Note dip in spectrum on Rev. 92 corresponding to quartz. Interpreted as indicating relatively high SiO_2 content (~60%) and differentiation of Mars. From Hanel *et al.* (1972).

consistent with an early or first differentiation. Such a crust would be significantly different from the lunar highland crust, probably because of the effect of water on igneous processes in Mars. In particular, magma generation in the presence of water tends to follow an andesitic trend (Fig. 6.10) (Yoder, 1969; Lowman, 1989). Since the Moon demonstrably lost most of its internal water – if it ever had any – at an early stage, magma generation should have followed

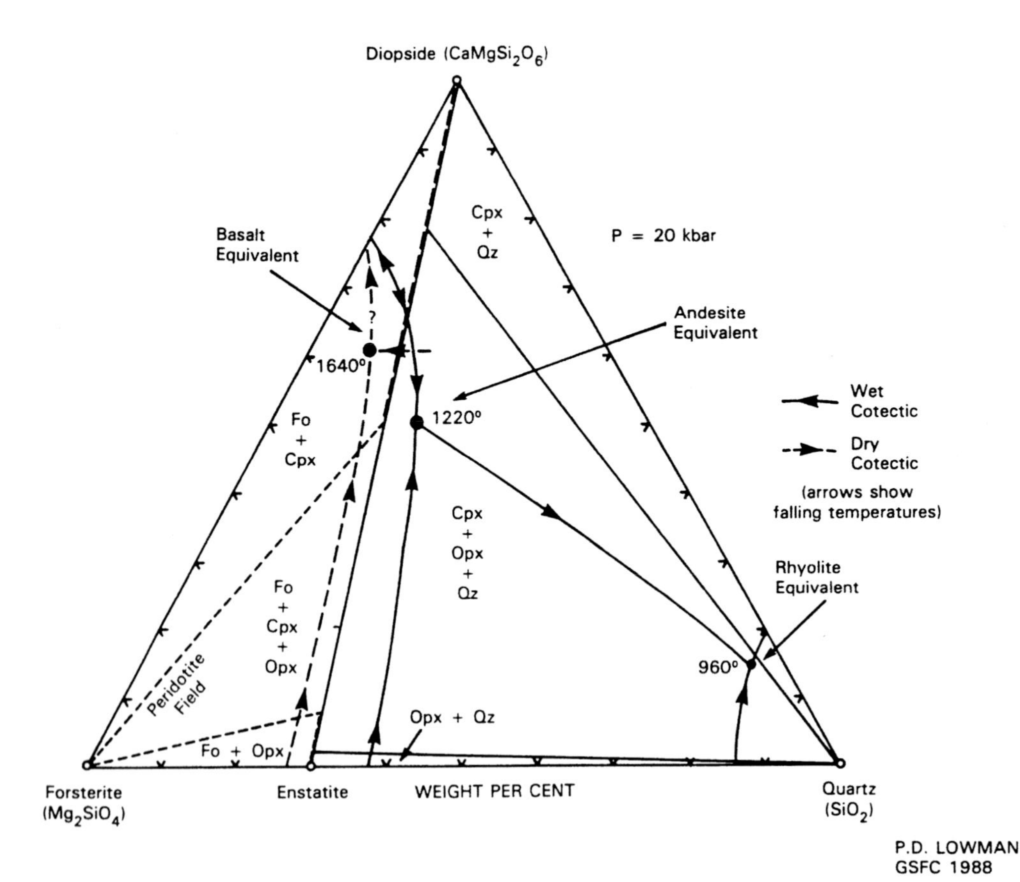

Fig. 6.10 Melting behavior of a rock-like material under wet and dry conditions. Temperatures in degrees Celsius, pressures in kilobars. "Andesite equivalent" line shows that melt produced under wet conditions is more siliceous, i.e., andesite-like, than the melt produced under dry conditions ("basalt equivalent"). Material shown does not represent an actual rock, aluminum being omitted.

a basaltic trend. The abundance of rocks formed by differentiation of basaltic magma, as well as the mare basalts, is quite consistent with this situation.

As the planet closest to the Earth in mass, Venus deserves special discussion. Because of its optically opaque cloud cover, the surface of Venus was not seen until radar methods were applied, first from Earth and then from space, with the *Venera*, *Pioneer Venus*, and especially *Magellan* missions. However, we now have a comprehensive picture of this once-invisible planet (Fig. 6.11). Results of the radar studies mentioned already fill volumes, but the main aspects of crustal evolution on Venus must be summarized at least briefly.

First, it appears that most of the Venusian crust is continental

Fig. 6.11 Mosaic of *Magellan* radar images of Venus (diameter 12,100 km); bright areas are features of high roughness.

in a physiographic sense. As discussed in Chapter 2, its topography is unimodal, in contrast to the Earth's bimodal (ocean basins and continents) distribution (Saunders and Carr, 1984). About 80% of the surface is within 500 m of the modal radius of 6051.1 km. Furthermore, the deviations from this rather flat terrain, or lowlands, are almost all positive, i.e., are elevations rather than depressions. It was thought, from the early low-resolution radar and altimetry data, that the Venusian lowlands might be oceanic crust, with higher areas such as Aphrodite representing zones of sea-floor spreading (Head and Crumpler, 1987). However, the *Magellan* radar imagery shows the topography to be clearly more similar to that of the Earth's continents than to its ocean basins (Solomon *et al.*, 1992). The crust displays broad, anastomosing fold belts (Suppe and Connors, 1992) (Fig. 6.12), remarkably similar in arrangement and apparently in structure to the greenstone belts of Archean terrains such as the Superior Province of the Canadian Shield (Fig. 6.13). Hamilton (1993) has specifically

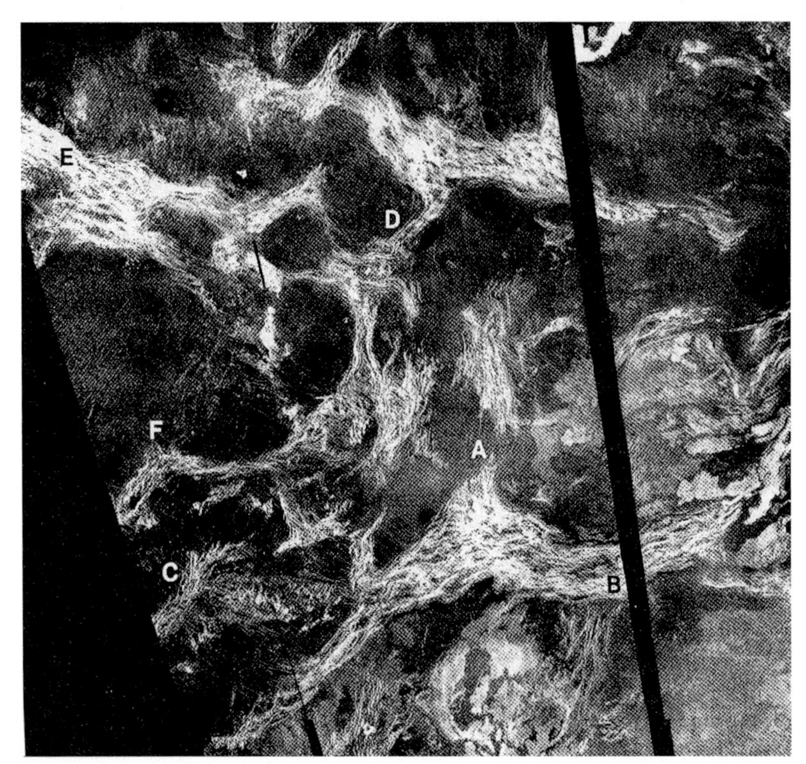

Fig. 6.12 *Magellan* radar image of fold belts in Lavinia Planitia; area 1843 km wide. Letters refer to discussion in Solomon *et al.*, 1992.

compared the tectonic style of Venus to that of the Canadian Shield, considering neither to have resulted from plate tectonic processes. Hamilton's view should be given special weight inasmuch as his study of the Indonesian region (Hamilton, 1979) is considered the authoritative application of plate tectonic theory to this area.

The "highlands" of Venus are extremely complex and interesting, and are not ancient cratered terrains analogous to the highlands of the Moon, Mercury, or Mars. The geochemical analyses carried out by the Soviet *Venera 8* (Surkov, 1983) indicated a composition suggestive of continental crust, although those from *Veneras 9* and *10* found basaltic rocks. This evidence, and the volcanic geomorphology, indicates that most of the exposed crust is basaltic, a "secondary" crust in the words of Solomon and Head (1991) rather than a "primary" one.

The issue is obviously still unresolved, but a reasonable interpretation is that Venus did form a primitive continent-like crust (Surkov, 1983) that has since been largely buried by later basalts – "resurfacing" (Head *et al.* 1992) or "overplating" (Lowman, 1989).

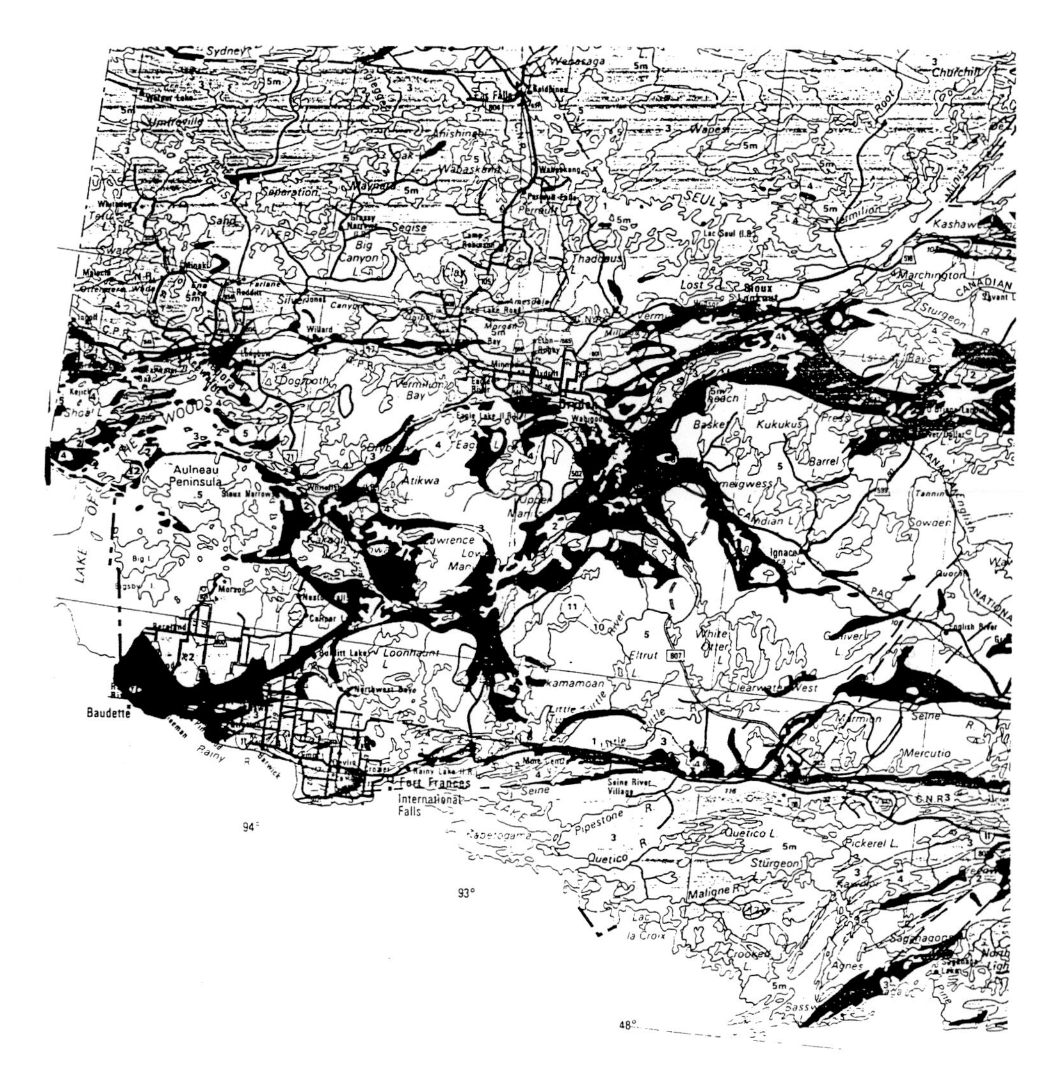

Fig. 6.13 Greenstone belts in western Ontario (dark pattern), intruded by granitoid rocks. From *Geological Highway Map*, Northern Ontario, Ministry of Northern Development and Mines (1986).

It has been pointed out by Fahrig and Wanless (1963) that the mafic dike swarms of the Canadian Shield (Fig. 6.14) and all other well-exposed shields were probably feeders for flood basalts, mostly removed by erosion. Venus might resemble the Canadian Shield had there been no erosion on the Earth, i.e., thoroughly overplated with basalt, or younger basalt plateaus such as those in the Pacific northwest, India, or Siberia. The *Magellan* radar has revealed many features interpreted as dike swarms (McKenzie *et al.*, 1992) that were specifically compared to those of the Canadian Shield. Collectively,

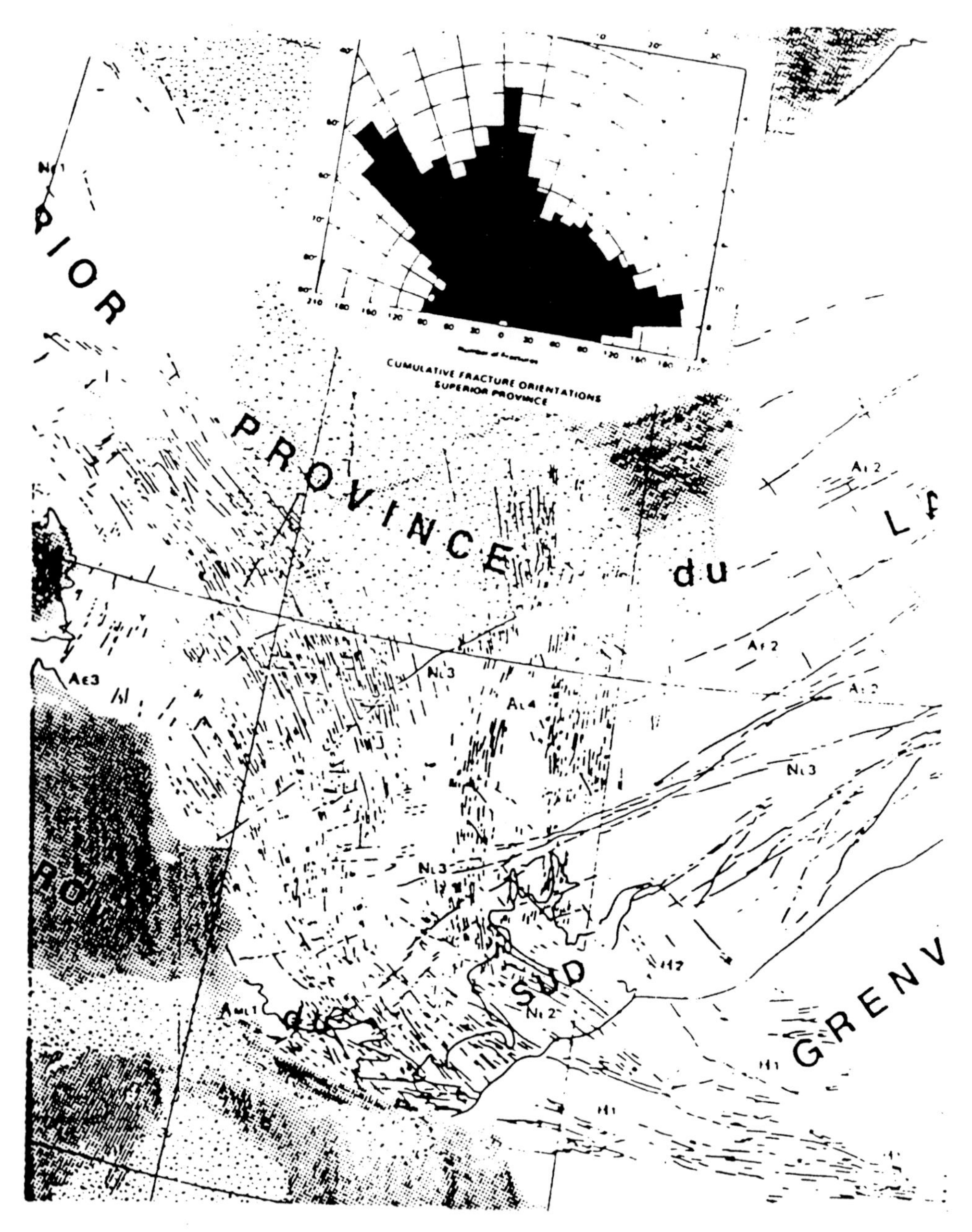

Fig. 6.14 Dike swarms and fractures on Canadian Shield, based on Fahrig and West (1986), with fractures from Lowman *et al.* (1992).

the chemical and geomorphic evidence suggests a primitive crust covered everywhere with basalt.

The *Magellan* radar images show a complex montage of impact, volcanic, and tectonic features, well preserved in the absence of water-dependent erosion and deposition. Most parts of Venus (Solomon *et al.*, 1992) express several superimposed periods of volcanism and tectonism. This characteristic is familiar to those acquainted with terrestrial continental geology. Any large crustal segment has undergone repeated episodes of magmatism, tectonism, and metamorphism, as previously illustrated by the geology around Sudbury, Ontario (see Figs. 6.4, 6.6). This history is only that recorded in exposed rocks, and there is nearly as much time to be accounted for by those not exposed, in the lower crust. This example should make it clear that the tectonic style of Venus is much closer to that of the Earth, in particular the continents (Kaula, 1990; Hamilton, 1993), than to that of the Moon, Mercury, or Mars.

In summary, the main features of the geology of Venus are more than "reminiscent" (Solomon *et al.*, 1992) of continental tectonism, and appear consistent with a first differentiation.

6.5.2 Late heavy bombardment

The next major event in the crustal histories of the Moon, Mercury, and Mars was a period of late heavy bombardment, during which the large impact craters known as mare basins on the Moon and analogous craters on Mercury and Mars were formed. Whether this was a discrete event or part of a declining impact history is still debated (Taylor, 1992, 2001). Nevertheless, radiometric dates and other information from lunar samples show clearly that there were several major impacts between 3800 and 4000 million years ago, forming Mare Imbrium and other circular mare basins. This event was covered in Chapter 5. Comparable impacts evidently occurred on Mars, producing the depressions occupied by the northern plains. The magnitude of such impacts suggests that they involved not random bodies from elsewhere in the solar system, but the actual planetesimals from which Mars was formed.

6.5.3 Second differentiation

The late heavy bombardment was followed, on the Moon and apparently on Mercury and Mars, by a prolonged period of lava eruptions. The lunar lavas, forming the maria, were basaltic, and most evidence indicates that the martian "smooth plains" are also largely basalt

(Christensen *et al.*, 2000) (see Fig. 6.8). The SNC (Shergotty, Nakla, Chassigny) meteorites are now considered samples of these martian basalts, with ages between 1000 and 2000 million years (McSween, 1985, 1999). (A useful discussion of basalt petrology on various planets was published by Walker *et al.*, 1979.) Reflectance spectroscopy of Mercury does not show evidence of basalt (Taylor, 1982), but the physiography of the mercurian plains is at least suggestive of the lunar maria and martian smooth plains. On Venus, there was extensive and perhaps global basaltic overplating, which as we have seen has buried a primordial crust and mare basins if there were any.

For the Moon and Mercury, the mare eruptions were essentially the end of their internal evolution (Walker *et al.*, 1979). Mars evolved further, with formation of immense shield volcanos, tectonic activity producing extensional features such as the Valles Marineris, and with a complex of surficial processes including fluvial erosion, mass wasting, and other Earth-like geomorphic events. The low ages, around 160 million years, of the Mars-derived shergottite meteorites, imply strongly that Mars is still internally active. This inference is supported by the finding of geomorphically young features, such as alluvial fans, produced by ground water (Malin *et al.*, 2000; Hartmann, 2001). The proposal of Sleep (1994), that the northern plains of Mars were produced by early plate tectonic activity, now seems unlikely, since the continuing activity of Mars would imply that sea-floor spreading would be still occurring. The northern plains do not show any geomorphic evidence of such processes.

Venus also is still internally active, judging from its impact crater population, freshness of topography, and other evidence summarized by Solomon *et al.* (1992). The landforms and the surface composition as measured by Soviet *Venera* landers all point to a dominantly basaltic nature for the presently-exposed terrain. In the absence of direct evidence for a "first differentiation" we can hardly call this basaltic magmatism a second differentiation. But it has all the earmarks of such.

6.5.4 Summary

To summarize the crustal evolution of silicate planets of the inner solar system, excluding for the moment our own, the main events seem to have been: *early intense heating*; a *first (felsic) differentiation*; a period of major *basin-forming impacts*; and a *second (basaltic) differentiation*. Mars may have gone beyond this, into incipient plate tectonics as shown by the Valles Marineris and other relatively young features (Hartmann, 2001). Venus is clearly still in an active period of tectonism and magmatism, but not "plate" tectonics.

Let us now turn to the question of whether we can see evidence of an analogous sequence of events in the geologic record on Earth, in particular the record preserved in continental rocks.

6.6 A model of continental crust

A few scientific problems are susceptible to something approaching pure thought. A classic example is the theory of special relativity, developed as a spare-time project by an isolated patent examiner in Switzerland, with no laboratory, limited library facilities, and no contact with professional physicists.

The origin of continental crust, in contrast, is a problem better suited to the descriptive approach. Many scientists, particularly physicists, are understandably sceptical of "butterfly collecting," as Rutherford put it, but in fact many problems in the geological sciences are well suited to this approach. The origin of continental crust is the origin of the rocks that make up this crust. To the extent that we understand the origin of granite, for example, we understand how the granitic part of the continental crust was formed. This example was deliberately chosen, because the "granite problem," a highly controversial issue until the mid-20th century, was largely solved by "butterfly collecting": field mapping, petrography, experimental petrology, and isotope geology (e.g., Best, 1982). The origin of continental crust is to a large degree a descriptive problem. If we had a really complete and accurate picture of the composition, structure, and age relationships of the crust, in particular the Precambrian crust, we would implicitly understand how it was formed.

Anything like a detailed description of the crust is obviously impossible here, but a generalized and highly focussed account is necessary. An important point also brought out in Wasserburg's (1961) seminal paper is that most continental crust was formed in the Precambrian Era, the nearly 4000 million years between accretion of the Earth and the beginning of the good fossil record in the Cambrian. The wide extent of Precambrian crust (Goodwin, 1990), underlying most continental areas, further implies that study of the origin of continental crust should concentrate on the Precambrian.

Precambrian rocks are best exposed in areas such as the Canadian Shield. Knowledge of the Precambrian has expanded greatly in the last decade or so, from field mapping with radiometric

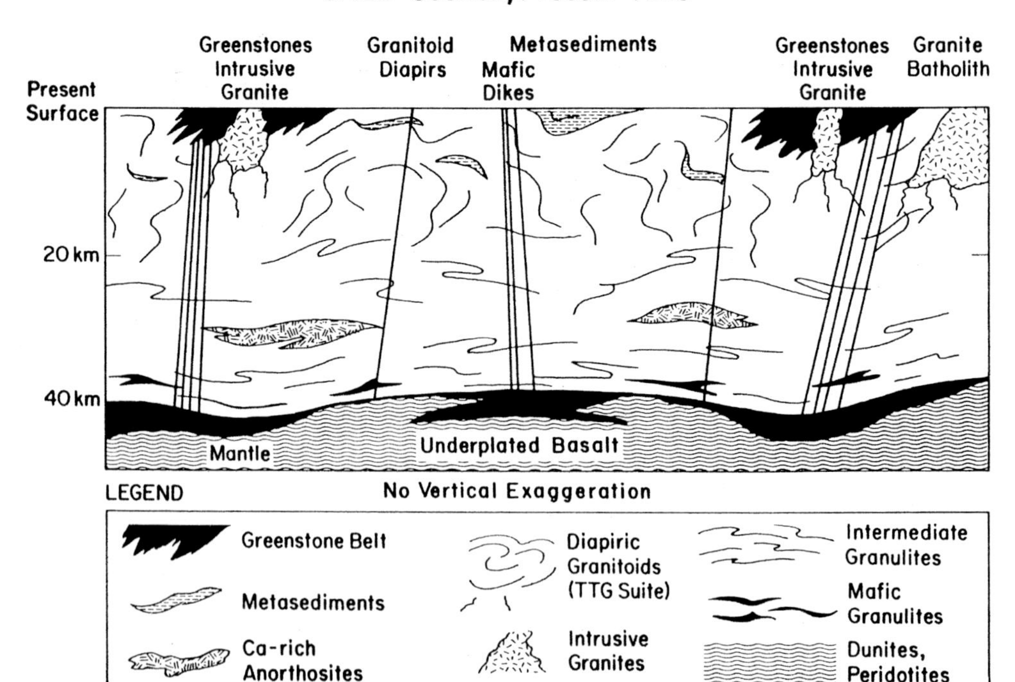

Fig. 6.15 Proposed crustal model for Archean granite–greenstone terrain, such as that under Superior Province of the Canadian Shield. From Lowman (1989).

dating and especially from seismic reflection profiling of the rarely exposed lower continental crust. Canada's LITHOPROBE has in particular been an epic exploration of the Earth's lower continental crust, the last frontier in a geological sense (Oliver, 1998), comparable to the *Apollo* program.

A composite cross section or model has been constructed (Lowman, 1989), showing the major rock types and structures of Archean rocks (older than 2500 million years) in a typical area such as the Superior Province of the Canadian Shield (Fig. 6.15). Such rocks comprise most Precambrian areas, and probably underlie Proterozoic fold belts. Accordingly, the Archean "granite–greenstone" terrains will be taken as the fundamental rock of the continental crust. (Lithologic terms are defined in the Glossary.)

Details of the model will not be discussed here, but a few general characteristics should be noted. The term "granite–greenstone" is used for convenience, being generally understood by geologists.

Fig. 6.16 Cliff at Scourie, northwest Scotland, showing deformed high-grade Archean granulites typical of lower crustal rocks in composition and structure. Cliff about 100 m high. From Lowman (1984).

However, both these rock types are generally confined to the upper continental crust, something only realized in recent years with the advent of reflection profiling. The lower continental crust is dominated by nearly horizontal layers (Fig. 6.16) of high-grade metamorphic rocks ("intermediate granulites") of intermediate silica content (Smithson, 1978; Clowes, 1993). These were probably formed largely from supracrustal protoliths, volcanics and sediments, representing a very early period of overplating (Lowman, 1984, 1989).

An important point brought out by Nisbet (1987) and Card (1990) is that these high-grade terranes are largely the highly-metamorphosed equivalents of upper-level rocks, not exotic material with a unique origin or provenance. Occasional layers of high-calcium anorthosite, comparable to those of the lunar crust

Fig. 6.17 *Landsat 1* view of area (160 km side) just south of Coronation Gulf, Canada, with low Sun elevation (9 deg.), emphasizing lineaments. Many with northwest trend are expression of Mackenzie dike swarm. *Landsat* scene 11206-18381-7, acquired 14 February 1973.

(Windley *et al.*, 1981) occur in the lower crust (Dymek, 1984). The crust becomes more mafic with depth, down to the Mohorovičič Discontinuity, grading into the mafic and ultramafic rocks of the mantle (Rudnick, 1995). In many areas, the crust has apparently been underplated, and intruded, by basaltic rocks (Fyfe, 1993), often expressed as regional magnetic anomalies discussed in Chapter 3.

Although a subordinate feature of the cross section shown here, basaltic dikes cut Precambrian shields on all continents (Fig. 6.17). Those of the Canadian Shield have been compiled by Fahrig and West (1986), who list dozens of radiometrically dated dike swarms. These features have only been studied in detail recently (e.g., Halls and Bates, 1990), but in fact represent a major class of continental magmatism.

6.7 Evolution of the continental crust

The general evolutionary sequence for silicate planets previously outlined can be applied to the Earth, and specifically to the origin of the continental crust, if it is reasonably consistent with the terrestrial geologic record. Is it?

An obvious first question that has been asked by many authors, such as Taylor (1982) is: "Is there any sign of a primitive crust analogous to that of the Moon?" Taylor's answer in 1982 was that "there is no isotopic or chemical evidence of the existence of such a crust." However, since then such evidence, admittedly fragmentary, has been found (Lowman, 1989). Detrital zircons 4300 million years old have been identified in Australia (Froude *et al.*, 1983; Compston and Pidgeon, 1986), suggesting a sialic source area since zircons are typically found in granites. Isotopic data, in particular positive neodymium epsilon values – a geochemical indicator for whose explanation the reader is referred to Faure (1986) – for even 3800-million-year-old rocks from India (Basu *et al.*, 1981) indicate derivation from mantle material *already* differentiated (Jacobsen and Dymek, 1988). Furthermore, this characteristic is now known to be "not unusual" for Archean rocks (Nisbet, 1987). In addition, negative neodymium values are also common in Precambrian rocks (Faure, 1986), implying derivation from or major mixing with pre-existing sialic crust. Bowring and Housh (1995) and Hofmann (1997) interpret the neodymium data as pointing to early differentiation of the Earth, followed by extensive reycling of much of the early crust.

A classic isotopic study of Precambrian basement rocks in the mid-continent region of North America (Muehlberger *et al.*, 1967) found that the distribution of radiometric dates showed that the continent had already attained at least 50% of its present area by 2500 million years ago. The authors concluded that lateral accretion was "of lesser importance" than usually supposed, and that a significant but unknown proportion of older basement had been incorporated in younger provinces. They also explicitly argued against what is today called terrane accretion for three separated Archean areas, roughly equivalent to the Superior, Wyoming, and Slave Provinces.

The distribution of Precambrian rocks along Pacific continental margins, which should represent continental accretion by plate tectonic processes if it occurs anywhere, argues strongly against such accretion as a major continent-forming process. In the Sierra Nevada and Peninsular Range batholiths of California, the isotopic evidence indicates formation on a continental margin 1800 million years old (DePaolo, 1980). Gehrels and Stewart (1998) have produced convincing evidence from detrital zircon U–Pb analyses that these minerals came from "widespread" source areas with ages between 1400 and 1800 million years in the southwest US and northwest Mexico, although allowing the possibility that some of these source areas were exotic "far traveled" terranes. Ernst (1988) presented isotopic and geologic evidence arguing against simple "amalgamation of preexisting continental fragments" as a mechanism for crustal growth in the western US. Rocks at the water's edge from Tierra del Fuego to northern Peru are Paleozoic to Proterozoic. Field evidence from Japan (Choi, 1984) points strongly to a now destroyed Precambrian landmass to the *east* of the Japanese islands, a strong argument against terrane accretion. Identification of Archean zircons from rocks in New Zealand (Ireland, 1992) suggests a nearby Precambrian source area, although there is no intact Precambrian crust in New Zealand itself.

Whole rock ages of 3960 million years have been found in the Canadian Shield by Bowring *et al.* (1990), and these rocks were formed when an earlier crust already existed. Furthermore, as shown in Fig. 6.15, even the oldest rocks exposed are generally underlain by 30–40 km of continental rock. The nearly level orientation of the layering in the lower crust (see Fig. 6.16), though partly deformational in origin (Clowes, 1993), suggests that if simple superposition relationships hold in general, these deeper rocks are older than the surface ones. There are possible structural complications obvious to any geologist, such as overthrusting, but as a rule, the deeper, the older. This interpretation suggests that the difficulty in finding truly primitive crust has been primarily an exposure effect. A traverse across the Colorado Plateau, for example, might lead a newly-landed martian geologist to think that the Earth's crust was not much older than Triassic (or the martian equivalent). But discovering the Grand Canyon, and going down to the Granite Gorge, would show him (or it) that there were rocks far older. Terrestrial geologists may have been misled in an analogous way; a primitive crust still exists, but only in the deepest levels, and in a highly metamorphosed and deformed form. To cite Wasserburg's (1961) seminal paper again: "From the apparent absence of this older event, it must not be concluded that it [*formation of a primitive crust*] did not occur."

Another possibility is that large volumes of continental crust were formed early in the Earth's history, but were recycled by processes such as subduction (Armstrong 1968, 1981) or delamination (Sylvester, 2000). Green *et al.* (2000) present evidence from the 3.5 Ga old Pilbara region of Australia indicating a pre-3.5 Ga continental crust at least 30 km thick. They suggest that the present apparent scarcity of such old crust is the result of "preservation and recycling."

To summarize the argument to this point, it appears that a primitive crust was formed on the early Earth, and reasonable explanations for its present restricted extent have been provided. Collectively, this and other evidence point to a general outline for the origin of continental crust, with the following main stages following the accretion of the Earth. It has been illustrated in the context of a diagram (Lowman, 1989) comparing the petrologic evolution of silicate planets (Fig. 6.18), although this evolutionary pattern can now be simplified, as will be shown.

6.7.1 Stage I: first differentiation

Following the earliest high-temperature stage, the Earth underwent a first, or felsic, differentiation, forming a global crust. This was a "primary" crust in the terms of Solomon *et al.* (1992) and Taylor (1992). Because of the presence of water, then being rapidly outgassed from the planet, the resulting volcanism formed a largely andesitic crust (Yoder, 1969, 1976; Hargraves, 1976; Shaw, 1980), whereas the dry conditions in the early Moon produced basalts and their magmatic differentiation products (anorthosites, troctolites, norites, and KREEP). The primordial Earth was extremely hot, and the volcanism was of an intensity and extent far greater than that of today. A possible physical (but not petrologic) analog of this early Earth is the jovian satellite Io (Fig. 6.19), continually erupting komatiitic lavas from dozens of volcanos as a result of the continued heat generation from body tides induced by Jupiter. The primordial Earth was probably rapidly overplated by andesites and minor basalts just as Io is visibly overplated today. The tectonic style of the Earth at this time was, like that of Io, probably dominated by closely-spaced mantle plumes, not spreading centers (Fyfe, 1978, 1993; Hamilton, 1993).

This earliest Archean crust, formed by volcanic overplating in the first few hundred million years, may survive today, as the intensely deformed originally supracrustal granulites of the lower continental crust (see Fig. 6.15). It was presumably intensely cratered and melted, with a megaregolith like that of the lunar highlands, but subsequent metamorphism and deformation have obliterated all evidence of this.

If a primitive crust does survive, is there any evidence that, like the lunar and Martian early crusts, it was global? We can, following

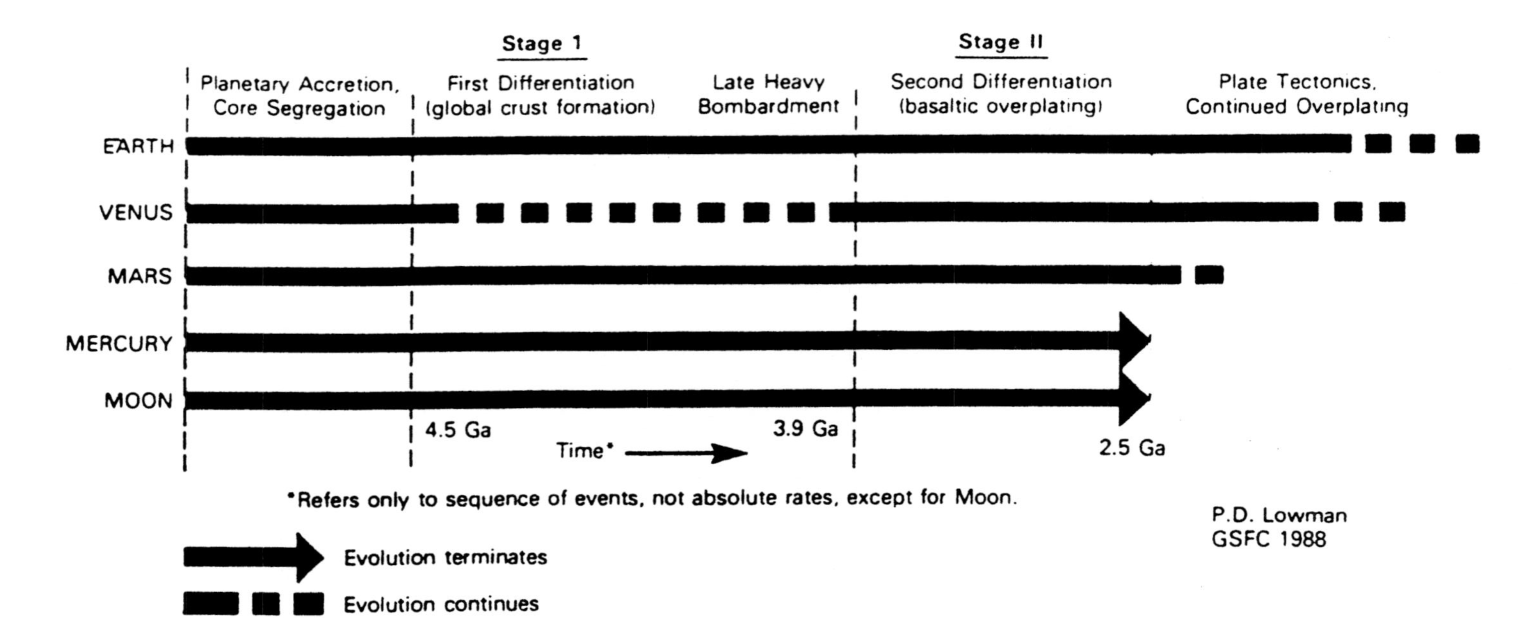

Fig. 6.18 Crustal evolution in silicate planets. True "plate" tectonics applicable only to Earth; term refers to one-plate tectonism on Venus, incipient plate tectonics on Mars.

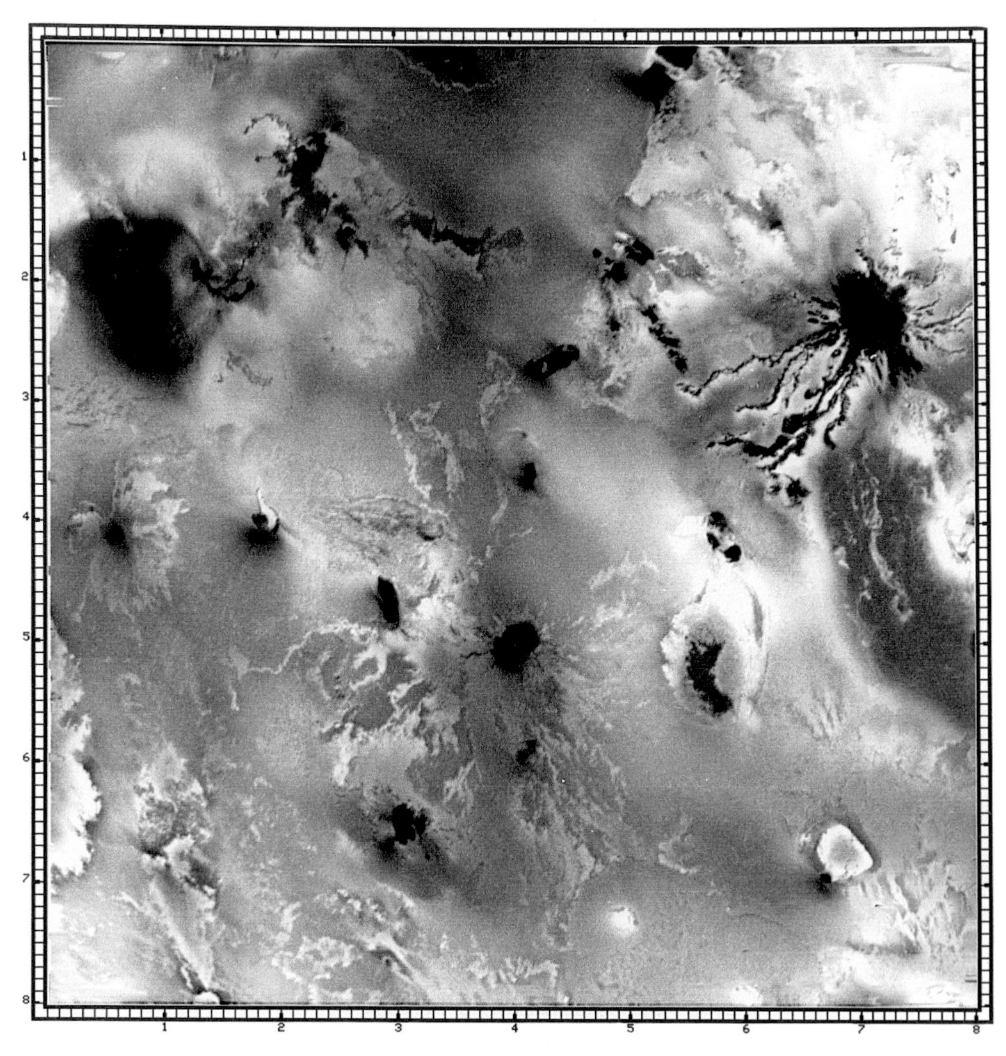

Fig. 6.19 *Voyager 1* image of limb area of Io (diameter 3640 km), showing calderas and overplating by sulfur-rich mafic or ultramafic volcanics. Elliptical caldera at lower right, Creidne Patera, 100×200 km.

Shaw (1976, 1980), apply Ockham's Razor and reason that since there is no evidence it was *not* global, we can assume it was. As we have seen, Venus does have a global crust that is much more nearly continental than oceanic. It is difficult to go beyond this argument for the Earth, because several hundred million years of sea-floor spreading and subduction appear to have destroyed such ancient crust in the ocean basins (but with significant exceptions: Meyerhoff *et al.*, 1992; Choi *et al.*, 1992). However, the fact that the terrestrial crust today is only one-third continental has been explained by Frey

(1980) as resulting from an early period of impact bombardment analgous to the one in which the lunar mare basins were formed. Such a bombardment, in Frey's interpretation, initiated sea-floor spreading some 4 Ga ago. This may account for the destruction, by subduction recycling, of what was once a global crust.

Another explanation for the present absence of some two-thirds of the supposed global primordial crust is also impact-related: removal by the giant impact, of a Mars-sized body, thought to have ejected the debris from which the Moon accreted. To explore this possibility, let us return to Venus, in particular to the apparent absence of structures formed by plate tectonic processes (e.g., Suppe and Connors, 1992), which had been expected (Phillips and Malin, 1984; Lowman, 1989). What explanation can be offered for this difference between "sister" planets?

Plunging into speculation, it may be suggested that the answer may be related to the fact that Venus has no satellite, while the Earth does (and a very large one at that). Given the evidence for venusian mantle convection discussed in Chapter 2, it appears that we may be seeing a planet in which Earth-like plate tectonics, i.e., ocean-basin tectonics, is muffled by an uninterrupted global crust. The fold belts, faults, and other features are suggestive of Kroner's (1985) "ensialic plate tectonics," a concept for orogeny and related phenomena involving mantle dynamics acting on continental crust. But suppose we could remove part of the global venusian crust. What sort of tectonic style would result?

This hypothetical crust removal may actually have happened on the primordial Earth, if the Moon was formed by impact of a Mars-sized planetesimal (Hartmann *et al.*, 1986). The origin of the Moon, a long-standing cosmological problem, must meet a number of constraints. The only mechanism that appears to do so is the "giant impact" hypothesis (Melosh, 1992), which has within the last few years met with general acceptance by default, so to speak. If the Moon was formed in this way, any original global crust may have been largely removed by the impact (along with much of the upper mantle). This concept resembles the 19th century proposal by G. H. Darwin (Melosh, 1992) that the Moon was formed by tidal fission, the Pacific Ocean being the scar of this event. Plate tectonic theory, with its implied unlimited crustal mobility and multi-cycle ocean basin formation, obviously does not support this idea. However, the unique size, structure, and geophysical character of the Pacific basin suggest that Darwin's concept may have been discarded too soon.

The giant-impact theory for formation of the Moon may account for the removal of much of the original terrestrial crust, as well as accounting for the differences between tectonic styles of the sister planets Earth and Venus. Removal of most of the Earth's original crust might have released mantle convection to produce the present plate tectonic style, a release that Venus has not had.

6.7.2 Stage II: second differentiation

The second or basaltic differentiation on the Moon was a long but essentially superficial event, expending the last of the Moon's internal energy to produce a relatively thin and limited veneer of mare basalts (Walker *et al.*, 1979; Head, 1976). On Mars, the second differentiation produced a similar veneer, but continued, perhaps to the present (McSween, 1999). This stage also produced the spectacular volcanos of the Tharsis Rise, dominating the planet's topography (Solomon and Head, 1982), but still superficial accumulations on a static lithosphere.

In contrast, the second differentiation of the Earth was a long and still active stage with profound influence on crustal evolution, petrologic *and* tectonic. Basaltic magmatism appears to have had the following specific effects over geologic time, starting at least 4000 million years ago.

Basaltic overplating This term covers eruption of continental flood basalts, typically tholeiitic in composition, comparable to the lavas of the Columbia Plateau. These are of course young – 10–20 million years old – and have been little altered by erosion, deposition, and metamorphism. Basaltic extrusions of the central eruptive type, i.e., volcanos, are restricted examples of overplating. As we go farther back in the geologic record, the basalts are more dispersed and covered by younger rocks. The greenstone belts previously described are thought by many authorities to be Precambrian flood basalts, now intensely deformed and metamorphosed. Their present areal extent is much less than originally because of the repeated diapiric intrusion of the "granites" making up most of the granite–greenstone terrains.

The dike swarms cutting all exposed Precambrian shields (Halls, 1982; Lowman *et al.*, 1992) were probably feeders for flood basalts since removed by erosion (Yoder, 1988). As discussed previously, Venus appears to have been overplated by basalts, apparently dike fed (McKenzie *et al.*, 1992). Venus has not undergone continual fluvial erosion as has the Earth, and the overplated basalts are still there. It has been suggested (Lowman, 1989) that in the absence of erosion, terrestrial continents would still be similarly covered with basalt. Even with erosion, the abundance of basaltic rocks is still impressive.

Basaltic underplating This relatively new term will be used here to cover basaltic composition rocks intruded under or into the continental crust, using "composition" to allow for different metamorphic grades up to eclogites. The importance of basaltic underplating has only recently been recognized (Newton *et al.*, 1980; McKenzie, 1984; Fyfe, 1986; Furlong and Fountain, 1986; and Lowman, 1989). It is becoming increasingly clear that mafic magmatism has played not just an important but a *dominant* role in the evolution of continental crust. The most direct aspect of this role, corresponding to the overplating just discussed, is simply the addition of material to this crust in subsurface, i.e., underplating and coeval intrusion, amounting to continental growth by vertical accretion. A wide range of geophysical and geochemical evidence indicates that the crust becomes more mafic with depth, specifically with increasingly abundant mafic granulites and eclogites (Rudnick and Fountain, 1995). The metamorphism that produced these rocks was probably promoted by the high temperatures resulting from basaltic underplating.

Remelting of underplated basalts may provide another way to meet the constraints of strontium isotopic evidence (Moorbath, 1975) and field data (McGregor, 1979) indicating that the dominant granitoid rocks of Precambrian shields, the "grey gneisses," form as juvenile additions from anatexis of mafic rocks. A plate tectonic mechanism would involve subduction zones, but it would appear that underplated basalts in other settings would also be suitable source rocks.

The heat added to the lower continental crust by basaltic underplating may be the most important second differentiation effect of all. It is becoming generally recognized (Brown *et al.*, 1995) that melting of the continental crust by mantle-derived basaltic magma has been a major and continuing process, accounting for many chemical and mineralogical characteristics of the crust. It was suggested by Lowman (1976) that partial melting had led to "redifferentiation" of the continental crust by generation of granites. As shown most recently by Raia and Spera (1997), such mafic magmas are probably "the major source of enthalpy that drives intracrustal differentiation" by anatexis. The rocks produced by this mechanism would be the "granites" of granite–greenstone terrains, actually granitoids of the tonalite–trondhjemite–granodiorite suite. These rocks would thus represent an extreme form of "reworking," previously discussed in reference to terrane accretion, not continental growth as generally believed. The intimate involvement of mantle-derived mafic magmas in their generation would be consistent with the indeterminate nature of isotopic evidence (Faure, 1986); both mantle and crustal sources would be involved in formation of any given granitoid rock.

Tectonic activity The global effects of the second differentiation are obvious, two-thirds of the Earth's crust consisting of mantle-derived mafic rocks formed by basaltic magmatism at spreading centers. It may be worth stressing the fundamental difference between terrestrial oceanic crust and the petrographically similar basaltic plains of the Moon, Mars, and perhaps Mercury. The latter are, as previously mentioned, relatively thin superficial coatings overlying, at least on the Moon, an older and very thick feldspathic crust formed in the first differentiation. However, the oceanic crust is a much more fundamental feature, resting directly on the mantle. Needless to say, there is no extraterrestrial counterpart to sea-floor spreading, even on Venus.

As discussed in previous chapters, ridge push may be a major driving force for sea-floor spreading, and has been invoked (Zoback, 1992) to explain the prevailing continent-wide compressional stress fields discovered in the *World Stress Map* Project. A generalized version of this map (Fig. 6.20) will illustrate this relationship. Richardson (1992) similarly concluded that ridge-push forces were largely responsible for intraplate compressional stress fields. Basaltic magmatism is in this interpretation not only a petrologic mechanism but a major cause of tectonic activity.

Another tectonic role for basaltic magmatism has been proposed by Meyerhoff *et al.* (1992), in the "surge tectonics" hypothesis. In this concept, interconnected magma chambers, or "surge channels," are the main driving mechanism for most terrestrial tectonic features including folded mountains, island arcs, and flood basalts. The mid-ocean ridges, spreading centers in plate tectonic theory, are considered "oceanic trunk channels." The surge tectonic hypothesis is too involved for detailed discussion here, but it demonstrates again the potential importance of basaltic magmatism as the driving force of plate tectonics.

The origin of mountain belts around the Mediterranean, although commonly ascribed to continental collision between Europe and Africa, may be partly or largely due instead to a fundamentally thermal process closely related to basaltic magmatism. As proposed most recently by Dewey (1988) under the title "Extensional collapse of orogens," the Mediterranean Sea may have been formed as it is today by diapiric upwelling from the mantle, accompanied by basaltic (and silicic) magmatism and sea-floor spreading. This hypothesis helps answer the anomaly brought out by the digital

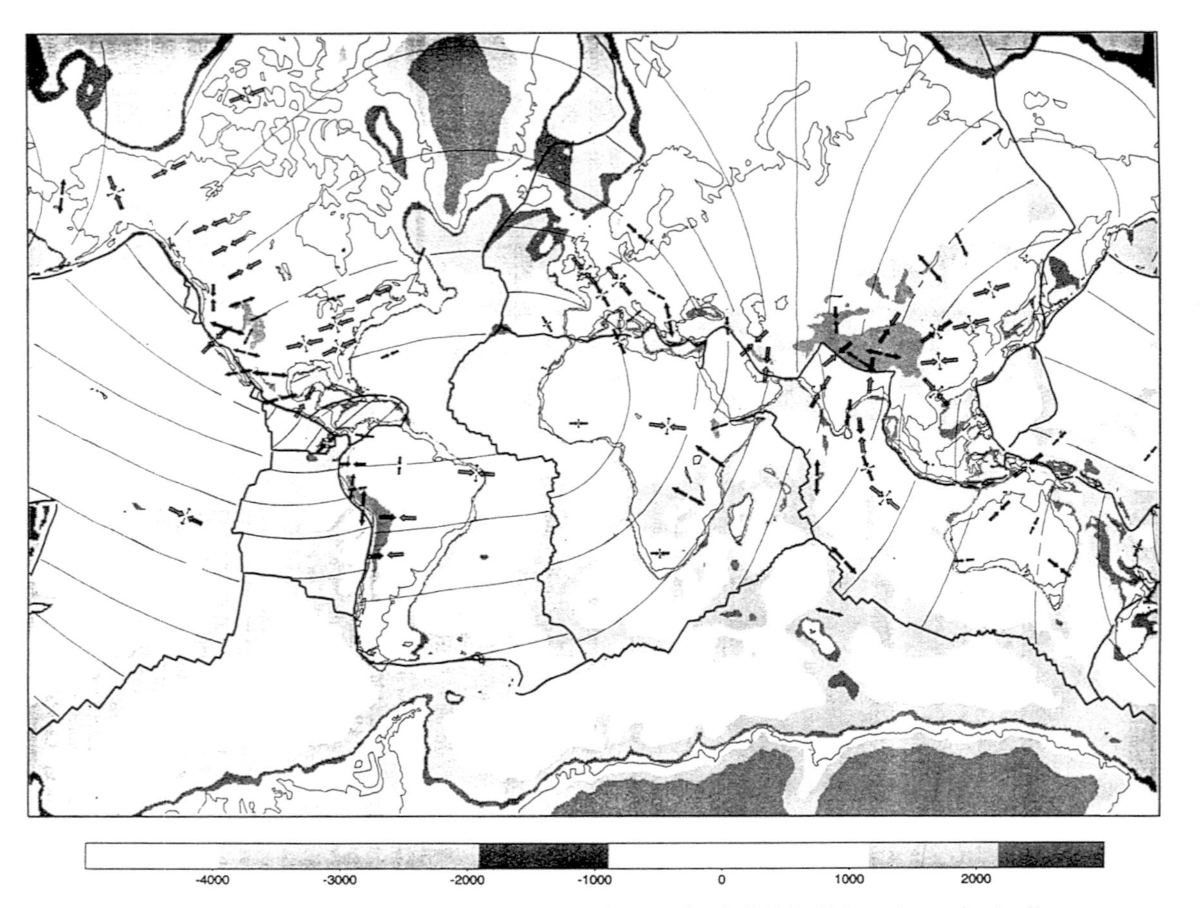

Fig. 6.20 Generalized version of the *World Stress Map*, from Zoback (1992). Values shown in shading are topographic elevations or depressions in meters above or below sea level, respectively. Refer to original paper for details.

tectonic activity map and its seismic source maps (Chapter 1). Africa and Eurasia are generally agreed to be the slowest-moving plates in all plate-motion models, yet the fold belt extending from Iran to the Atlantic Ocean is intensely active today, arguing for some mechanism other than simple plate collision.

In earlier expositions of the crustal evolution theory presented here (Lowman, 1976, 1989), the second differentiation was followed by a third stage, labeled "plate tectonics" or simply "tectonic." However, the second differentiation now appears to be far more important than previously appreciated. It clearly accounts (by sea-floor spreading) for formation of some two-thirds of the Earth's crust. It provides a driving force for sea-floor spreading and indirectly for subduction and attendant tectonism. It produces the newly-discovered intracontinental stress fields and probably many intracontinental tectonic features illustrated in the digital tectonic activity map (Chapter 1). If conventional "plate tectonics and continental drift" theory should be completely valid, sea-floor spreading was responsible throughout geologic time for practically all internally-caused geologic features on the planet – a true "master plan" in Hamblin's (1978) term. In Precambrian granite–greenstone terrains, the second differentiation produced the greenstone belts directly, and the "granite" indirectly by anatexis – "redifferentiation" (Lowman, 1976) or "intracrustal differentiation" (Dewey and Windley, 1981).

In view of the foregoing interpretation, a third stage of crustal evolution appears unnecessary; virtually all terrestrial geologic history has been dominated, in any plausible theory of global tectonics, by basaltic magmatism, the Stage II "second differentiation."

6.8 Petrologic evolution of the Earth

The petrologic evolution of the Earth can be divided into the same major stages as the evolution of other silicate planets: a first differentiation, accompanied by or followed by heavy impact bombardment (perhaps initiated by the ultimate "impact bombardment" that formed the Moon), and a second differentiation. This essentially two-stage history differs from that of the Moon, Mercury, and Mars in that the second differentiation has continued in full strength to the present, whereas in the smaller bodies it either stopped completely or slowed down greatly. Venus has apparently undergone a second differentiation and may also be still active, but having no plates (or only one plate) has a tectonic style very different from that of the Earth.

The continents are interpreted in conventional theory as the result of a long-term process of lateral growth around continental nuclei, dominated by plate tectonic phenomena. The primordial Earth in this view had no continents, the primordial crust being basaltic or komatiitic (Engel, 1963; Condie, 1981, 1989a). The interpretation derived here, from comparative planetology and terrestrial geology, is fundamentally different: that the present continents are the remnants, greatly deformed and metamorphosed, of *an originally global andesitic crust formed largely in the first few hundred million years of the Earth's history in the first, or felsic, differentiation.* Most magmas separated from the mantle since then have been basaltic, though they have often given rise to massive intracrustal melting that generated the granitoid rocks of the upper crust in Precambrian shields. The primordial crust still survives under continents as the intensely deformed and metamorphosed granulitic gneisses, the high-grade terrains of Windley (1984).

The continental crust as we know it is of course radically different from the only other crust we know much about, the lunar highlands. However, these differences can be understood fairly easily. Compared to the Moon, the Earth has greater "endurance" (Walker *et al.*, 1979), remaining intensely active tectonically and magmatically, leading to frequent reworking and remelting of the original crust. The abundance of granites in the continental crust, in contrast to their scarcity on the Moon, is one result of this reworking. Another major difference between crustal evolution in Moon and Earth stems from the relative abundance of water. The Moon evidently lost all its water, at least from the outer few hundred kilometers, early in its history. Not only are lunar rocks totally anhydrous, their chemistry is highly reduced, implying very dry magmas. In the Earth, however, water has continually played a major petrologic role. Magma generation under hydrous conditions tends to produce andesites or rhyolites, but in the absence of water, basalts or their differentiation products (Yoder, 1969; Kushiro, 1972). In view of these considerations, it is reasonable to consider the continental crust of the Earth analogous to the crust of the Moon, and to the primitive crusts of other planets, especially to the demonstrably global crust of Venus.

A scientific theory must be testable or falsifiable to be of any value. The one proposed here has passed its first test: the *Pathfinder* discovery of andesites on Mars, as previously discussed, supported by orbital remote sensing from the *Mars Global Surveyor*. It was proposed (Lowman, 1989) that the "most significant test" of this theory for the origin of continental crust would be "determination of the composition and age of the martian highland crust." Because of the retention of water in Mars, it was predicted that the "first differentiation should have followed an andesitic trend." There are many controversial issues surrounding the *Pathfinder* results (McSween, 1998), such as the source of the analyzed rocks, the degree to which they are representative of the martian highland crust, and their age. However, the predicted "andesitic trend" was a specific one based on extensive petrologic data (Yoder, 1969). A supposed Chinese proverb says, more or less, that "Prediction is difficult, especially with regard to the future." Andesite on Mars was predicted before the *Pathfinder* mission.

The research on which this chapter is based was originally focussed on the continental crust, but it has led to a much broader and radically different view of the Earth's crustal evolution as a whole. The continents in this broad view are subordinate features of the planet's crust, the repeatedly reworked and deformed remnants of a primordial sialic layer. The overwhelmingly dominant process, petrologically *and* tectonically, for roughly the last 3000 to 4000 million years has been **basaltic magmatism**. In various forms, this magmatism – the "second differentiation" in planetary terms – has generated most of today's crust, as indeed recognized in plate tectonic theory. But beyond that, it has remelted, metamorphosed, and deformed the original continental crust after the catastrophic disruption of the Moon-forming and other impacts. It has repeatedly overplated the continents by repeated fissure eruptions. Plume-initiated sea-floor spreading (Burke and Dewey, 1973), the dominant expression of basaltic magmatism today, produces the "ridge push" responsible for most of today's intraplate tectonic activity, and of course the complementary subduction zones responsible for volcanism and seismicity over them. The evolution of the Earth has thus been thermally driven (Hamilton, 1993), plate tectonic activity being primarily the expression of heat rejection from the interior.

This view of crustal evolution is radically different from that of conventional theory. It is a minority view, it needs further testing, and it may be wrong. But its very novelty illustrates, if nothing else, the impact that space exploration has had on one of the great problems of geology, whose solution is the "ultimate goal" of this science (Poldervaart, 1955).

There is one more aspect of crustal evolution in the Earth that has not yet been covered: the effect of life. This topic deserves a chapter of its own, to follow.

CHAPTER 7

Geology and biology: the influence of life on terrestrial geology

7.1 Introduction

One of the most important results of space exploration is a growing appreciation of the fundamental impact of life on the Earth's geology. "Geology" as a separate discipline is rapidly expanding into "earth system science," which includes the study of feedback interactions among the crust, oceans, atmosphere, and especially the biosphere (Jacobson *et al.*, 2000). This new link between geology and biology originated in the Gaia hypothesis.

First proposed under that name by James Lovelock (1979, 1990), this hypothesis holds that **surface conditions on the Earth have been for most of geologic history regulated by life.** This simple statement is directly counter to traditional biological theory and to common sense, which would indicate that life is controlled by its environment. Obviously, locally and temporarily this is true. But Lovelock has proposed that from a planetary perspective, and on geologic time-scales, the opposite is true: life has controlled its environment.

The Gaia hypothesis – to be explained shortly – was directly stimulated by space exploration, as recounted by Lovelock (1979). He was invited by NASA in the early-1960s, first to help plan lunar missions, and then to identify methods for detecting life on Mars, at the Jet Propulsion Laboratory. He rapidly became immersed in fundamental questions about the nature of life in general, and about the origin of the Earth's life-supporting environment. The atmosphere of Mars was even then known to consist largely of carbon dioxide, a chemically stable and non-reactive gas. Lovelock realized that the Earth's atmosphere, in contrast, is out of chemical equilibrium. Gases such as oxygen and methane co-exist, although they should rapidly react with each other. Similar examples of disequilibrium soon came to light. This led him, with various collaborators such as

Lynn Margulis, to propose that atmospheric composition is regulated by biological processes. Further study gave birth to the Gaia concept, after the Greek Earth goddess. This concept, or hypothesis, has been compared to Darwin's theory of evolution in importance by an authority on Archean geology, Euan Nisbet (1987). However, another equally qualified Precambrian geologist, Paul Hoffman, described it in a 1997 lecture as an "interesting but untestable metaphor." In any event, the Gaia concept leads to a new perspective on crustal evolution of the Earth, and to a unified biogenic theory of the Earth's crustal evolution.

7.2 Gaia

The concept of "feedback" is familiar to most people, though not necessarily under that name. It amounts to self-regulation, or automatic control, much as a household thermostat regulates temperature. Another term, more directly relevant to the present discussion, is "homeostasis," referring to the physiological mechanisms such as sweating (or, in season, shivering) by which humans compensate for changes in their external environment. The Gaia hypothesis holds that the collective life on Earth is capable of such homeostasis, or self-regulation of the global environment. Useful summaries are those by Margulis and Lovelock (1974) and Margulis and West (1993).

The concept is well illustrated by the "Daisyworld" of Watson and Lovelock (1983) (Fig. 7.1). This charming hypothetical planet is carpeted entirely with black and white daisies. As time goes on, its hypothetical sun becomes brighter and *Daisyworld* becomes hotter. This will favor an increasing proportion of heat-reflecting white daisies. If the sun becomes dimmer, the proportion of heat-absorbing black daisies will increase. Life thus keeps the planetary temperature within livable limits. Lovelock's discussion is more elaborate than this, and he has produced a modified model, but the feedback principle should be clear with this simplified example. The *Daisyworld* analogy has been further used by Lenton (1998) to show how such "planetary self-regulation" can arise from natural selection, thus unifying Darwinian evolution with the Gaia concept.

A much less hypothetical example is the probability that the actual Earth's carbon dioxide atmospheric content is to a degree regulated by plants (especially oceanic phytoplankton), which should tend to increase with increasing CO_2 and metabolize the excess (Falkowski *et al.*, 1998). There are many other mechanisms, biological and otherwise, that could contribute to this "geophysiology" in Lovelock's term.

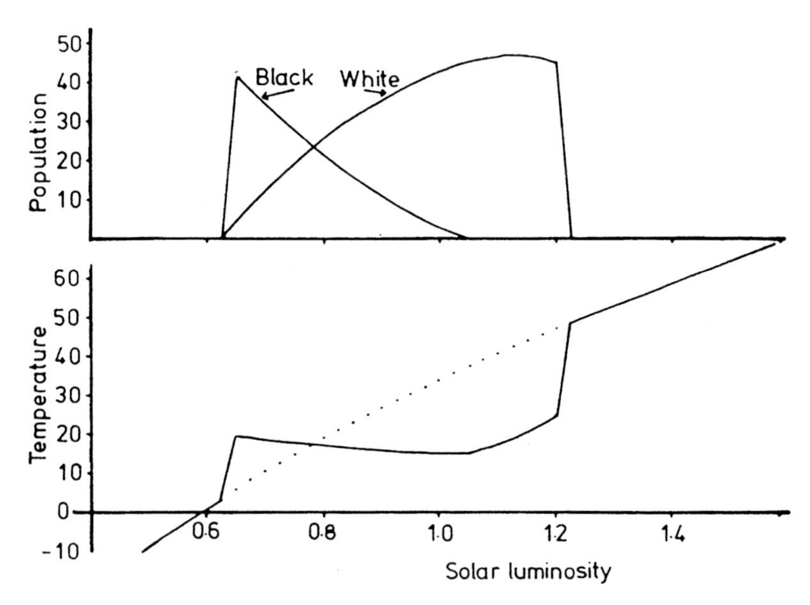

Fig. 7.1 *Daisyworld* planet model of Watson and Lovelock (1983). *Lower diagram*: temperature vs. solar luminosity; dotted line shows temperature on a lifeless planet, solid line temperature regulated by daisy population. *Upper diagram*: relative proportions of black and white daisies.

The discussion to this point should make it clear that Gaia can be simply defined as "the physiologically regulated Earth" (Margulis, 1998). The theoretical soundness of the Gaia concept is undeniable, despite the religious overtones it has unfortunately acquired. Before discussing possible weaknesses it will be helpful to give further background. A specific problem has been the near-certainty that the Sun's luminosity in the early Precambrian was substantially less than it is now, to the point that water on Earth would have been frozen. Yet there is indisputable proof that there has been abundant liquid water as far back (almost 4000 million years) as the rock record goes. One reason for proposing the Gaia concept has been to account for this anomaly, the "faint early Sun paradox" (Gilliland, 1989). Lovelock has shown that, in principle, early life – methane-producing decomposition organisms – might have regulated global temperature by increasing CH_4 output, the resulting greenhouse effect thus retaining heat. There is good fossil evidence that oceanic phytoplankton existed as early as 3.5 billion years ago, and could have helped regulate planetary temperature by generation of CO_2 much as they do today (Falkowski *et al.*, 1998). There is a broad if indirect base of evidence in the geologic record for the Gaia hypothesis.

There has been, predictably, considerable debate about the Gaia hypothesis. The opposing view is, in brief, that non-biological feedback mechanisms can regulate terrestrial environments. A recent popular review of the controversy has been published by Allegre and Schneider (1995). A technical treatment illustrating the inorganic feedback school of thought is that of Kasting (1989), discussing the "long-term stability of the Earth's climate." Citing the agreed-upon geologic evidence for the continued (though perhaps intermittent, interrupted by glaciations) existence of liquid water through geologic time, Kasting shows that several non-biological mechanisms could control global temperatures, directly or indirectly, through controlling atmospheric composition (specifically greenhouse gases). For example, excess carbon dioxide could be compensated for by weathering of silicate minerals, followed by absorption in the oceans by $CaCO_3$ (limestone) deposition (Walker *et al.*, 1981). If the global temperatures fell sharply, weathering rates would decrease, increasing the CO_2 content of the atmosphere (contributed by volcanism) and hence the greenhouse warming. The actual situation is far more complicated than this simplified example, which is intended only to show that there are sound arguments against the Gaia concept.

A possible problem with the Gaia hypothesis goes back to the question that originally inspired Lovelock's thinking: life on Mars. Lovelock concluded rapidly, long before any Mars landings, that the CO_2 atmosphere of Mars alone ruled out life, which on Earth is detectable by the non-equilibrium composition of our atmosphere. The *Viking* landers of 1976 are thought to have vindicated Lovelock's pessimistic conclusion, although there is still disagreement about this (e.g., DiGregorio, 1997). The celebrated (and disputed) finding of possible nannofossils in the Allan Hills martian meteorite 84001 by McKay *et al.* (1996) raised hope that there might after all be subsurface life, perhaps single-celled, like that whose existence deep within the Earth has now been demonstrated (Frederickson and Onstott, 1996). The climate of opinion of life on Mars may be changing, one reason being the now-abundant evidence for large quantities of liquid water on and in the planet. The finding of Mars-derived meteorites (shergottites) with radiometric ages of 160 million years implies magmatic activity at a relatively recent time, perhaps persisting to the present (McSween, 1999). Furthermore, the predicted discovery of andesites implies that the early petrologic evolution of Mars and Earth was grossly similar, as discussed in the previous chapter. Since life evidently arose on the Earth almost as far back as the rock record goes, not more than about 500 million years after the Earth itself accreted, the burden of proof has shifted slightly toward those arguing that Mars is totally lifeless. Further speculation at this point is not useful, but it should be remembered that the Gaia concept was originally based largely on Lovelock's initial conclusion that there was no life on Mars. Whether Lovelock's arguments rule out a subsurface biosphere of prokaryotic non-photosynthesizing organisms is not clear.

The Gaia controversy cannot be adequately covered here, much less resolved. Such resolution may actually be impossible in principle. Chamberlain's Principle of Multiple Working Hypotheses, often misunderstood, warns us that in sciences such as geology, a given effect will frequently be the result of multiple causes, not just one. As applied to Gaia, it implies that *both* biological and non-biological mechanisms are probably involved in producing the well-documented effect of environmental stability. Furthermore, Jacobson *et al.* (2000) have shown that there are several well-studied biogeochemical cycles operating on the Earth, controlling not only temperature but pH, redox state, and ocean salinity. Nevertheless, for present purposes, let us assume that biological feedback mechanisms are dominant, and discuss the implications for terrestrial crustal evolution on that assumption.

7.3 The geologic role of water

It was pointed out in the previous chapter that water has had major petrologic effects on terrestrial geology. However, there is much more to the story.

The surficial effects of liquid water on Earth are obvious, and there are very few landforms outside young lava flows that do not reflect erosion and deposition. The physiography of Venus is fascinating for that reason, in that we see an Earth-size planet, with terrestrial tectonics, that has apparently never had appreciable liquid water, an uneroded Earth in effect.

The petrologic effects of water, previously discussed, are profound. The nature of magma generation can be totally changed by the presence of water in the parent rock, as brought out in the discussion of andesite formation. However, on the other end of petrologic evolution, the nature of magmatic differentiation is also strongly water-dependent. Water-rich magmas, other factors equal, produce greater quantities of silicic differentiates, ranging from granites and granite pegmatites to veins of pure quartz. It has been shown (Campbell and Taylor, 1983) that water on a planet may be essential to the formation of granites, although as pointed out in the previous chapter a prolonged period of tectonic activity is also necessary to promote redifferentiation. A whole class of petrologic processes, hydrothermal alteration, is by definition water-dependent, as are chemical weathering processes. Regional metamorphism, responsible for formation of gneisses and schists, is now recognized as being strongly influenced by water pressure. It was in fact argued by Yoder (1955) that various metamorphic facies could reflect water pressure rather than temperature and total pressure alone. Retrograde metamorphism is generally agreed to be largely controlled by re-introduction of water after peak metamorphism to the rock involved.

Rock deformation under normal crustal conditions (excluding the extremes of shock metamorphism) is now realized to be strongly affected by water. Simple plastic deformation is promoted by small amounts of water in the crystal lattice, replacing some Si—O bonds with Si—OH bonds (Seyfert, 1987a). Regional scale structures, in particular overthrust faulting, were shown by Hubbert and Rubey to depend on high fluid (high water) pressures, permitting overthrust sheets to, in effect, float on their own interstitial fluids. (Summaries of this topic are given by Seyfert, 1987b, and Lowman, 1996.)

The largest structures of all, lithospheric plates, may owe their movement partly to water. An especially prominent aspect of this is the cooling of oceanic plate segments by ocean water, and by hydrothermal circulation in oceanic crust (Stein and Stein, 1992). Cooling of oceanic lithosphere away from spreading centers is generally recognized as a significant contributor to plate motion, added to ridge push and slab pull in subduction zones. Given the existence of the continent–ocean dichotomy, and spreading centers, plate motion might occur anyway, but in the absence of oceans would almost certainly be much slower. Another possibility, suggested by Anderson (1984), is that the absence of active plate tectonics on Venus may reflect the absence of life. He points out that if [biogenic] limestone could form on Venus, it might raise the depth of the basalt–eclogite transition and induce sea-floor spreading. Although only a thought experiment, Anderson's suggestion brings out the possible tectonic importance of life on the Earth.

In summary, the effects of abundant water on and within the Earth are second only to the effects of gross heat flow and subsequent magma generation previously discussed. Rocks of the crust are either water-deposited, water-modified, or formed from hydrous magmas. The only major exceptions are fresh basaltic rocks from subaerial volcanos or fissures, but even these have a hydrous component, in contrast to the totally anhydrous basalts of the Moon.

This brings us to the central question, well phrased by Lovelock (1990): How have we kept our oceans?

7.4 Gaia and geology

The conclusion to which the foregoing discussion points has been reached by others, that life is "the architect of our planet" in the words of Nisbet (1987). To quote Lovelock (1988): ". . .the Earth's crust, oceans, and air [are] either directly the product of living things or else massively modified by their presence." This view is of course constrained by comparative planetology. It was shown in the previous chapter that planets almost surely lifeless – the Moon, Mercury, and Venus – have undergone sequences of early crustal evolution fundamentally similar to those of the Earth. Furthermore, the geologic effects of impact owe little to biology. If the impact origin of the Earth's ocean basins is correct, the basic crustal divisions of the Earth were originally formed without the intervention of life. Lovelock's analogy between the Earth's crust and the inner no-longer-living part of a tree is at best questionable, since the inner layers of a tree trunk were once alive, while very few rocks – notably reef limestones, coquinas, and similar types – were actually part of living beings or colonies.

A compromise view is that the major concentric layers of the Earth – core, mantle, and crust – were formed by petrologic processes, and that the main crustal dichotomy of the Earth – ocean basins and continents – was externally initiated by major impacts. But from then on – that is, for the last 3 billion years if not longer – terrestrial geology has been dominated by the presence of life, through its ability to regulate surface temperature and thus retain liquid water continuously. This view could be summarized as biogenically-maintained and regulated crustal evolution, but this summary is something of an understatement, as the following discussion will make clear.

7.5 A biogenic theory of tectonic evolution

Comparative planetology, viewed in the light of the Gaia concept, has led the author to a radically different view of the geologic evolution of the Earth, presented here as a unified **biogenic theory for the tectonic evolution of the Earth**. In this theory, as described in previous chapters, the Earth underwent early global differentiation, perhaps even as it accreted, forming a global crust of andesitic composition. But about 4.5 billion years ago, an event – apparently unique in the solar system – occurred: the catastrophic impact of a Mars-sized body on the primordial Earth. Such an impact, and subsequent ones, may have destroyed or at least dispersed much of the initial global crust, and triggered mantle upwelling, basaltic magmatism, and similar phenomena – the second differentiation, including plate tectonic processes. The early stages of this evolution are apparently similar to those of the Moon, Mercury, Mars, and possibly Venus. The main differences in the crustal evolution of the Earth are probably due to the presence of life for some four billion years.

In the unified theory proposed here, the broad aspects of the Earth's geology as it is now – continents, ocean basins, the oceans themselves, sea-floor spreading and related processes – are the product of fundamentally biogenic processes, acting on a crustal dichotomy formed by several enormous impacts on the primordial Earth. The main stages, starting when the Earth attained essentially its present size, can be summarized in tabular form (Table 7.1).

This outline obviously does not include surficial water-dependent processes of weathering, erosion and sedimentation, which are easily understood in the context of a water-rich planet. However, it should be pointed out that the biogenic processes listed, in particular plate tectonics, have apparently affected not just the outer layers of the Earth – the lithosphere – but the deep structure of the planet as far down as the core–mantle boundary. The basis for this statement is the nature of subduction and more specifically the fate of subducted slabs of oceanic crust. It is now becoming clear, from seismic tomography, geochemistry, and isotope geology (Kellogg *et al.*, 1999), that these slabs go all the way down to the base of the mantle. The structure and convective behaviour of the mantle are strongly affected by subduction (van der Hilst and Karason, 1999).

The fundamental structure of the Earth, not just its exterior and outer layers, thus appears to have been dominated by water-dependent – and thus life-dependent – plate tectonic processes. It is for this reason that the theory is termed biogenic *tectonic* evolution, rather than simply crustal evolution.

7.6 Summary

This final chapter can be summarized by describing the Earth as it might be reported – in free translation – by a departing interstellar planetologist, whom we will assume to see the same colors we do, and to have made a few surreptitious rock-collecting landings before returning home.

This spectacular blue and white planet is unique in the solar system

Table 7.1 *Unified biogenic theory: tectonic evolution of the Earth*

Early global differentiation (4.55 Ga ago; early segregation of the core; massive degassing and hydrous anatexis of primordial mantle; global crust, largely andesitic, formed)

Impact of a Mars-sized planetesimal (4.5 Ga ago; extensive melting; dispersal of much of initial crust; basic crustal dichotomy formed)

Impacts of several more planetesimals (4 Ga ago: further melting; destruction of more differentiated crust; ***prokaryotic*** life arises; oceans form)

Sea-floor spreading localized by impacts (4 Ga ago; basaltic volcanism at spreading centers, accompanied by subduction, transform faulting, movement of small crustal fragments, local terrane accretion; recycling of sialic crust in subduction zones)

Complex *eukaryotic* life arises and expands (2.5 Ga ago; atmospheric composition becomes oxidizing; biological homeostasis, i.e., Gaia, becomes dominant)

Continued intermittent generation of basaltic magma by partial melting of the mantle (4 Ga ago to present; formation of large igneous provinces by dike-fed fissure eruptions; subcrustal gabbroic intrusions; anatexis and metamorphism of sialic crust by basaltic underplating; crust becomes significantly more stable and rigid about 2.5 Ga ago. Sea-floor spreading, subduction, and transform faulting control ocean-basin and adjacent tectonism and volcanism. Local terrane accretion leads to continental growth in some areas; subduction erosion reduces continental area in others)

in several ways, largely stemming from its distance from the Sun, its greater mass compared to other silicate planets, and the presence of life for several billion years. The most striking characteristic of the Earth is its abundant water: colloidally suspended in the atmosphere; covering two-thirds of its surface; coating, falling on, and flowing over the remaining one-third; and infiltrating the crust and mantle. It retains this water partly because of the planet's surface temperature, but also because the Earth behaves like a living organism that maintains this temperature by a wide variety of feedback mechanisms, many of which are caused by life itself. The Earth is the only silicate planet with abundant life, perhaps the only one with any life at all. Life on the Earth, every variety of which contains DNA, apparently arose because the presence of abundant water permitted formation, probably over four billion years ago, of RNA and DNA, which in turn promoted the formation of the complex carbon compounds that comprise terrestrial life. The Earth thus provides an example of self-regulation on a planetary scale: water permitted life to arise, and this life in turn regulated the surface environment to preserve conditions under which the water survives as liquid. Terrestrial geology is dominated by water, which promotes magmatic, metamorphic, and tectonic processes, in addition to sculpturing the surface of the Earth into a greater variety of landforms than displayed by any other body in the solar system. Most rocks, structures, and landforms of the continental crust are fundamentally biogenic, in that water has been essential for their origin. Although the Earth has followed a crustal evolution sequence whose initial stages are common to other silicate planets, it has evolved and continues to evolve geologically much further. It is unique, a uniqueness that may be due primarily to its life.

AFTERWORD

This book is not an autobiography, but there is an autobiographical aspect to it that can serve as a summing-up. The book's subtitle, *New Understanding of the Earth from Space Research,* might more accurately be *The Author's New Understanding* (etc.), because it reflects the evolution of my own concepts of terrestrial geology over some four decades. A personal account of this evolution may help the reader assimilate the broad range of topics covered in this book.

By the time I joined NASA, in 1959, I had acquired a solid and comprehensive geologic education, seasoned with field work. My intellectual frame of reference at the beginning of the Space Age was therefore essentially that of most geologists of the day. But it will be obvious to the reader that my frame of reference in 2002 is very different in several respects from that of today's geological community.

Terrestrial geologic thought is dominated today by plate tectonic theory, a master plan generally believed to have solved most of the major geologic problems of the early-20th century: the origin of continents, mountain belts, ocean basins, volcanic fields, and many others. Being preoccupied with space research, I followed development of this "master plan" through the 1960s as an interested outsider. I knew many of its founders personally: Hess, Dietz, Wilson, Crowell, Hamilton, and others, and hold them in high regard. Yet as the reader will know by now, I think plate tectonic theory has been over-extended, beyond its very real strengths. My view is, in brief, that it describes very well the tectonics of ocean basins, and some adjacent mountain belts, such as the Andes and the Aleutians. But I consider it of little relevance for describing and explaining continental geology in general. Turning to a fundamental problem, the origin of continental crust, I think that plate tectonic theory falls far short of "master plan" status, except for the few areas such as Alaska where terrane accretion has been convincingly demonstrated.

My tectonic views may suggest legal phraseology: *John Doe vs. The United States*. To put it explicitly: How can I have the confidence – or if you will, arrogance – to defy almost the entire geological community? The answer to this is essentially autobiographical, a brief summary of the development of my thinking since joining NASA.

Perhaps the most important aspect of this development is the fact that, by coincidence of several factors, I was unusually lucky to get involved in both externally-oriented space exploration and at the same time study of the Earth from space. Not more than a few dozen geologists and geophysicists have been able to follow such a dual career. This has permitted me, first, to compare the Earth in detail with other planets (including the Moon), and secondly to survey the Earth in its entirety to a degree impossible before the achievement of space flight. In the widest terms, what have I learned about the Earth?

First, I conclude that the Earth's early geologic evolution was, very broadly speaking, essentially analogous to that of other silicate bodies: the Moon, Mars, Mercury, Venus, and perhaps the larger asteroids. After an intensely hot formation process, these bodies all underwent similar early evolution, forming differentiated crusts. But the Earth is unique in its crustal dichtomy; no other body has true ocean basins with sea-floor spreading and its complementary processes. The Earth is also unique in its large satellite, the Moon, and I think the two unique characteristics are linked through the origin of the Moon. Venus has no moon, and no crustal dichotomy. Mars has a crustal dichotomy, but the younger segment – the northern plains – is not truly analogous to the Earth's active and geologically ephemeral oceanic crust. The same of true of the lunar maria. Plate tectonics, then, is uniquely terrestrial.

Despite the absence of plate tectonics, other planets have developed differentiated crusts. What information we now have about the crust of Mars indicates it to be partly of andesitic composition, similar to the bulk composition of terrestrial continental crust. Other planets have been partly covered by flood basalts, apparently dike-fed on Venus, similar to the plateau basalts of Earth. Mars and Venus have rift valleys that may be analogous to the African Rift Valleys and their counterparts on other continents.

The geology of Venus, close to the Earth in size and density, is unusually relevant to terrestrial tectonic theory. Venus is a no-plate, or one-plate planet; whatever its crustal composition, there is no two-fold crustal division like that of the Earth. There are no well-developed plate boundaries, if any at all, and it is agreed by all who have studied Venus that plate tectonics has not been effective there. Nevertheless, Venus has developed spectacular fold-and-thrust belts strikingly similar to those of the Earth, differing largely in being expressed directly as tectonic landforms, not by differential erosion.

In my view, this demonstrates that folded mountain belts can be formed by mechanisms not involving plate collisions, since Venus has no plates. Just what these mechanisms are, on the Earth, is a question for geologists of this new century. However, they may find themselves reviving theories of the last one, perhaps the last two. The contraction theory was for many years the only plausible explanation for terrestrial mountain belts. More recent possibilities include "A-subduction," subduction occuring within a continent, or "surge tectonics," involving lateral transport of magma in the upper mantle. "Delamination," detachment of a mafic lower part of continental crust, may play a role in mountain belt formation. In any event, Earth's "sister planet," in the familiar phrase, has developed Earth-like mountains without plate tectonics.

Turning now to the earth-oriented phase of my dual career in space research, let me summarize briefly four decades of orbital remote sensing and space geodesy in relation to tectonic theory.

First, space geodesy has confirmed beyond any reasonable doubt the reality of plate tectonic mechanisms, by demonstrating plate rigidity and predicted plate movement in the Pacific Basin. The measured plate movements are astonishingly close to those estimated, from the spacing of dated magnetic anomalies in the oceanic crust, for the last three million years. Sea-floor spreading thus appears to be a very steady process. Pacific plate movements are not possible without sea-floor spreading and complementary subduction, both long supported by geophysical data such as focal mechanism determinations. Transform faulting was confirmed almost immediately by focal mechanisms along such faults.

So far, so good: space research has fully supported the core of plate tectonic theory. Furthermore, the movement of some blocks of continental crust, certainly Baja California and perhaps Australia, also appears supported by space geodesy. However, remote sensing from orbit has demonstrated that continental crust as a whole is far too complex and active to be described realistically in terms of the usual twelve standard plates. This finding is most dramatic in central Asia, where *Landsat*, *SPOT*, and other satellites have shown the crust far from the nominal plate boundaries to be riddled with active faults, a finding fully confirmed by seismology. A Eurasian Plate can be identified, but its southern boundaries are extremely broad and ill defined. A similarly broad boundary has long been known between the North American and Pacific Plates; the San Andreas fault is a small part of this boundary.

Has continental drift been confirmed by space geodesy? Many feel it has. Space geodesy stations in western Europe appear to be moving in agreement with the NUVEL-1 model, but this model shows only plate motions relative to another plate, in this case the Pacific Plate. True plate movement, over the mantle, should be at right angles to the NUVEL-1 (Pacific Plate fixed) model. The space geodesy site motions relative to a fixed North American Plate are consistent with push from the Mid-Atlantic Ridge as shown by the *World Stress Map*. However, even these motions can not realistically be considered to represent movement of the Eurasian Plate as a whole. Such movement (an Eulerian rotation) can only be demonstrated by many decades of inter- and intraplate measurements, of which those done in the 20th century are only the beginning.

This autobiographical essay has up to now been essentially a critique of plate tectonic theory in the light of the findings of space research. In his 1920s essay "The cult of hope," Mencken argued that valid criticism is justified by itself – that we should not confuse the function of criticism with that of reform. However, I prefer to go beyond criticism, and offer an alternative theory of the Earth's geologic evolution and its tectonic behavior.

The Earth's continental crust is in this alternative concept the greatly modified remnant of an originally global differentiated crust, formed largely by water-catalyzed andesitic volcanism very early in the planet's history. This global crust was catastrophically disrupted even as it formed, about 4.5 billion years ago, by the impact of a Mars-sized object on the primordial Earth. Debris from this event, mixed with material from the impacting object, re-accreted from the ejected cloud to form the Moon. The crustal dichotomy of the Earth, unique in the solar system, dates from this period, but was accentuated by further impacts analogous to those that formed the lunar mare basins and similar basins on Mars and Mercury. Mantle up-welling and sea-floor spreading, the start of plate tectonics, were first triggered and localized by the terrestrial mare basins, so to speak. The unimodal crust of Venus reflects the absence of a catastrophic satellite-forming event. Mars, Mercury, and the Moon were too small to sustain post-impact sea-floor spreading.

Basaltic volcanism and sea-floor spreading are obviously linked. But in my view, basaltic magmatism in general has dominated the Earth's petrologic *and* tectonic evolution, oceanic and continental, for at least four billion years. The oceanic crust is obviously largely basalt, generated by reasonably well-understood processes. However, continental basaltic magmatism, expressed as flood basalts, dike swarms, and intracrustal intrusions, has played a major role in formation of the "granite" part of the granite–greenstone terrains, by promoting intracrustal melting. The greenstone belts themselves, dominantly basaltic, represent a form of vertical continental growth. Dike swarms on all continents represent basaltic overplating, although in most areas the overplated basalt has been removed by erosion. Regional metamorphism has been promoted by heat added from underplated or intruded basaltic magma. The continents may be dominantly granitic, in the broadest sense, but they have been repeatedly modified, directly and indirectly, by the generation of basaltic magma.

The Earth's tectonic evolution has been comparably affected by basaltic magmatism, through the mechanism of sea-floor spreading and ridge push. The *World Stress Map*, viewed with the results of space geodesy, has shown that ridge push is a major contributor to continental crust deformation and possibly to continental drift. Movement of purely oceanic plates is further promoted by gravity-driven slab pull. In summary, the petrologic and tectonic activity of the Earth has been dominated by generation of basaltic magma, traceable back at least four billion years and obviously continuing today. However, there is a critical aspect not yet covered: the effects of life on terrestrial geology.

By some four billion years ago, according to my theory, the Earth had reached essentially its present style of tectonism: actively spreading basaltic oceanic crust and relatively passive but deforming continental crust. This style is unique to the Earth, partly because of the Earth's mass and greater internal energy. However, another major reason for this uniqueness is the continued existence of large bodies of water on the Earth's surface. It is here that life began to play a critical tectonic role, expressed as the Gaia concept, essentially a biological feedback theory.

Prokaryotic life arose initially, perhaps in the oceans, some four billion years ago, through mechanisms not yet understood. This life was apparently similar to the marine phytoplankton of today's oceans, and may have begun to regulate the atmospheric composition of the Earth at an early date. The rise of eukaryotic algae, probably about 2.5 billion years ago, increased the oxygen content of the atmosphere to something like that of the present. The atmosphere became oxidizing and, more importantly, it became thermally stabilized such that oceans could survive, despite occasional "snow-ball earth" periods of widespread glaciation.

The existence of oceans is of course essential for deposition of many sedimentary rocks, and most erosion is the result of fluvial or marine processes. But the influence of the oceans is far more fundamental, for their cooling effect promotes sea-floor spreading and subduction. Subducted crust is now known to descend deep into the mantle, perhaps to the core–mantle boundary. The main controls on terrestrial geology since the early Archean have thus been **life**, which by regulating surface temperatures permits oceans to survive, and **oceans**, which permit oceanic plate tectonic processes to continue. Simplifying even further, I conclude that the dominant single factor in the last two billion years of terrestrial geology has been life, which has not only controlled crustal geology but indirectly the deep structure of the Earth. I therefore label this concept a "biogenic theory of tectonic evolution."

I can do no better, in summarizing *New Understanding of the Earth*, than quote my imaginary interstellar planetologist: "**It is unique, a uniqueness that may be due primarily to its life.**"

APPENDIX A

Essentials of physical geology

Geology: The Study of the Earth and Earth-like bodies (Moon, Mars, etc.)

Physical: Minerals, rocks, and structures. **Tectonics** concerns regional and global strucure. **Plate tectonics:** Explains mountain-building, volcanism, & earthquakes in terms of **ridges** (spreading centers), **trenches** (subduction zones), and **transform faults**.

Historical: History of Earth and inhabitants thereof (dinosaurs, etc.)

Essentials: Elements Minerals Rocks Structures

Elements: Order of abundance in Earth's *crust*:

Oxygen	Silicon	Aluminum	Iron	Calcium	Sodium	Potassium	Magnesium
O	Si	Al	Fe	Ca	Na	K	Mg

Minerals: Most minerals are **silicates**; others: oxides, carbonates, sulfides

Quartz: (SiO_2) – Most abundant simple mineral in Earth's crust

Feldspars: (Al silicates of K, Na, Ca in solid solution) – Main rock-forming minerals (in granites, basalts, gabbros, etc.)

Calcite: ($CaCO_3$) – Main mineral of limestone, chalk, marble

Olivine, Pyroxene, Amphibole: Main Fe, Mg silicates in basalt, peridotite; olivine main component of upper mantle

Clay minerals: Formed chiefly by weathering of silicates

Rocks: Composed of one or more minerals

Igneous – Formed from lava (extrusive) or magma (intrusive). Most common types: **basalt, granite, gabbro, rhyolite, andesite, anorthosite**

Sedimentary – Deposited by water, wind, or ice, or evaporation of water. Most common types **sandstone, shale, limestone, conglomerate**

Metamorphic – Chiefly formed by solid-state recrystallization of pre-existing rocks; e.g., marble is recrystallized limestone. Shock metamorphism may destroy crystal structure. Most common types: **gneiss, schist, marble, amphibolite, serpentinite**

Structures:

Rock layers – sedimentary bedding, metamorphic foliation, lava flows, volcanic ash deposits 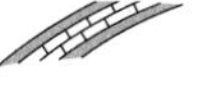

Folds – anticlines, arch-like upfolds
synclines, trough-like downfolds 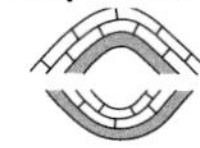

Faults: fractures along which there has been movement

normal (tensional)

reverse or thrust (compressional) 

strike-slip or wrench (horizontal shear)
(San Andreas is this type)

Joints: simple fractures, movement away from fracture

APPENDIX B

Lunar missions, 1958 to 1994 (NASA History Office)

SPACECRAFT	MISSION	LAUNCH DATE	ARRIVAL DATE	REMARKS
Pioneer 1 USA	Lunar Orbit	Oct 11, 1958		Did not achieve lunar trajectory; launch vehicle second and third stages did not separate evenly. Returned data on Van Allen Belt and other phenomena before reentering on October 12, 1958.
Pioneer 2 USA	Lunar Orbit	Nov 8, 1958		Third stage of launch vehicle failed to ignite. Returned data that indicated the Earth's equatorial region has higher flux and energy levels than previously believed. Did not achieve orbit.
Pioneer 3 USA	Lunar Probe	Dec 6, 1958		First stage of launch vehicle cut off prematurely; transmitted data on dual bands of radiation around Earth. Reentered December 7, 1958.
Luna 1 USSR	Lunar Impact	Jan 2, 1959		Intended to impact the Moon; carried instruments to measure radiation. Passed the Moon and went into solar orbit.
Pioneer 4 USA	Lunar Probe	Mar 3, 1959	Mar 4, 1959	Passed within 37,300 miles from the Moon; returned excellent data on radiation. Entered solar orbit.
Luna 2 USSR	Lunar Impact	Sep 12, 1959	Sep 15, 1959	First spacecraft to reach another celestial body. Impacted east of the Sea of Serenity; carried USSR pennants.
Luna 3 USSR	Lunar Probe	Oct 4, 1959		First spacecraft to pass behind Moon and send back pictures of far side. Equipped with a TV processing and transmission system, returned pictures of far side including composite full view of far side. Reentered Apr 29, 1960.
Pioneer P-3 USSR	Lunar Orbit	Nov 26, 1959		Payload shroud broke away 45 seconds after liftoff. Did not achieve orbit.
Ranger 1 USA	Lunar Probe	Aug 23, 1961		Flight test of lunar spacecraft carrying experiments to collect data on solar plasma, particles, magnetic fields, and cosmic rays. Launch vehicle failed to restart resulting in low Earth Orbit. Reentered August 30, 1961.

SPACECRAFT	MISSION	LAUNCH DATE	ARRIVAL DATE	REMARKS
Ranger 2 USA	Lunar Probe	Nov 18, 1961		Flight test of spacecraft systems for future lunar and interplanetary missions. Launch vehicle altitude control system failed, resulting in low Earth orbit. Reentered November 20, 1961.
Ranger 3 USA	Lunar Landing	Jan 26, 1962		Launch vehicle malfunction resulted in spacecraft missing the Moon by 22,862 miles. Spectrometer data on radiation were received. Entered solar orbit.
Ranger 4 USA	Lunar Landing	Apr 23, 1962	Apr 26, 1962	Failure of central computer and sequencer system rendered experiments useless. No telemetry received. Impacted on far side of the Moon.
Ranger 5 USA	Lunar Landing	Oct 18, 1962		Power failure rendered all systems and experiments useless; 4 hours of data received from gamma ray experiment before battery depletion. Passed within 450 miles of the Moon. Entered solar orbit.
Sputnik 25 USSR	Lunar Probe	Jan 4, 1963		Unsuccessful lunar attempt.
Luna 4 USSR	Lunar Orbiter	Apr 2, 1963		Attempt to solve problems of landing instrument containers. Contact lost as it passed the Moon. Barycentric orbit.
Ranger 6 USA	Lunar Photo	Jan 30, 1964	Feb 2, 1964	TV cameras failed; no data returned. Impacted in the Sea of Tranquility area.
Ranger 7 USA	Lunar Photo	Jul 28, 1964	Jul 31, 1964	Transmitted high quality photographs, man's first close-up lunar views, before impacting in the Sea of Clouds area.
Ranger 8 USA	Lunar Photo	Feb 17, 1965	Feb 20, 1965	Transmitted high quality photographs before impacting in the Sea of Tranquility area.
Ranger 9 USA	Lunar Photo	Mar 21, 1965	Mar 24, 1965	Transmitted high quality photographs before impacting in the Crater of Alphonsus. Almost 200 pictures were shown live via commercial television in the first TV spectacular from the Moon.

SPACECRAFT	MISSION	LAUNCH DATE	ARRIVAL DATE	REMARKS
Luna 5 USSR	Lunar Lander	May 9, 1965	May 12, 1965	First soft landing attempt. Retrorocket malfunctioned; spacecraft impacted in the Sea of Clouds.
Luna 6 USSR	Lunar Lander	Jun 8, 1965		During midcourse correction maneuver, engine failed to switch off. Spacecraft missed Moon and entered solar orbit.
Zond 3 USSR	Lunar Probe	Jul 18, 1965		Photographed lunar far side and transmitted photos to Earth 9 days later. Entered solar orbit.
Luna 7 USSR	Lunar Lander	Oct 4, 1965	Oct 7, 1965	Retrorockets fired early; crashed in Ocean of Storms.
Luna 8 USSR	Lunar Lander	Dec 3, 1965	Dec 6, 1965	Retrorockets fired late; crashed in Ocean of Storms.
Luna 9 USSR	Lunar Lander	Jan 31, 1966	Feb 3, 1966	First successful soft landing; first TV transmission from lunar surface. Three panoramas of the lunar landscape were transmitted from the eastern edge of the Ocean of Storms.
Cosmos 111 USSR	Lunar Probe	Mar 11, 1966		Unsuccessful lunar attempt. Reentered March 16, 1966.
Luna 10 USSR	Lunar Orbiter	Mar 31, 1966		First lunar satellite. Studied lunar surface radiation and magnetic field intensity; monitored strength and variation of lunar gravitation. Selenocentric orbit.
Surveyor 1 USA	Lunar Lander	May 30, 1966	Jun 2, 1966	First U.S. spacecraft to make a fully controlled soft landing on the Moon; landed in the Ocean of Storms area. Returned high quality images, from horizon views of mountains to close-ups of its own mirrors, and selenological data.
Lunar Orbiter 1 USA	Lunar Orbiter	Aug 10, 1966	Aun 14, 1966	Photographed over 2 million square miles of the Moon's surface. Took first photo of Earth from lunar distance. Impacted on the far side of the Moon on October 29, 1966.
Luna 11 USSR	Lunar Orbiter	Aug 24, 1966		Second lunar satellite. Data received during 277 orbits. Selenocentric orbit.

SPACECRAFT	MISSION	LAUNCH DATE	ARRIVAL DATE	REMARKS
Surveyor 2 USA	Lunar Lander	Sep 20, 1966	Sep 22, 1966	Spacecraft crashed onto the lunar surface southeast of the crater Copernicus when one of its three vernier engines failed to ignite during a mid-course maneuver.
Luna 12 USSR	Lunar Orbiter	Oct 22, 1966		TV system transmitted large-scale pictures of Sea of Rains and Crater Aristarchus areas. Tested electric motor for Lunokhod's wheels. Selenocentric orbit.
Lunar Orbiter 2 USA	Lunar Orbiter	Nov 6, 1966	Nov 10, 1966	Photographed landing sites, including the Ranger 8 landing point, and surface debris tossed out at impact. Impacted the Moon on October 11, 1967.
Luna 13 USSR	Lunar Lander	Dec 21, 1966	Dec 24, 1966	Soft landed in Ocean of Storms and sent back panoramic views. Two arms were extended to measure soil density and surface radioactivity.
Lunar Orbiter 3 USA	Lunar Orbiter	Feb 4, 1967	Feb 8, 1967	Photographed lunar landing sites; provided gravitational field and lunar environment data. Impacted the Moon on October 9, 1967.
Surveyor 3 USA	Lunar Lander	Apr 17, 1967	Apr 19, 1967	Vernier engines failed to cut off as planned and the spacecraft bounced twice before landing in the Ocean of Storms. Returned images, including a picture of the Earth during lunar eclipse, and used a scoop to make the first excavation and bearing test on an extraterrestrial body. Returned data on a soil sample. Visual range of TV cameras was extended by using two flat mirrors.
Lunar Orbiter 4 USA	Lunar Orbiter	May 4, 1967	May 8, 1967	Provided the first pictures of the lunar south pole. Impacted the Moon on Oct 6, 1967.
Surveyor 4 USA	Lunar Lander	Jul 14, 1967	Jul 17, 1967	Radio contact was lost 2-1/2 minutes before touchdown when the signal was abruptly lost. Impacted in Sinus Medii.
Lunar Orbiter 5 USA	Lunar Orbiter	Aug 1, 1967	Aug 5, 1967	Increased lunar photographic coverage to better than 99%. Used in orbit as a tracking target. Impacted the Moon on January 31, 1968.

SPACECRAFT	MISSION	LAUNCH DATE	ARRIVAL DATE	REMARKS
Surveyor 5 USA	Lunar Lander	Sep 8, 1967	Sep 10, 1967	Technical problems were successfully solved by tests and maneuvers during flight. Soft-landed in the Sea of Tranquility. Returned images and obtained data on lunar surface radar and thermal reflectivity. Performed first on-site chemical soil analysis.
Surveyor 6 USA	Lunar Lander	Nov 7, 1967	Nov 9, 1967	Soft-landed in the Sinus Medii area. Returned images of the lunar surface, Earth, Jupiter, and several stars. Spacecraft engines were restarted, lifting the spacecraft about 10 feet from the surface and landing it 8 feet from the original site.
Surveyor 7 USA	Lunar Lander	Jan 7, 1968	Jan 9, 1968	Landed near the crater Tycho. Returned stereo pictures of the surface and of rocks that were of special interest. Provided first observation of artificial light from Earth.
Luna 14 USSR	Lunar Orbiter	Apr 7, 1968		Studied gravitational field and "stability of radio signals sent to spacecraft at different locations in respect to the Moon." Made further tests of geared electric motor for Lunokhod's wheels. Selenocentric orbit.
Zond 5 USSR	Circumlunar	Sep 15, 1968		First spacecraft to circumnavigate the Moon and return to Earth. Took photographs of the Earth. Capsule was recovered from the Indian Ocean on September 21, 1968. Russia's first sea recovery.
Zond 6 USSR	Circumlunar	Nov 10, 1968		Second spacecraft to circumnavigate the Moon and return to Earth "to perfect the automatic functioning of a manned spaceship that will be sent to the Moon." Photographed lunar far side. Reentry made by skip-glide technique; capsule was recovered on land inside the Soviet Union on November 17, 1968.
Luna 15 USSR	Lunar Sample Return	Jul 13, 1969	Jul 21, 1969	First lunar sample return attempt. Began descent maneuvers on its 52nd revolution. Spacecraft crashed at the end of a 4 minute descent in the Sea of Crises.
Zond 7 USSR	Circumlunar	Aug 7, 1969		Third circumlunar flight. Far side of Moon photographed. Color pictures of Earth and Moon brought back. Reentry by skip-glide technique on August 14, 1969.

SPACECRAFT	MISSION	LAUNCH DATE	ARRIVAL DATE	REMARKS
Cosmos 300 USSR	Lunar Probe	Sep 23, 1969		Unsuccessful lunar attempt. Reentered September 27, 1969.
Cosmos 305 USSR	Lunar Probe	Oct 22, 1969		Unsuccessful lunar attempt. Reentered October 24, 1969.
Luna 16 USSR	Lunar Sample Return	Sep 12, 1970	Sep 20, 1970	First recovery of lunar soil by an automatic spacecraft. Controlled landing achieved in Sea of Fertility; automatic drilling rig deployed; samples collected from lunar surface and returned to Earth on September 24, 1970.
Zond 8 USSR	Circumlunar	Oct 20, 1970		Fourth circumlunar flight. Color pictures taken of Earth and Moon. Russia's second sea recovery occurred on October 27, 1970, in the Indian Ocean.
Luna 17 USSR	Lunar Rover	Nov 10, 1970	Nov 17, 1970	Carrying the first Moon robot, soft landed in Sea of Rains. Lunokhod 1, driven by 5-man team on Earth, traveled over the lunar surface for 11 days; transmitted photos and analyzed soil samples.
Luna 18 USSR	Lunar Lander	Sep 2, 1971		Attempted to land in Sea of Fertility on September 11, 1971. Communications ceased shortly after command was given to start descent engine.
Luna 19 USSR	Lunar Orbiter	Sep 28, 1971		From lunar orbit, studied Moon's gravitational field; transmitted TV pictures of the surface. Selenocentric orbit.

SPACECRAFT	MISSION	LAUNCH DATE	ARRIVAL DATE	REMARKS
Luna 20	Lunar Sample Return	Feb 14, 1972		Soft landed in Sea of Crises. Used "photo-telemetric device" to relay pictures of surface. A rotary-percussion drill was used to drill into rock; samples were lifted into a capsule on ascent stage and returned to Earth on Feb 25, 1972.
Luna 21	Lunar Rover	Jan 8, 1973	Jan 15, 1973	Carried improved equipment and additional instruments; second Lunokhod rover soft landed near the Sea of Serenity. Lunar surface pictures were transmitted and experiments were performed. Ceased operating on the 5th lunar day.
Luna 22	Lunar Orbiter	May 29, 1974	Jun 2, 1974	Placed in circular lunar orbit then lowered to obtain TV panoramas of high quality and good resolution. Altimeter readings were taken and chemical rock composition was determined by gamma radiation. Selenocentric orbit.
Luna 23	Lunar Sample Return	Oct 28, 1974		Landed on the southern part of the Sea of Crises on November 6, 1974. Device for taking samples was damaged; no drilling or sample collection possible.
Luna 24	Lunar Sample Return	Aug 9, 1976	Aug 14, 1976	Landed in Sea of Crises on August 18, 1976. Carried larger soil carrier. Core samples were drilled and returned. U.S. and British scientists were given samples for analyses.
Clementine USA	Lunar Flyby	Jan 25, 1994		Carrying ultraviolet/visible and near-infrared cameras, mineralogical mapping of the moon will enhance the scientific knowledge of the surface for future exploration. The mission failed in its attempt to flyby the asteroid Geographos.

APPENDIX C

Planetary missions, 1961 to 1992 (NASA History Office)

SPACECRAFT	MISSION	LAUNCH DATE	ARRIVAL DATE	REMARKS
Pioneer 11 USA	Jupiter/Saturn Flyby	Apr 5, 1973	Dec 2, 1974 (Jupiter) Sep 1, 1979 (Saturn)	The successful encounter of Jupiter by Pioneer 10 permitted Pioneer 11 to be retargeted in flight to fly by Jupiter and encounter Saturn. Still operating in the outer Solar System.
Mars 4 & 5 USSR	Mars Orbiters and Landers	Jul 21, 1973 Jul 25, 1973	Feb 10, 1974 Feb 12, 1974	Pair of spacecraft launched to Mars. Mars 4 retro rockets failed to fire, preventing orbit insertion. As it passed the planet, Mars 4 returned one swath of pictures and some radio occultation data. Mars 5 was successfully placed in orbit, but operated only a few days, returning photographs of a small portion of southern hemisphere of Mars.
Mars 6 & 7 USSR	Mars Orbiters and Landers	Aug 5, 1973 Aug 9, 1973	Mar 12, 1974 Mar 9, 1974	Second pair of spacecraft launched to Mars. Mars 6 lander module transmitted data during descent, but transmissions abruptly ceased when the landing rockets were fired. Mars 7 descent module was separated from the main spacecraft due to a problem in the operation of one of the onboard systems, and passed by the planet.
Mariner 10 USA	Venus/Mercury Flyby	Nov 3, 1973	Feb 5, 1974 (Venus) Mar 29, 1974 (Mercury) Sep 21, 1974 (Mercury) Mar 16, 1975 (Mercury)	First dual-planet mission. Used gravity of Venus to attain Mercury encounter. Provided first ultraviolet photographs of Venus; returned close-up photographs and detailed data of Mercury. Transmitter was turned off March 24, 1975, when attitude control gas was depleted. Spacecraft is inoperable in solar orbit.
Venera 9 USSR	Venus Orbiter and Lander	Jun 8, 1975	Oct 22, 1975	First spacecraft to transmit a picture from the surface of another planet. The lander's signals were transmitted to Earth via the orbiter. Utilized a new parachute system, consisting of six chutes. Signals continued from the surface for nearly 2 hrs 53 mins.
Venera 10 USSR	Venus Orbiter and Lander	Jun 14, 1975	Oct 25, 1975	During descent, atmospheric measurements and details of physical and chemical contents were transmitted via the orbiter. Transmitted pictures from the surface of Venus.

SPACECRAFT	MISSION	LAUNCH DATE	ARRIVAL DATE	REMARKS
Viking 1 USA	Mars Orbiter and Lander	Aug 20, 1975	Jul 19, 1976 (in orbit) Jul 20, 1976 (landed)	First U.S. attempt to soft land a spacecraft on another planet. Landed on the Plain of Chryse. Photographs showed an orange-red plain strewn with rocks and sand dunes. Both Orbiters took a total of 52,000 images during their mission; approximately 97% percent of the surafce was imaged. Orbiter 1 operated until August 7, 1980, when it used the last of its attitude control gas. Lander 1 ceased operating on Nov 13, 1983.
Viking 2 USA	Mars Orbiter and Lander	Sep 9, 1975	Aug 7, 1976 (in orbit) Sep 3, 1976 (landed)	Landed on the Plain of Utopia. Discovered water frost on the surface at the end of the Martian winter. The two Landers took 4,500 images of the surface and provided over 3 million weather reports. Orbiter 2 stopped operating on July 24, 1978, when its attitude control gas was depleted because of a leak. Lander 2 operated until April 12, 1980, when it was shut down due to battery degeneration.
Voyager 2 USA	Tour of the Outer Planets	Aug 20, 1977	Jul 9, 1979 (Jupiter) Aug 25, 1981 (Saturn) Jan 24, 1986 (Uranus) Aug 25, 1989 (Neptune)	Investigated the Jupiter, Saturn and Uranus planetary systems. Provided first close-up photographs of Uranus and its moons. Used gravity-assist at Uranus to continue on to Neptune. Swept within 1280 km of Neptune on August 25, 1989. The spacecraft will continue into interstellar space.
Voyager 1 USA	Tour of Jupiter and Saturn	Sep 5, 1977	Mar 5, 1979 (Jupiter) Nov 12, 1980 (Saturn)	Investigated the Jupiter and Saturn planetary systems. Returned spectacular photographs and provided evidence of a ring encircling Jupiter. Continues to return data enroute toward interstellar space.
Pioneer Venus 1 USA	Venus Orbiter	May 20, 1978	Dec 4, 1978	Mapped Venus' surface by radar, imaged its cloud systems, explored its magnetic environment and observed interactions of the solar wind with a planet that has no intrinsic magnetic field. Provided radar altimetry maps for nearly all of the surface of Venus, resolving features down to about 50 miles across. Still operating in orbit around Venus.
Pioneer Venus 2 USA	Venus Probe	Aug 8, 1978	Dec 9, 1978	Dispatched heat-resisting probes to penetrate the atmosphere at widely separated locations and measured temperature, pressure, and density down to the planet's surface. Probes impacted on the surface.

SPACECRAFT	MISSION	LAUNCH DATE	ARRIVAL DATE	REMARKS
Venera 1 USSR	Venus Probe	Feb 12, 1961		First Soviet planetary flight; launched from Sputnik 8. Radio contact was lost during flight; spacecraft was not operating when it passed Venus.
Mariner 1 USA	Venus Flyby	Jul 22, 1962		Destroyed shortly after launch when vehicle veered off course.
Sputnik 19 USSR	Venus Probe	Aug 25, 1962		Unsuccessful Venus attempt.
Mariner 2 USA	Venus Flyby	Aug 27, 1962	Dec 14, 1962	First successful planetary flyby. Provided instrument scanning data. Entered solar orbit.
Sputnik 20 USSR	Venus Probe	Sep 1, 1962		Unsuccessful Venus attempt.
Sputnik 21 USSR	Venus Probe	Sep 12, 1962		Unsuccessful Venus attempt.
Sputnik 22 USSR	Mars Probe	Oct 24, 1962		Spacecraft and final rocket stage blew up when accelerated to escape velocity.
Mars 1 USSR	Mars Probe	Nov 1, 1962		Contact was lost when the spacecraft antenna could no longer be pointed towards Earth.
Sputnik 24 USSR	Mars Probe	Nov 4, 1962		Disintegrated during an attempt at Mars trajectory from Earth parking orbit.
Zond 1 USSR	Venus Probe	Apr 2, 1964		Communications lost. Spacecraft went into solar orbit.
Mariner 3 USA	Mars Flyby	Nov 5, 1964		Shroud failed to jettison properly; Sun and Canpous not acquired; spacecraft did not encounter Mars. Transmissions ceased 9 hours after launch. Entered solar orbit.
Mariner 4 USA	Mars Flyby	Nov 28, 1964	Jul 14, 1965	Provided first close-range images of Mars, confirming the existence of surface craters. Entered solar orbit.
Zond 2 USSR	Mars Probe	Nov 30, 1964		Passed by Mars; failed to return data. Went into solar orbit.

SPACECRAFT	MISSION	LAUNCH DATE	ARRIVAL DATE	REMARKS
Venera 2 USSR	Venus Probe	Nov 12, 1965	Feb 27, 1966	Passed by Venus, but failed to return data.
Venera 3 USSR	Venus Probe	Nov 16, 1965	Mar 1, 1966	Impacted on Venus, becoming the first spacecraft to reach another planet. Failed to return data.
Venera 4 USSR	Venus Probe	Jun 12, 1967	Oct 18, 1967	Descent capsule transmitted data during parachute descent. Sent measurements of pressure, density, and chemical composition of the atmosphere before transmissions ceased.
Mariner 5 USA	Venus Flyby	Jun 14, 1967	Oct 19, 1967	Advanced instruments returned data on Venus' surface temperature, atmosphere, and magnetic field environment. Entered solar orbit.
Venera 5 USSR	Venus Probe	Jan 5, 1969	Mar 16, 1969	Entry velocity reduced by atmospheric braking before main parachute was deployed. Capsule entered atmosphere on planet's dark side; transmitted data for 53 minutes while traveling into the atmosphere before being crushed.
Venera 6 USSR	Venus Probe	Jan 10, 1969	Mar 17, 1969	Descent capsule entered the atmosphere on the planet's dark side; transmitted data for 51 minutes while traveling into the atmosphere before being crushed.
Mariner 6 USA	Mars Flyby	Feb 24, 1969	Jul 31, 1969	Provided high-resolution photos of Martian surface, concentrating on equatorial region. Entered solar orbit.
Mariner 7 USA	Mars Flyby	Mar 27, 1969	Aug 5, 1969	Provided high-resolution photos of Martian surface, concentrating on southern hemisphere. Entered solar orbit.
Venera 7 USSR	Venus Lander	Aug 17, 1970	Dec 15, 1970	Entry velocity was reduced aerodynamically before parachute deployed. After fast descent through upper layers, the parachute canpoy opened fully, slowing descent to allow fuller study of lower layers. Gradually increasing temperatures were transmitted. Returned data for 23 minutes after landing.

SPACECRAFT	MISSION	LAUNCH DATE	ARRIVAL DATE	REMARKS
Cosmos 359 USSR	Venus Lander	Aug 22, 1970		Unsuccessful Venus attempt; failed to achieve escape velocity.
Mariner 8 USA	Mars Orbiter	May 8, 1971		Centaur stage malfunctioned shortly after launch.
Cosmos 419 USSR	Mars Probe	May 10, 1971		First use of Proton launcher for a planetary mission. Placed in Earth orbit but failed to separate from fourth stage.
Mars 2 USSR	Mars Orbiter and Lander	May 19, 1971	Nov 27, 1971	Landing capsule separated from spacecraft and made first, unsuccessful attempt to soft land. Lander carried USSR pennant. Orbiter continued to transmit data.
Mars 3 USSR	Mars Orbiter and Lander	May 28, 1971	Dec 2, 1971	Landing capsule separated from spacecraft and landed in the southern hemisphere. Onboard camera operated for only 20 seconds, transmitting a small panoramic view. Orbiter transmitted for 3 months.
Mariner 9 USA	Mars Orbiter	May 30, 1971	Nov 13, 1971	First interplanetary probe to orbit another planet. During nearly a year of operations, obtained detailed photographs of the Martian moons, Phobos and Deimos, and mapped 100 percent of the Martian surface. Spacecraft is inoperable in Mars orbit.
Pioneer 10 USA	Jupiter Flyby	Mar 2, 1972	Dec 3, 1973	First spacecraft to penetrate the Asteroid Belt. Obtained first close-up images of Jupiter, investigated its magnetosphere, atmosphere and internal structure. Still operating in the outer Solar System.
Venera 8 USSR	Venus Lander	Mar 27, 1972	Jul 22, 1972	As the spacecraft entered the upper atmosphere, the descent module separated while the service module burned up in the atmosphere. Entry speed was reduced by aerodynamic braking before parachute deployment. During descent, a refrigeration system was used to offset high temperatures. Returned data on temperature, pressure, light levels, and descent rates. Transmitted from surface for about 1 hour.
Cosmos 482 USSR	Venus Lander	Mar 31, 1972		Unsuccessful Venus probe; escape stage misfired leaving craft in Earth orbit.

SPACECRAFT	MISSION	LAUNCH DATE	ARRIVAL DATE	REMARKS
Venera 11 USSR	Venus Orbiter and Lander	Sep 9, 1978	Dec 25, 1978	Arrived at Venus 4 days after Venera 12. The two landers took nine samples of the atmosphere at varying heights and confirmed the basic components. Imaging system failed; did not return photos. Operated for 95 minutes.
Venera 12 USSR	Venus Orbiter and Lander	Sep 14, 1978	Dec 21, 1978	A transit module was positioned to relay the lander's data from behind the planet. Returned data on atmospheric pressure and components. Did not return photos; imaging system failed. Operated for 110 minutes.
Venera 13 USSR	Venus Orbiter and Lander	Oct 31, 1981	Mar 1, 1982	Provided first soil analysis from Venusian surface. Transmitted eight color pictures via orbiter. Measured atmospheric chemical and isotopic composition, electric discharges, and cloud structure. Operated for 57 minutes.
Venera 14 USSR	Venus Orbiter and Lander	Nov 4, 1981	Mar 3, 1982	Transmitted details of the atmosphere and clouds during descent; soil sample taken. Operated for 57 minutes.
Venera 15 USSR	Venus Orbiter	Jun 2, 1983	Oct 10, 1983	Obtained first high-resolution pictures of polar area. Compiled thermal map of almost entire northern hemisphere.
Venera 16 USSR	Venus Orbiter	Jun 7, 1983	Oct 16, 1983	Provided computer mosiac images of a strip of the northern continent. Soviet and U.S. geologists cooperated in studying and interpreting these images.
Vega 1 & 2 USSR	Venus/Halley	Dec 15, 1984 Dec 21, 1984	Jun 11, 1985 (Venus) Mar 6, 1986 (Halley) Jun 15, 1985 (Venus) Mar 9, 1986 (Halley)	International two-spacecraft project using Venusian gravity to send them on to Halley's Comet after dropping the Venusian probes. The Venus landers studied the atmosphere and acquired a surface soil sample for analysis. Each lander released a helium-filled instrumented balloon to measure cloud properties. The other half of the Vega payloads, carrying cameras and instruments, continued on to encounter Comet Halley.

SPACECRAFT	MISSION	LAUNCH DATE	ARRIVAL DATE	REMARKS
Phobos 1 & 2 USSR	Mars/Phobos	Jul 7, 1988 Jul 12, 1988	Jan 1989 (Mars) Jan 1989 (Mars)	International two-spacecraft project to study Mars and its moon Phobos. Phobos 1 was disabled by a ground control error. Phobos 2 was successfully inserted into Martian orbit in January 1989 to study the Martian surface, atmosphere, and magnetic field. On March 27, 1989, communications with Phobos 2 were lost and efforts to contact the spacecraft were unsuccessful.
Magellan USA	Venus Radar Mapping	May 4, 1989	Aug 1990	Returned radar images that showed geological features unlike anything seen on Earth. One area scientists called crater farms; another area was covered by a checkered pattern of closely spaced fault lines running at right angles. Most intriguing were indications that Venus still may be geologically active. Will continue to map the entire surface and observe evidence of volcanic eruption into 1991.
Galileo* USA	Jupiter Orbiter and Probe	Oct 18, 1989	Dec 8, 1990 (Earth) Feb 1991 (Venus)	A sophisticated two-part spacecraft; an Orbiter will be inserted into orbit around Jupiter to remotely sense the planet, its satellites and the Jovian magnetosphere and a Probe will descend into the atmosphere of Jupiter to make in situ measurements of its nature. Galileo flew by Venus, conducting the first infrared imagery and spectroscopy below the planet's cloud deck and used the Earth's gravity to speed it on its way to Jupiter.
Mars Observer USA	Mars Orbiter	Sep 25, 1992		Communication was lost with the Mars Observer on August 21, 1993, 3 days before the orbit insertion burn.

**Galileo* went into orbit around Jupiter in December 1995, deploying an atmospheric probe.

GLOSSARY OF GEOLOGIC TERMS

This glossary provides brief, informal definitions of technical terms not necessarily defined in the text. Readers are cautioned that these definitions may not be complete, and may wish to consult publications such as the American Geological Institute's *Glossary of Geology*, edited by R. L. Bates and J. A. Jackson (1987). *Webster's Ninth New Collegiate Dictionary*, Merriam–Webster (1986) includes many technical terms. For geophysical terms, R. E. Sheriff's *Encyclopedic Dictionary of Exploration Geophysics* published by the Society of Exploration Geophysicists (Tulsa) (1999) is most useful.

Terms not defined here or in the text may be found in Appendix A.

achondrite A stony meteorite with a crystalline texture and no chondrules; equivalent to an igneous rock; dominantly pyroxene and plagioclase.

anatexis Partial or complete melting of a pre-existing rock.

andesite A volcanic rock, generally fine-grained, with roughly 60% SiO_2; chemically referred to as "intermediate." Typically found around the Pacific Ocean, in areas such as the Andes.

anorthosite An igneous rock, frequently light-colored and usually coarse-grained, consisting largely of plagioclase. Found in the lower crust of the Earth and in the lunar highlands.

basalt A dark-colored volcanic rock, generally fine-grained, with roughly 50% SiO_2, relatively rich in Fe and Mg. Extremely widespread on continents, and forming much of the oceanic crust. A coarse-grained equivalent is gabbro.

biogenic Produced by living organisms.

breccia A rock composed of fragments, generally angular, of pre-existing rocks or minerals. In impact craters, may include glass or other melt rock.

Bouguer anomaly A gravity anomaly, or residual value, remaining after correction has been made for attraction of the rock between the measurement site and the reference surface (as well as for elevation and latitude).

chondrite A meteorite composed of roughly spheroidal grains, or chondrules; roughly ultramafic in composition.

crust In geology, the chemically distinct outer major layer of the Earth overlying the mantle; broadly, basaltic in ocean basins, granitic in continents. Lower boundary taken as the Mohorovičič Discontinuity, or Moho. Generally around 40 km thick under continents, around 10 km under ocean basins, but variable.

differentiation As used in planetology, the processes by which a planet or satellite becomes zoned into concentric shells (core, mantle, crust). *See* **magmatic differentiation**.

dike In structural geology, a tabular igneous intrusion cutting across the bedding of foliation of the enclosing rock.

diorite A medium- to coarse-grained igneous rock, composed of plagioclase, amphibole, pyroxene, and minor amounts of quartz. The deep-seated or intrusive equivalent of andesite.

earthquake magnitude A numerical measure of the strength of an earthquake, first developed by Charles Richter for local California earthquakes. Except for such earthquakes, the Richter magnitude has been replaced by better measures. **Ms** is the surface-wave magnitude, roughly equal to Richter magnitude for values up to about 6. For very large earthquakes, the moment magnitude **Mw** is now used, taking into account factors such as fault displacement at the earthquake focus. **Mw** values are roughly equal to Ms values up to about 7.5, but can be substantially higher for larger earthquakes.

eukaryotic Referring to cells with nuclei. *See* **prokaryotic**.

facies As used in metamorphic petrology (metamorphic facies), a characteristic mineral assemblage indicating a definite range of pressure, temperature, water pressure, and other variables.

fault A fracture in the Earth's crust along which there has been movement, horizontal, vertical, or oblique. *See* Appendix A.

felsic As used for igneous rocks, referring to a rock, generally light-colored, rich in feldspar and quartz. The complement of **mafic**. Granite is a typical felsic rock.

field As used in physics, a continuous volume of space in which the effects of gravity, magnetism, or electricity are measurable.

free-air anomaly A gravity anomaly remaining after corrections have been made for elevation of the measurement site above the reference surface, but not for attraction of the intervening rock. Satellite free-air anomalies generally reflect topography; mountains or high plateaus are generally shown as positive anomalies.

graben An extensional feature in the Earth's crust, essentially a down-dropped block (sometimes expressed as a valley) bounded by normal faults. *See* Appendix A.

granite A light-colored, frequently pink, igneous rock, medium to coarse-grained, composed largely of potassium feldspar, plagioclase, quartz, and mica. Generally with about 70% SiO_2.

granulite As used in metamorphic petrology, a rock formed by very high-pressure and high-temperature recrystallization, leading to loss of water. Typically composed of pyroxenes, plagioclase, and various mafic minerals.

isostasy A condition, analogous to floating, of equilibrium of lithospheric units such as the Tibetan Plateau. Two isostatic concepts are Airy (constant crustal density with different lithospheric thickness) and Pratt (constant lithospheric thickness, but with differing crustal densities).

lithosphere The rigid upper part of the Earth, including crust and mantle, underlain by a less-rigid zone, the asthenosphere. In plate tectonic theory, divided into spherical caps, i.e., plates.

mafic As used for igneous rocks, referring to a rock, generally dark-colored, rich in ferromagnesian minerals, i.e., relatively rich in magnesium (Mg) and iron (Fe), from which the term is derived. Basalt is a typical mafic rock.

magma A naturally-occurring rock melt, below the surface of the Earth or similar body, forming igneous rocks when solidified. The subsurface equivalent of lava.

magmatic differentiation The various processes, such as fractional crystallization, by which a magma can form a variety of igneous rocks on solidifying. Most common in basaltic magmas.

nucleosynthesis In nature, the process or processes by which elements are built up from lighter ones, as in stellar nuclear fusion or in supernovae.

plate In geology, a torsionally-rigid block of crust and mantle, i.e., the lithosphere, of regional extent, bounded by some combination of spreading centers (ridges), transform faults, and subduction zones.

plate tectonics A theory of crustal dynamics holding that most geologic phenomena such as mountain building, earthquakes, and volcanic activity are caused by movement of **plates**: convergence, divergence, or shear.

potential field As used in geophysics, a continuous volume of space in which the effects of gravity or magnetism are measurable; for each point in the field there is a unique value of gravity or magnetism.

prokaryotic Referring to cells without nuclei, notably bacteria.

regional metamorphism Solid-state recrystallization of pre-existing rock under conditions of high temperature and pressure, generally during periods of deformation; typical metamorphic rocks include gneiss, schist, marble, and quartzite.

redifferentiation A term proposed by Lowman (1976) for processes such as metamorphism and anatexis by which the continental crust of the Earth has been zoned into a felsic upper region and a mafic granulitic lower region. Equivalent to "intracrustal differentiation" of Dewey and Windley (1981).

right lateral As used for faults, displacement along a fault resulting in offset of the opposite side of the fault to the right (left, in left-lateral movement).

scalar As used in physics, referring to a quantity, such as mass or time, with magnitude but not direction. *See* **vector**.

secular As used in earth sciences, referring to a long period of indefinite duration.

shergottite An achondritic stony meteorite, largely composed of pyroxene and plagioclase or glassy equivalent; formed by crystallization of a basaltic magma.

slip vector Motion on a fault during an earthquake.

strike In structural geology, the direction or bearing of a horizontal line in the plane (stratum, lava flow, fault, joint, vein, dike) in question.

strike-slip fault A fault along which the dominant movement is horizontal. **Transform faults** are generally strike-slip faults, but most strike-slip faults are local features and not transform faults.

suture In plate tectonic theory, a surface or zone along which crustal blocks, or **terranes**, have been joined by plate movements. Generally marked by a fault or by ophiolite remnants, marking the site of an ocean closure.

tectonic Referring to the broad structure of the outer part of the Earth; structural geology is more localized, involving individual folds, faults, or intrusions.

terrane A fault-bounded block of crust of regional extent, whose geologic history is distinct from that of adjacent blocks. In plate tectonic theory, thought to have been carried significant distance by sea-floor spreading and sutured to a land mass.

transform fault A major fault in the Earth's crust, with dominantly horizontal displacement, generally offsetting segments of a mid-ocean ridge or connecting such a ridge with a trench or island arc. Actual movement in earthquakes may be opposite to the apparent offset.

ultramafic As used in petrology, referring to a rock high in magnesium and iron, generally dark-colored, with about 40% SiO_2. Typical ultramafic rocks are peridotite, serpentinite, and dunite.

vector As used in physics, referring to a quantity, such as stress, with magnitude and direction. See **scalar**.

SELECTED BIBLIOGRAPHY

Chapter 1

Burke, K. C., The African Plate, 24th Alex du Toit Memorial Lecture, *South African Journal of Geology*, **99**, 341–409, 1996.

Burke, K. C., and Wilson, J. T., Is the African plate stationary?, *Nature*, **239**, 387–390, 1972.

Canup, R. M., and Righter, K. (Editors), *Origin of the Earth and Moon*, University of Arizona Press, Tucson, and Lunar and Planetary Institute, Houston, 555 pp., 2000.

Chenoweth, P. A., Comparisons of the ocean floor with the lunar surface, *Geological Society of America Bulletin*, **73**, 199–210, 1962.

Dietz, R. S., Continent and ocean basin evolution by spreading of the sea floor, *Nature*, **190**, 854–857, 1961.

Frey, H. V., MAGSAT and POGO anomalies over the Lord Howe Rise: Evidence against a simple continental crustal structure, *Journal of Geophysical Research*, **90**, 2631–2639, 1985.

Hamblin, W. K., *Earth's Dynamic Systems*, Burgess, Minneapolis, 1978.

Hess, H. H., History of ocean basins, in *Petrologic Studies: A Volume to Honor A. F. Buddington*, A. E. J. Engel, H. L. James and B. F. Leonard, Editors, Geological Society of America, Boulder, pp. 599–620, 1962.

Langel, R. A., Study of the crust and mantle using magnetic surveys by Magsat and other satellites, *Proceedings of the Indian Academy of Sciences*, **99**, 581–618, 1990.

Lowman, P. D., Jr., A more realistic view of global tectonism, *Journal of Geological Education*, **30**, 97–107, 1982.

Lowman, P. D., Jr., Landsat and Apollo: The forgotten legacy, *Photogrammetric Engineering and Remote Sensing*, **65**, 1143–1147, 1999.

Lowman, P. D., Jr., Yates, J., Masuoka, P., Montgomery, B., O'Leary, J., and Salisbury, D., A digital tectonic activity map of the Earth, *Journal of Geoscience Education*, **437**, 430–433, 1999.

Press, F., and Siever, R., *Understanding Earth* (Third Edition), W. H. Freeman, New York, 2001.

Robbins, J. W., Smith, D. E., and Ma, C., Horizontal crustal deformation and large scale plate motions inferred from space geodetic techniques, in *Contributions of Space Geodesy to Geodynamics: Crustal Dynamics*, D. E. Smith and D. L. Turcotte, Editors, American Geophysical Union, **23**, 21–36, 1993.

Sandwell, D. T., Geophysical applications of satellite altimetry, *Reviews of Geophysics, Supplement*, 132–137, 1991.

Skinner, B. J., and Porter, S. C., *The Blue Planet*, John Wiley & Sons, New York, 1995.

Smith, W. H. F., and Sandwell, D. T., Global sea floor topography from satellite altimetry and ship depth soundings, *Science*, **277**, 1956–1962, 1997.

Thatcher, W., Microplate versus continuum descriptions of active tectonic deformation, *Journal of Geophysical Research*, **100**, B3, 3885–3894, 1995.

Wyllie, P. J., *The Way the Earth Works*, John Wiley, New York, 1976.

Yates, J., Montgomery, B. C., and Lowman, P. D., Jr., An integrative technique to create a Digital Tectonic Activity Map (DTAM) of the Earth, *Towards Digital Earth: Proceedings of the International Symposium on Digital Earth*, Beijing, December 1999, G. Xu and Y. Chen, Editors, Science Press, Beijing, China, pp. 783–787, 1999.

Chapter 2

Agocs, W. B., Meyerhoff, A. A., and Kis, K., Reykjanes Ridge: quantitative determinations from magnetic anomalies, in *New Concepts in Global Tectonics*, S. Chatterjee and N. Hotton, Editors, Texas Tech University Press, Lubbock, TX, pp. 221–238, 1992.

Allenby, R., *Crustal Dynamics Project Site Selection Criteria*, NASA Technical Memorandum 85084, Goddard Space Flight Center, Greenbelt, MD, 28 pp., 1983.

Argus, D. F., and Gordon, R. F., Tests of the rigid-plate hypothesis and bounds on intraplate deformation using geodetic data from very long baseline interferometry, *Journal of Geophysical Research*, **101**, 13,555–13,572, 1996.

Arkani-Hamed, J., Viscosity of the Moon, *The Moon*, **6**, 100–111, 1973.

Arnadottir, T., and Segall, P., The 1989 Loma Prieta earthquake imaged from inversion of geodetic data, *Journal of Geophysical Research*, **99**, B11, 21,835–21,855, 1994.

Atwater, T., Implications of plate tectonics for the Cenozoic tectonic evolution of western North America, *Geological Society of America Bulletin*, **81**, 3513–3536, 1970.

Baldwin, R. B., *The Measure of the Moon*, University of Chicago Press, Chicago, 1963.

Bender, P. L., and 12 others, The lunar laser ranging experiment, *Science*, **182**, 229–238, 1973.

Bills, B. G., and Ferrari, A. J., Mars topography harmonics and geophysical implications, *Journal of Geophysical Research*, **83**, B7, 3497–3508, 1978.

Bills, B. G., and Ferrari, A. J., A harmonic analysis of lunar gravity, *Journal of Geophysical Research*, **85**, B2, 1013–1025, 1980.

Bills, B. G., and Kobrick, M., Venus topography: A harmonic analysis, *Journal of Geophysical Research*, **90**, B1, 827–836, 1985.

Bills, B. G., Kiefer, W. S., and Jones, R. L., Venus gravity: A harmonic analysis, *Journal of Geophysical Research*, **92**, B10, 10,335–10,351, 1987.

Binder, A. B., Lunar Prospector: Overview, *Science*, **281**, 1475–1476, 1998.

Bird, P., and Li, Y., Interpolation of principal stress directions by nonparametric statistics: Global maps with confidence limits, *Journal of Geophysical Research*, **101**, B3, 5435–5443, 1996.

Blewitt, G., Advances in Global Positioning System technology for geodynamics investigations: 1978–1992, in *Contributions of Space Geodesy to Geodynamics*: *Technology*, D. E. Smith and D. L. Turcotte, Editors, American Geophysical Union, **25**, 195–213, 1993.

Bonneville, A., Barriot, J. P., and Bayer, R., Evidence from geoid data of a hotspot origin for the southern Mascarene Plateau and Mascarene Islands (Indian Ocean), *Journal of Geophysical Research*, **93**, B5, 4199–4212, 1988.

Bostrom, R. C., Subsurface exploration via satellite: Structure visible in Seasat images of North Sea, Atlantic continental margin, and Australia, *American Association of Petroleum Geologists Bulletin*, **73**, 1053–1064, 1989.

Bratt, S. R., Solomon, S. C., Head, J. W., and Thurber, C. H., The deep structure of lunar basins: Implications for basin formation and modification, *Journal of Geophysical Research*, **90**, B4, 3049–3064, 1985.

Buchar, R., Motion of the nodal line of the second Russian earth satellite (1957b) and flattening of the earth, *Nature*, **182**, 98–199, 1958.

Burgmann, R., Segall, P., Losowski, M., and Svarc, J. P., Rapid aseismic slip on the Berrocal fault zone following the Loma Prieta earthquake (abstract), *EOS Transactions*, American Geophysical Union, **73**(43), Fall Meeting Supplement, 119, 1992.

Burns, J. O. MERI: an ultra-long baseline Moon–Earth interferometer, in *Future Astronomical Observatories on the Moon*, NASA Conference Publication 2489, J. O. Burns and W. W. Mendell, Editors, pp. 97–103.

Cameron, A. G. W. The origin of the Moon and the single impact hypothesis V, *Icarus*, **126**, 126–137, 1997.

Carey, S. W., *The Expanding Earth*, Elsevier, Amsterdam, 1976.

Carey, S. W., (Editor), *The Expanding Earth: A Symposium*, University of Tasmania, Sydney, 1981a.

Carey, S. W., Earth expansion and the null universe, in *The Expanding Earth: A Symposium*, S. W. Carey, Editor, University of Tasmania, Sydney, pp. 365–372, 1981b.

Carr, M. H. *The Surface of Mars*, Yale University Press, New Haven, 1982.

Carter, W. E., and Robertson, D. S., Studying the Earth by very long baseline interferometry, *Scientific American*, **11**, 46–54, 1986.

Cazenave, A., and Dominh, K., Global relationship between oceanic geoid and seafloor depth: New results, *Geophysical Research Letters*, **14**, 1–4, 1987.

Cazenave, A., and 6 others, Geodetic results from Lageos 1 and Doris satellite data, in *Contributions of Space Geodesy to Geodynamics: Crustal Dynamics*, D. E. Smith and D. L. Turcotte, Editors, American Geophysical Union, **23**, 81–98, 1993.

Chao, B. F., The geoid and earth rotation, in *The Geoid and Its Geophysical Interpretation*, P. Vanicek and N. Christou, Editors, CRC Press, Boca Raton, Florida, pp. 285–298, 1994.

Chase, C. G., Plate kinematics: The Americas, east Africa, and the rest of the world, *Earth and Planetary Science Letters*, **37**, 355–368, 1978.

Christodoulidis, D. C., Smith, D. E., Kolenkiewicz, R., Klosko, S. M., Torrence, M. H., and Dunn, P. J., Observing tectonic plate motions and deformations from satellite laser ranging, *Journal of Geophysical Research*, **90**, B11, 9249–9263, 1985.

Cohen, S. C., Time-dependent uplift of the Kenai Peninsula and adjacent regions of south central Alaska since the 1964 Prince William Sound earthquake, *Journal of Geophysical Research*, **101**, B4, 8595–8604, 1996.

Cohen, S. C., and Freymuller, J. T., Deformation of the Kenai Peninsula, Alaska, *Journal of Geophysical Research*, **102**, B9, 20,479–20,487, 1997.

Cohen, S. C., and Smith, D. E., LAGEOS scientific results: Introduction, *Journal of Geophysical Research*, **90**, B11, 9217–9220, 1985.

Cohen, S. C., Holdhal, S., Caprette, D., Hilla, S., Safford, R., and Schultz, D., Uplift of the Kenai Peninsula, Alaska, since the 1964 Prince William Sound earthquake, *Journal of Geophysical Research*, **100**, 2031–2038, 1995.

Colombo, O., and Watkins, M. M., Satellite positioning, *Reviews of Geophysics, Supplement*, **29**, 138–147, 1991.

Craig, C. H., and Sandwell, D. T., Global distribution of seamounts from Seasat profiles, *Journal of Geophysical Research*, **93**, B9, 10,408–10,420, 1988.

Daly, P., An introduction to the USSR GLONASS navigation satellites, *Journal of the British Interplanetary Society*, **46**, 385–390, 1993.

Daly, R. A., *Strength and Structure of the Earth*, Prentice-Hall, Englewood Cliffs, NJ, 1940.

Davis, R. E., and Foote, F. S., *Surveying: Theory and Practice* (Third Edition), McGraw-Hill Book Company, New York, 1940.

Degnan, J. J., Millimeter accuracy satellite laser ranging: A review, in *Contributions of Space Geodesy to Geodynamics: Technology*, D. E.

Smith and D. L. Turcotte, Editors, American Geophysical Union, pp. 134–162, 1993.

DeMets, C., Gordon, R. G., Argus, D. F., and Stein, S., Current plate motions, *Geophysical Journal International*, **101**, 425–478, 1990.

Dewey, J. F., Plate tectonics, *Reviews of Geophysics and Space Physics*, **13**, 326–332, 1975.

Dewey, J. F., and Sengor, C. A. M., Aegean and surrounding regions: Complex multiplate and continuum tectonics in a convergent zone, *Geological Society of America Bulletin*, **90**, 84–92, 1979.

Dickey, J. O. and Eubanks, T. M., Earth rotation and polar motion: Measurements and implications, *IEEE Transactions of Geoscience and Remote Sensing*, **GE-23**, 373–384, 1985.

Dickey, J. O. and 11 others, Lunar laser ranging: A continuing legacy of the Apollo Program, *Science*, **265**, 482–490, 1993.

Donellan, A., Hager, B. H., King, R. W., and Herring, T. A., Deformation of the Ventura Basin region, southern California, *Journal of Geophysical Research*, **98**, B12, 21,727–21,739, 1993.

Douglas, B. C., and Cheney, R. E., Geosat: Beginning a new era in satellite oceanography, *Journal of Geophysical Research*, **95**, C3, 2833–2836, 1990.

Dvořák, J., and Phillips, R. J., The nature of the gravity anomalies associated with large young lunar craters, *Geophysical Research Letters*, **4**, 380–382, 1977.

Dvořák, J., and Phillips, R. J., Lunar Bouguer gravity anomalies : Imbrian age craters, *Proceedings of the Ninth Lunar and Planetary Science Conference*, Pergamon Press, New York, pp. 3651–3668, 1978.

England, P., and Jackson, J., Active deformation of the continents, *Annual Reviews of Earth and Planetary Science*, **17**, 197–226, 1989.

Esposito, P. B., and 7 others, Gravity and topography, in *Mars*, H. H. Kieffer *et al.*, Editors, University of Arizona Press, Tucson, AZ, pp. 209–248, 1992.

Feigl, K. L., and 14 others, Space geodetic measurements of crustal deformation in central and southern California, 1984–1992, *Journal of Geophysical Research*, **98**, B12, 21,677–21,712, 1993.

Ferrari, A. J., and Bills, B. G., Planetary geodesy, *Reviews of Geophysics and Space Physics*, **17**, 1663–1678, 1979.

Ferrari, A. J., Nelson, D. L., Sjogren, W. L., and Phillips, R. J., The isostatic state of the lunar Apennines and regional surroundings, *Journal of Geophysical Research*, **83**, B6, 2863–2871, 1978.

Filmer, P. E., and McNutt, M. K., Geoid anomalies over the Canary Island group, *Marine Geophysical Researches*, **11**, 77–87, 1989.

Flack, C. and Warner, M., Three-dimensional mapping of seismic reflections from the crust and upper mantle, northwest of Scotland, *Tectonophysics*, **173**, 469–481, 1990.

Flinn, E. A., Application of space technology to geodynamics, *Science*, **213**, 89–96, 1981.

Flynn, L. P., Harris, A. J. L., Rothery, D. A., and Oppenheimer, C., High-spatial-resolution thermal remote sensing of active volcanic features using Landsat and hyperspectral data, in *Remote Sensing of Active Volcanism*, P. J. Mouginis-Mark, J. A. Crisp, and J. H. Fink, Editors, Geophysical Monograph 116, American Geophysical Union, pp. 161–177, 2000.

Frey, H. V., and Schultz, R.A., Large impact basins and the mega-impact theory for the crustal dichotomy on Mars, *Geophysical Research Letters*, **15**, 229–232, 1988.

Frey, H. V., Roark, J. H., and Sakimoto, S. E. H., Impact craters and basins at the boundary: Windows into the evolution of the martian dichotomy (Abstract), 2000 Spring Meeting, American Geophysical Union, p. S292, 2000.

Gaposchkin, E. M., and Lambeck, K., Earth's gravity field to the sixteenth degree and station coordinates from satellite and terrestrial data, *Journal of Geophysical Research*, **76**, 4855–4883, 1971.

Garland, G. D., *The Earth's Shape and Gravity*, Pergamon Press, New York, 1965.

Glasstone, S., *Sourcebook on the Space Sciences*, D. Van Nostrand, Princeton, 1965.

Golombek, M. P., Structural analysis of lunar grabens and the shallow crustal structure of the Moon, *Journal of Geophysical Research*, **84**, 4657–4666, 1979.

Gordon, R. G., Plate motion, *Reviews of Geophysics*, *Supplement*, 748–758, April 1991.

Gordon, R. G., and Stein, S., Global tectonics and space geodesy, *Science*, **256**, 333–342, 1992.

Hager, B. H., King, R. W., and Murray, M. H., Measurement of crustal deformation using the Global Positioning System, *Annual Reviews of Earth and Planetary Sciences*, **19**, 351–382, 1991.

Haley, A. G., and Rosen, M. W., On the utility of an artificial unmanned earth satellite, *Jet Propulsion*, **25**, 71–78, 1955.

Hall, S. S., *Mapping the Next Millenium*, Random House, New York, 1992.

Hallam, A., *Great Geological Controversies*, Oxford University Press, New York, 1983.

Haxby, W. F., *Gravity Field of the World's Oceans*, National Geophysical Data Center, National Oceanic and Atmospheric Adminstration, 1987.

Haxby, W. F., and Weissel, J. K., Evidence for small-scale mantle convection from Seasat altimeter data, *Journal of Geophysical Research*, **91**, B3, 3507–3520, 1986.

Head, J. W., Hiesinger, H., Ivanov, M. A., Kreslavsky, M. A., Pratt, S., and Thompson, B. J., Testing for oceans on Mars. Evidence for configuration and recession (Abstract), 2000 Spring Meeting, American Geophysical Union, p. S292, 2000.

Heiskanen, W. A., and Vening Meinesz, F. A., *The Earth and Its Gravity Field*, McGraw-Hill, New York 1958.

Herrick, R. R., and Phillips, R. J., Geological correlations with the interior density structure of Venus, *Journal of Geophysical Research*, **97**, E10, 16,017–16,034, 1992.

Jeffreys, H., *The Earth* (Fourth Edition), Cambridge University Press, Cambridge, 1962.

Jordan, T. H., and Minster, J. B., Beyond plate tectonics: Looking at plate deformation with space geodesy, in *The Impact of VLBI on Astrophysics and Geophysics*, M. J. Reid and J. M. Moran, Editors, International Astronomical Union, pp. 341–350, 1988.

Kahn, W. D., and Bryan, J. W., Use of a spacecraft borne altimeter for determining the mean sea surface and the geopotential, X-550-72-19, 29 pp., Goddard Space Flight Center, 1972.

Kaula, W. M., *An Introduction to Planetary Physics*, John Wiley & Sons, New York, 1968.

Kaula, W. M., Interpretation of lunar mass concentrations, *Physics of the Earth and Planetary Interiors*, **2**, 123–137, 1969.

Kaula, W. M., Global gravity and tectonics, in *The Nature of the Solid Earth*, E. C. Robertson, J. F. Hays, and L. Knopoff, Editors, McGraw-Hill, New York, pp. 385–402, 1972.

Kaula, W. M., The gravity and shape of the Moon, *EOS*, *Transactions*, *American Geophysical Union*, **56**, 309–316, 1975.

Kaula, W. M., Gravity fields: Implications for planetary interiors, in *The Encyclopedia of Solid Earth Geophysics*, D. E. James, Editor, Van Nostrand Reinhold, New York, pp. 622–627, 1989.

Kaula, W. M., Venus: A contrast in evolution to Earth, *Science*, **247**, 1191–1196, 1990.

Kerr, R. A. Continental drift nearing certain detection, *Science*, **229**, 953–955, 1985.

Kiefer, W. S., Richards, M. A., Hager, B. H., and Bills, B. G., A dynamical model of Venus's gravity field, *Geophysical Research Letters*, **13**, 14–17, 1986.

Kieffer, H. H., Jakosky, B. M., Snyder, C. W., and Matthews, M. S. (Editors), *Mars*, University of Arizona Press, Tucson, AZ, 1992.

King, N. E., and 7 others, Continuous GPS across the Hayward fault (Abstract), *EOS, Transactions*, *American Geophysical Union*, **73**(43), Fall Meeting Supplement, 121, 1992.

King-Hele, D., The shape of the Earth, *Science*, **192**, 1293–1300, 1976.

King-Hele, D., *Observing Earth Satellites*, Van Nostrand Reinhold, New York, 1983.

King-Hele, D., *A Tapestry of Orbits*, Cambridge University Press, Cambridge, 1992.

Knopoff, L., The thickness of the lithosphere from the dispersion of surface waves, *Geophysical Journal of the Royal Astronomical Society*, **74**, 55–81, 1983.

Kolenkiewicz, R., Ryan, J., and Torrence, M. H., A comparison between LAGEOS laser ranging and very long baseline interferometry determined baseline lengths, *Journal of Geophysical Research*, **90**, B11, 9265–9274, 1985.

Kopal, Z., Moments of inertia of the lunar globe, and their bearing on chemical differentiation of its outer layers, *The Moon*, **3**, 28–33, 1971.

Konopliv, A. S., Binder, A. B., Hood, L. L., Kuscinskas, A. B., Sjogren, W. L., and Williams, J. G., Improved gravity field of the Moon from Lunar Prospector, *Science*, **281**, 1476–1480, 1998.

Lambeck, K., *The Earth's Variable Rotation: Geophysical Causes and Consequences*, Cambridge University Press, Cambridge, 1980.

Lambeck, K., *Geophysical Geodesy*, Clarendon Press, Oxford, 1988.

Larson, K. M., Freymuller, J. T., and Philipsen, S., Global plate velocities from the Global Positioning System, *Journal of Geophysical Research*, **102**, B5, 9961–9981, 1997.

Lemoine, E. F., Pavlis, N. K., Kenyon, S. C., Rapp, R. H., Pavlis, E. C., and Chao, B. F., New high-resolution model developed for Earth's gravitational field, *EOS*, *Transactions*, *American Geophysical Union*, **79**, 113–118, 1998a.

Lemoine, F.G., and 14 others, *The Development of the Joint NASA GSFC and the National Imagery and Mapping Agency (NIMA) Geopotential Model EGM 96*, NASA/TP-1998–206861, 575 pp., 1998b.

Le Pichon, X., Camot-Rooke, N., Lallemant, S., Noomen, R., and Veis, G., Geodetic determination of the kinematics of central Greece with respect to Europe: Implications for eastern Mediterranean tectonics, *Journal of Geophysical Research*, **100**, B7, 12,675–12,690, 1995.

Lowman, P. D., Jr., Mechanical obstacles to the movement of continent-bearing plates, *Geophysical Research Letters*, **12**, 223–225, 1985.

Lowman, P. D., Jr., Plate tectonics with fixed continents: a testable hypothesis, I and II, *Journal of Petroleum Geology*, **8**, 373–388, 1985, and **9**, 71–88, 1986.

Lowman, P. D., Jr., Eastern North America as a convergent plate boundary: A suggested factor in eastern earthquakes (Abstract), *1991 Abstracts with Programs*, Geological Society of America, **23** (1), 60, 1991.

Lowman, P. D., Jr., Plate tectonics and continental drift: A skeptic's view, in *The Blue Planet, An Introduction to Earth System Science*, B. J. Skinner and S. C. Porter, Editors, John Wiley & Sons, New York, pp. 187–189, 1995.

Lowman, P. D., Jr., T plus twenty five years: A defense of the Apollo Program, *Journal of the British Interplanetary Society*, **49**, 71–79, 1996.

Lowman, P. D., Jr., A global tectonic activity map of the Earth: Implications for plate tectonics (Abstract), *EOS Transactions, American Geophysical Union, Supplement*, p. S407, May 9, 2000.

Lowman, P. D., Jr., and Frey, H. V., *A Geophysical Atlas for Interpretation of Satellite-derived Data*, Technical Memorandum 79722, Goddard Space Flight Center, Greenbelt, MD, 54 pp., 1979.

Lowman, P. D., Jr., Allenby, R. A., and Frey, H. V., *Proposed Satellite Laser Ranging and Very Long Baseline Interferometry Sites for Crustal Dynamics Observations*, NASA Technical Memorandum 80563, Goddard Space Flight Centre, Greenbelt, MD, 64 pp. 1979.

Ma, C., Ryan, J. W., and Caprette, D. S., *Crustal Dynamics Project Data Analysis – 1991*, NASA Technical Memorandum 104552, Goddard Space Flight Center, Greenbelt, MD, 424 pp., 1992.

Ma, C., Sauber, J. M., Bell, L. J., Clark, T. A., Gordon, D., Himwich, W. E., and Ryan, J. W., Measurement of horizontal motions in Alaska using very long baseline interferometry, *Journal of Geophysical Research*, **95**, B13, 21,991–22,011, 1990.

MacDonald, G. J. F., The internal constitutions of the inner planets and the Moon, *Space Science Reviews*, **2**, 473–557, 1963.

McKenzie, D. P., Speculations on the consequence and causes of plate motions, *Geophysical Journal of the Royal Astronomical Society*, **18**, 1–32, 1969.

McKenzie, D. P., Active tectonics of the Mediterranean region, *Geophysical Journal of the Royal Astronomical Society*, **30**, 109–185, 1972.

McNutt, M. K., and Judge, A. V., The superswell and mantle dynamics beneath the South Pacific, *Science*, **248**, 969–975, 1990.

Manchester, W., *A World Lit Only By Fire*, Little, Brown, and Company, Boston, 1988.

Marsh, J. G., Satellite-derived gravity maps, in *A Geophysical Atlas for Interpretation of Satellite-derived Data*, P. D. Lowman Jr. and H. V. Frey, Editors, NASA Technical Memorandum 79722, Goddard Space Flight Center, Greenbelt, MD, pp. 9–14, 1979.

Marsh, J. G., and Chang, E. S. 5′ Detailed gravimetric geoid in the northwestern Atlantic Ocean, *Marine Geodesy*, **1**, 253–261,1985.

Marsh, J. G., Lerch, F. J., and Williamson, R. G., Precision geodesy and geodynamics using Starlette laser ranging, *Journal of Geophysical Research*, **90**, B11, 9335–9345, 1985.

Martin, B. D., Constraints to major right-lateral movements, San Andreas fault system, central and northern California, USA, in *New Concepts in Global Tectonics*, S. Chatterjee and N. Hotton III, Editors, Texas Tech University Press, Lubbock, TX, pp. 131–148, 1992.

Massey, H. S. W., and Boyd, R. L. F., Scientific observations of the artificial earth satellites and their analysis, *Nature*, **181**, 78–80, 1958.

Masursky, H., Eliason, E., Ford, P. G., McGill, G. E., and Pettengill, G. H., Pioneer Venus radar results: geology from images and altimetry, *Journal of Geophysical Research*, **85**, 8322–8360, 1980.

Melosh, H. J., The tectonics of mascon loading, *Proceedings of the Ninth Lunar and Planetary Science Conference*, pp. 3513–3525, 1978.

Merson, R. H., and King-Hele, D. G., Use of artificial satellites to explore the Earth's gravitational field: results from Sputnik 2, *Nature*, **182**, 640–641, 1958.

Meyerhoff, A. A., and Meyerhoff, H. A., The new global tectonics: Major inconsistencies, *American Association of Petroleum Geologists Bulletin*, **56**, 269–336, 1972.

Meyerhoff, A. A., Taner, I., Morris, A. E. L., Martin, B. D., Agocs, W. B., and Meyerhoff, H. A., Surge tectonics: A new hypothesis of Earth dynamics, in *New Concepts in Global Tectonics*, S. Chatterjee and N. Hotton III, Editors, Texas Tech University Press, Lubbock, TX, pp. 309–409, 1992.

Michael, W. H., and Blackshear, W. T., Recent results on the mass, gravitational field, and moments of inertia of the Moon, *The Moon*, **3**, 388–402, 1972.

Minster, J. B., and Jordan, T. H., Present-day plate motions, *Journal of Geophysical Research*, **83**, 5531–5554, 1978.

Molnar, P., Continental tectonics in the aftermath of plate tectonics, *Nature*, **335**, 131–137, 1988.

Molnar, P., and Atwater, T., Relative motion of hot spots in the mantle, *Nature*, **246**, 288–291, 1973.

Muller, P. M., and Sjogren, W. L., Mascons: Lunar mass concentrations, *Science*, **161**, 680–684, 1968.

Murray, B. C., The artificial earth satellite – A new geodetic tool, *ARS Journal*, July, 924–931, 1961.

Munk, W. H., and MacDonald, G. J. F., *The Rotation of the Earth*, Cambridge University Press, 1960.

NASA, *The National Geodynamics Program: An Overview*, NASA Technical Paper 2147, National Aeronautics and Space Administration, Washington, 125 pp., 1983.

NASA, *NASA Geodynamics Program Summary Report*: *1979–1987*, NASA Technical Memorandum 4065, National Aeronautics and Space Administration, Washington, 144 pp., 1988.

National Research Council, *Current Problems in Geodesy*, National Academy Press, Washington, 19 pp., 1987.

Nerem, R. S., Bills, B. G., and McNamee, J. B., A high resolution gravity model for Venus: GVM-1, *Geophysical Research Letters*, **20**, 599–602, 1993.

Newell, H. E., *Beyond the Atmosphere: Early Years of Space Science*, NASA Special Publication 4211, National Aeronautics and Space Administration, Washington, 497 pp., 1980.

Nozette, S., and 33 others, The Clementine mission to the Moon: Scientific overview, *Science*, **266**, 1835–1839, 1994.

O'Keefe, J. A., The geodetic significance of an artificial satellite, *Jet Propulsion*, **25**, 75–76, 1955.

O'Keefe, J. A., Satellite methods in geodesy, *Surveying and Mapping*, **XVIII**, 418–422, 1958.

O'Keefe, J. A., Zonal harmonics of the earth's gravitational field and the basic hypothesis of geodesy, *Journal of Geophysical Research*, **64**, 2389–2392, 1959.

O'Keefe, J. A., The equilibrium shape of the earth in the light of recent discoveries in space science, *Space Mathematics, Pt. II*, pp. 119–154, 1966.

O'Keefe, J. A., Inclination of the moon's orbit, *Irish Astronomical Journal*, **10**, 241–250, 1972.

O'Keefe, J. A., and Batchlor, C. D., Perturbations of a close satellite by the equatorial ellipticity of the earth, *Astronomical Journal*, **62**, 183–185, 1957.

O'Keefe, J. A., Eckels, A., and Squires, R. K., Vanguard measurements give pear-shaped component of earth's figure, *Science*, **129**, 565–566, 1959.

Owen, H. G., Ocean-floor spreading: Evidence of Earth expansion, in *The Expanding Earth*, S. W. Carey, Editor, University of Tasmania, Sydney, pp. 31–58. 1981.

Peltier, W. R., Mantle convection and viscoelasticity, *Annual Reviews of Fluid Mechanics*, **17**, 561–608, 1985.

Phillips, R. J., and Lambeck, K., Gravity fields of the terrestrial planets: Long-wavelength anomalies and tectonics, *Reviews of Geophysics and Space Physics*, **18**, 27–76, 1980.

Phillips, R. J., and Malin, M. C., Tectonics of Venus, *Annual Reviews of Earth and Planetary Sciences*, **12**, 411–443, 1984.

Proverbio, E., and Quesada, V., Secular variations in latitudes and longitudes and continental drift, *Journal of Geophysical Research*, **79**, 4941–4943, 1974.

Rapp, R. H., The determination of geoid undulations and gravity anomalies from Seasat altimeter data, *Journal of Geophysical Research*, **88**, 1552–1562, 1983.

Rapp, R. H., The decay of the spectrum of the gravitational potential and the topography for the Earth, *Geophysical Journal International*, **99**, 449–455, 1989.

Reasenberg, R. D., Shapiro, I. I., and White, R. D., The gravity field of Mars, *Geophysical Research Letters*, **2**, 89–92, 1975.

Robbins, J. S., Smith, D. E., and Ma, C., Horizontal crustal deformation and large scale plate motions inferred from space geodetic techniques, in *Contributions of Space Geodesy to Geodynamics: Crustal Dynamics*, D. E. Smith and D. L. Turcotte, Editors, American Geophysical Union, **23**, pp. 21–36, 1993.

Royer, J-Y., and Sandwell, D. T., Evolution of the eastern Indian Ocean since the Late Cretaceous: Constraints from Geosat altimetry, *Journal of Geophysical Research*, **94**, B10, 13,755–13,782, 1989.

Rubincam, D. P., Tidal friction and the early history of the moon's orbit, *Journal of Geophysical Research*, **80**, 1537–1548, 1975.

Rubincam, D. P., Information theory lateral density distribution for Earth inferred from global gravity field, *Journal of Geophysical Research*, **87**, 5541–5552, 1982.

Rubincam, D. P., Postglacial rebound observed by Lageos and the effective viscosity of the lower mantle, *Journal of Geophysical Research*, **89**, 1077–1087, 1984.

Runcorn, S. K., Convection in the Moon and the existence of a lunar core, *Proceedings of the Royal Society, Series A*, No. 1446, **296**, 1967.

Runcorn, S. K., Geophysical tests of the Earth expansion hypothesis, in *The Expanding Earth*, S. W. Carey, Editor, University of Tasmania, Sydney, p. 327, 1981.

Ryan, J. W., Ma, C., and Caprette, D. S., *NASA Space Geodesy Program – GSFC Data Analysis – 1992, Final Report of the Crustal Dynamics Project VLBI Geodetic Results 1979–91*, NASA Technical Memorandum 10457, Goddard Space Flight Center, Greenbelt, MD, 471 pp., 1993.

Sandwell, D. T., Geophysical applications of satellite altimetry, *Reviews of Geophysics*, Supplement, 132–137, 1991.

Sandwell, D. T., and Renkin, M. L., Compensation of swells and plateaus in the North Pacific: No direct evidence for mantle convection, *Journal of Geophysical Research*, **93**, B4, 2775–2783, 1988.

Satalich, J., On shifting ground: GPS control networks in California, *GPS World*, **4**, 9, 36–42, 1993.

Sato, K., Tectonic plate motion and deformation inferred from very long baseline interferometry, *Tectonophysics*, **220**, 69–87, 1993.

Sauber, J., Thatcher, W., and Solomon, S. C., Geodetic measurement of deformation in the central Mojave Desert, California, *Journal of Geophysical Research*, **91**, B12, 12,683–12,693, 1986.

Sauber, J., Thatcher, W., Solomon, S. C., and Lisowski, M., Geodetic slip rate for the eastern California shear zone and the recurrence time of Mojave desert earthquakes, *Nature*, **367**, 264–266, 1994.

Sauber, J., McClusky, S., and King, R., Relation of ongoing deformation rates to the subduction zone process in southern Alaska, *Geophysical Research Letters*, **24**, 2853–2856, 1997.

Schmid, H. H., Worldwide geometric satellite triangulation, *Journal of Geophysical Research*, **84**, 7599–7615, 1974.

Schmidt, P. W., and Embleton, B. J. J., Recognition of common Precambrian polar wandering reveals a conflict with plate tectonics, *Nature*, **282**, 705–708, 1979.

Silver, P. G., Carlson, R. W., and Olson P., Deep slabs, geochemical heterogeneity, and the large-scale structure of mantle convection: Investigation of an enduring paradox, *Annual Reviews of Earth and Planetary Sciences*, **16**, 477–541, 1988.

Simpson, G. G., Holoarctic mammalian faunas and continental relationships during the Cenozoic, *Geological Society of America Bulletin*, **58**, 613–688, 1947.

Siry, J. W., Satellite orbits and atmospheric densities at altitudes up to 750 km obtained from the Vanguard orbit determination program, *Planetary and Space Science*, **1**, 184–192, 1959.

Sleep, N. H., Martian plate tectonics, *Journal of Geophysical Research*, **99**, 5639–5655, 1994.
Smith, D. E., Determination of the earth's gravitational potential from satellite orbits, *Planetary and Space Science*, **8**, 43–48, 1961.
Smith, D. E., and 9 others, Tectonic motion and deformation from satellite laser ranging to LAGEOS, *Journal of Geophysical Research*, **95**, B13, 22,013–22,041, 1990.
Smith, D.E., and 9 others, *LAGEOS Geodetic Analysis – SL7.1*, NASA Technical Memorandum 104549, Goddard Space Flight Center, Greenbelt, MD, 251 pp., 1991.
Smith, D. E., and 8 others, Contemporary global horizontal crustal motion. *Geophysical Journal International*, **119**, 511–520, 1994.
Smith, D. E., and 11 others, Topography of the northern hemisphere of Mars from the Mars Orbital Laser Altimeter, *Science*, **279**, 1686–1692, 1998.
Smith, W. H. F., and Sandwell, D. T., Bathymetric prediction from dense satellite altimetry and sparse shipboard bathymetry, *Journal of Geophysical Research*, **99**, B11, 21,803–21,824, 1994.
Smith, W. H. F., and Sandwell, D. T., Global sea floor topography from satellite altimetry and ship depth soundings, *Science*, **277**, 1956–1962, 1997.
Solomon, S. C., The nature of isostasy on the moon; How big a Pratt-fall for Airy models? *Proceedings of the Ninth Lunar and Planetary Science Conference*, pp. 3499–3511, 1978.
Solomon, S. C., and Head, J. W., Vertical movement in mare basins: Relation to mare emplacement, basin tectonics, and lunar thermal history, *Journal of Geophysical Research*, **84**, 1667–1682, 1979.
Solomon, S. C., and 10 others, Venus tectonics: An overview of Magellan observations, *Journal of Geophysical Research*, **97**, E8, 13,199–13,255, 1992.
Spudis, P. D., Reisse, R. A., and Gillis, J. J., Ancient multiring basins on the Moon revealed by Clementine laser altimetry, *Science*, **266**, 1848–1851, 1994.
Stein, C. A., Cloetingh, S., and Wortel, R., Seasat-derived gravity constraints on stress and deformation in the northeastern Indian Ocean, *Geophysical Research Letters*, **16**, 823–826, 1989.
Stein, S., Space geodesy and plate motions, in *Contributions of Space Geodesy to Geodynamics: Crustal Dynamics*, D. E. Smith and D. L. Turcotte, Editors, American Geophysical Union, pp. 5–20, 1993.
Struve, H., Bestimmung der Abplattung und des Aequators von Mars, *Astronomische Nachrichten*, **138**, 3302, 218–228, 1895.
Tapley, B. D., Schutz, B. E., Eanes, R. J., Ries, J. C., and Watkins, M. M., Lageos laser ranging contributions to geodynamics, geodesy, and orbital dynamics, in *Contributions of Space Geodesy to Geodynamics: Earth Dynamics*, D. E. Smith and D. L. Turcotte, Editors, American Geophysical Union, **24**, pp. 147–173, 1993.

Thatcher, W., Microplate versus continuum descriptions of active tectonic deformation, *Journal of Geophysical Research*, **100**, B3, 3885–3894, 1995.

Thurber, C. H., and Solomon, S. C., An assessment of crustal thickness variations on the lunar near side: models, uncertainties, and implications for crustal differentiation, *Proceedings of the Ninth Lunar and Planetary Science Conference*, pp. 3481–3497, 1978.

UNAVCO (University NAVSTAR Consortium), Online Brochure, 1998 www.unavco.ucar.edu/community/brochure

Urey, H. C., *The Planets*, Yale University Press, New Haven, CT, 1952.

Wegener, A., *The Origin of Continents and Ocean Basins* (English translation by J. Biram), Dover Publications, New York, 1966 (original publication 1928).

Wessel, P., and Haxby, W. F., Geoid anomalies at fracture zones and thermal models for the oceanic lithosphere, *Geophysical Research Letters*, **16**, 827–830, 1989.

Wieczorek, M. A., and Phillips, R. J., The structure and compensation of the lunar highland crust, *Journal of Geophysical Research*, **102**, E5, 10,993–10,943, 1997.

Wilford, J. N., *The Mapmakers*, Alfred A. Knopf, New York, 1981.

Williams, J. G., Newhall, X. X., and Dickey, J. O., Lunar laser ranging: Geophysical results and reference frames, in *Contributions of Space Geodesy to Geodynamics: Earth Dynamics*, D. E. Smith and D. L. Turcotte, Editors, American Geophysical Union, **24**, pp. 83–88, 1993.

Wise, D. U., and Yates, M. T., Mascons as structural relief on a lunar "Moho," *Journal of Geophysical Research*, **75**, 261–268, 1970.

Wyllie, P. J., *The Dynamic Earth*, John Wiley, New York, 1971.

Yunck, T. P., Melbourne, W. G., and Thornton, C. L., GPS-based satellite tracking system for precise positioning, *IEEE Transactions on Geoscience and Remote Sensing*, **GE-23**, 450–457, 1985.

Zoback, M. L., First- and second-order patterns of stress in the lithosphere: The World Stress Map Project, *Journal of Geophysical Research*, **97**, B8, 11,703–11,728, 1992.

Zuber, M. T., Smith, D. E., Lemoine, F. G., and Neumann, G. A., The shape and internal structure of the Moon from the Clementine mission, *Science*, **266**, 1839–1843, 1994.

Zuber, M. T., and 20 others, Observations of the north polar region of Mars from the Mars Orbiter Laser Altimeter, *Science*, **282**, 2053–2060, 1998.

Zuber, M. T., and 14 others, Internal structure and early thermal evolution of Mars from Mars Global Surveyor topography and gravity, *Science*, **287**, 1788–1793, 2000.

Chapter 3

Acuña, M. H., and 19 others, Magnetic field and plasma observations at Mars: Initial results of the Mars Global Surveyor Mission, *Science*, **279**, 1676–1680, 1998.

Arkani-Hamed, J., and Strangway, D. W., An interpretation of magnetic signatures of subduction zones detected by Magsat, *Tectonophysics*, **133**, 45–55, 1987.

Arkani-Hamed, J., Urquhart, W. E. S., and Strangway, D. W., Scalar magnetic anomalies of Canada and northern United States derived from Magsat data, *Journal of Geophysical Research*, **90**, B3, 2599–2608, 1985.

Arkani-Hamed, J., Langel, R. A., and Purucker, M., Scalar magnetic anomaly maps of Earth derived from POGO and Magsat data, *Journal of Geophysical Research*, **98**, 24,075–24,090, 1994.

Bagenal, F., Introduction to the special section: Magnetospheres of the outer planets, *Journal of Geophysical Research*, **103**, E9, 19,841–19,842, 1998.

Baldwin, R. T., and Frey, H. V., Defining the crustal fields in Magsat data over Africa (Abstract), *EOS, Transactions, American Geophysical Union*, **69**, 336, 1988.

Bates, C. C., Gaskell, T. F., and Rice, R. B., *Geophysics in the Affairs of Man*, Pergamon Press, New York, 1982.

Bloxham, J., The steady part of the secular variation of the Earth's magnetic field, *Journal of Geophysical Research*, **97**, B13, 19,565–19,579, 1992.

Bloxham, J., and Jackson, A., Time-dependent mapping of the magnetic field at the core–mantle boundary, *Journal of Geophysical Research*, **97**, B13, 19,537–19,563, 1992.

Bullard, E., The interior of the Earth, in *The Earth as a Planet*, G. P. Kuiper, Editor, University of Chicago Press, Chicago, pp. 57–137, 1954.

Butler, R. F., *Paleomagnetism*, Blackwell Scientific Publications, Boston, 1992.

Cain, J. C., Structure and secular change of the geomagnetic field, *Reviews of Geophysics and Space Physics*, **13**, 203–206, 1975.

Cain, J. C., Schmitz, D. R., and Muth, L., Small-scale features in the Earth's magnetic field observed by Magsat, *Journal of Geophysical Research*, **89**, B2, 1070–1076, 1984.

Cisowski, S. M., and Fuller, M., The effect of shock on the magnetism of terrestrial rocks, *Journal of Geophysical Research*, **83**, 3441–3458, 1978.

Clark, S. C., Frey, H., and Thomas, H. H., Satellite magnetic anomalies over subduction zones: The Aleutian Arc anomaly, *Geophysical Research Letters*, **12**, 41–44, 1985.

Cohen, Y., and Achache, J., New global vector magnetic anomaly maps derived from satellite data, *Journal of Geophysical Research*, **95**, B12, 10,783–10,800, 1990.

Coles, R. L., Magsat scalar magnetic anomalies at northern high latitudes, *Journal of Geophysical Research*, **90**, B3, 2576–2582, 1985.

Connerney, J. E. P., The magnetospheres of Jupiter, Saturn, and Uranus, *Reviews of Geophysics*, **25**, 615–638, 1987.

Connerney, J. E. P., and 9 others, Magnetic lineations in the ancient crust of Mars, *Science*, **284**, 794–798, 1999.

Crary, F. J., and Bagenal, F., Remanent ferromagnetism and the interior structure of Ganymede, *Journal of Geophysical Research*, **103**, E11, 25,757–26,773, 1998.

Dyal, P., Planetary Magnetism, Origin, in *The Astronomy and Astrophysics Encyclopedia*, S. P. Maran, Editor, Van Nostrand Reinhold, New York, 1992.

Dyson, F., *Infinite in All Directions*, Princeton University Press, 1989.

Einstein, A., On the electrodynamics of moving bodies (original in German), *Annalen der Physik*, **17**, 1905.

Frey, H. V., Magsat scalar anomalies and major tectonic boundaries in Asia, *Geophysical Research Letters*, **16**, 299–302, 1982.

Frey, H. V., Magsat and POGO magnetic anomalies over the Lord Howe Rise: Evidence against a simple continental crustal structure, *Journal of Geophysical Research*, **90**, B3, 2631–2639, 1985.

Frey, H. V., Langel, R., Mead, G., and Brown, K., POGO and Paneaea, *Tectonophysics*, **95**, 181–189, 1983.

Fullerton, L. G., Frey, H. V., Roark, J. H., and Thomas, H. H., Evidence for a remanent contribution in Magsat data from the Cretaceous Quiet Zone in the South Atlantic, *Geophysical Research Letters*, **16**, 1085–1088, 1989.

Girdler, R. W., Taylor, P. T., and Frawley, J. J., A possible impact origin for the Bangui magnetic anomaly (Central Africa), *Tectonophysics*, **212**, 45–48, 1992.

Grant, A. C., Intracratonic tectonism: key to the mechanism of diastrophism, in *New Concepts in Global Tectonics*, S. Chatterjee and N. Hotton III, Editors, Texas Tech University Press, Lubbock, pp. 65–73, 1992.

Grieve, R. A. F., Stöffler, D., and Deutsch, A., The Sudbury Structure: Controversial or misunderstood, *Journal of Geophysical Research*, **96**, E5, 22,753–22,764, 1991.

Haggerty, S. E., and Toft, P. B., Native iron in the continental lower crust: Petrological and geophysical implications, *Science*, **229**, 647–649, 1985.

Heirtzler, J. R., The change in the magnetic-anomaly pattern at the ocean–continent boundary, in *The Utility of Regional Gravity and Magnetic Anomaly Maps*, W. J. Hinze, Editor, Society of Exploration Geophysicists, Tulsa, OK, pp. 339–346, 1985.

Heirtzler, J. R., *The Geomagnetic Field and Radiation in Near-Earth Orbits*, Technical Memorandum 1999–22094881, 36 pp., Goddard Space Flight Center, Greenbelt, MD, 1999.

Hinze, W. J., Von Frese, R. R. B., Longacre, M. B., Braile, L. W., Lidiak, E. G., and Keller, G. R., Regional magnetic and gravity anomalies of South America, *Geophysical Research Letters*, **9**, 314–317, 1982.

Hood, L. L., and Hartdegen, K., A crustal magnetization model for the

magnetic field of Mars: A preliminary study of the Tharsis Region, *Geophysical Research Letters*, **24**, 727–730, 1997.
Hood, L. L., Russell, C. T., and Coleman, P. J., Jr., Contour maps of lunar remanent magnetic fields, *Journal of Geophysical Research*, **86**, B2, 1055–1069, 1981.
Hood, P. J., McGrath, P. H., and Teskey, D. J., Evolution of Geological Survey of Canada magnetic-anomaly maps: A Canadian perspective, in *The Utility of Regional Gravity and Magnetic Anomaly Maps*, W. J. Hinze, Editor, Society of Exploration Geophysicists, Tulsa, OK, pp. 62–68, 1985.
Jacobs, J. A., *Deep Interior of the Earth*, Chapman & Hall, London, 1992.
Jeanloz, R., The Earth's core, *Scientific American*, **249**, 59–65, 1983.
Kaula, W. M., The shape of the Earth, in *Introduction to Space Science* (Seond Edition), W. N. Hess and G. D. Mead, Gordon and Breach, New York, pp. 315–339, 1968.
Khurana, K. K., and 6 others, Induced magnetic fields as evidence for subsurface oceans in Europa and Callisto, *Nature*, **395**, 777–780, 1998.
Kivelson, M. G., Serendipitous science from flybys of secondary targets: Galileo at Venus, Earth, and asteroids; Ulysses at Jupiter, US National Report to IUGG, *Reviews of Geophysics*, **33**, Supplement, 1995.
Kivelson, M. G., Discovery of Ganymede's magnetic field by the Galileo spacecraft, *Nature*, **384**, 537–541, 1996.
Kivelson, M. G., Khurana, K. K., Means, J. D., Russell, C. T., and Snare, R. C., The Galileo magnetic field investigation, *Space Science Reviews*, **60**, 357–383, 1992.
Kletetschka, G., Wasilewski, P. J., and Taylor, P. T., Mineralogy of the sources for the magnetic anomalies on Mars, *Meteoritics and Planetary Science*, Sept. 2000.
LaBrecque, J. L., and Raymond, C. A., Seafloor spreading anomalies in the Magsat field of the North Atlantic, *Journal of Geophysical Research*, **90**, B3, 2565–2575, 1985.
LaBrecque, J. L., Cande, S. C., and Jarrard, R. D., Intermediate-wavelength magnetic anomaly field of the North Pacific and possible source distributions, *Journal of Geophysical Research*, **90**, B3, 2549–2564, 1985.
Langel, R. A., Introduction to the special issue: A perspective on Magsat results, *Journal of Geophysical Research*, **90**, B3, 2441–2444, 1985.
Langel, R. A., The main geomagnetic field, in *Geomagnetism*, J. A. Jacobs, Editor, Academic Press, New York, pp. 249–512, 1987.
Langel, R. A., Study of the crust and mantle using magnetic surveys by Magsat and other satellites, *Proceedings of the Indian Academy of Sciences* (*Earth and Planetary Sciences*), **99**, 581–618, 1990a.
Langel, R. A., Global magnetic anomaly maps derived from POGO spacecraft data, *Physics of the Earth and Planetary Interiors*, **62**, 208–602, 1990b.

Langel, R. A., International Geophysical Reference Field: The sixth generation, *Journal of Geomagnetism and Geoelectricity*, **44**, 679–707, 1992.

Langel, R. A., The use of low altitude satellite data bases for modeling of core and crustal fields and the separation of external and internal fields, *Surveys in Geophysics*, **14**, 31–87, 1993.

Langel, R. A., and Hinze, W. J., *The Magnetic Field of the Earth's Lithosphere: The Satellite Perspective*, Cambridge University Press, Cambridge, UK, 1998.

Lanzerotti, L. J., Langel, R. A., and Chave, A. D., Geoelectromagnetism, in *Encyclopedia of Applied Physics*, V. 2, VCH Publsihers, Inc., pp. 109–123, 1993.

Levy, E. H., Generation of planetary magnetic fields, *Annual Review of Earth and Planetary Sciences*, **4**, 159–185, 1976.

Lin, R. P., and 8 others, Lunar surface magnetic fields and their interaction with the solar wind: Results from Lunar Prospector, *Science*, **281**, 1480–1484, 1998.

Lowman, P. D., Jr., Global tectonic and volcanic activity maps, in *A Geophysical Atlas for Interpretation of Satellite-derived Data*, NASA Technical Memorandum 79722, P. D. Lowman, Jr., and H. V. Frey, Editors, Goddard Space Flight Center, Greenbelt, MD, pp. 35–42, 1979.

Lowman, P. D., Jr., Formation of the earliest continental crust: Inferences from the Scourian Complex of northwest Scotland and geophysical models of the lower continental crust, *Precambrian Research*, **24**, 199–215, 1984.

Lowman, P. D., Jr., Comparative planetology and the origin of continental crust, *Precambrian Research*, **44**, 171–195, 1989.

Lowman, P. D., Jr., The Sudbury Structure as a terrestrial mare basin, *Reviews of Geophysics*, **30**, 227–243, 1992.

Lowman, P. D., Jr., Harris, J., Masuoka, P. M., Singhroy, V. H., and Slaney, V. R., Shuttle Imaging Radar (SIR-B) investigations of the Canadian Shield: Initial report, *IEEE Transactions on Geoscience and Remote Sensing*, **GE-25**, 55–66, 1987.

Marcus, A., *Electricity for Technicians*, Prentice-Hall, Englewood Cliffs, NJ, 1968.

Mayhew, M. A., Curie isotherm surfaces inferred from high-altitude satellite magnetic anomalies, *Journal of Geophysical Research*, **90**, B3, 2647–2654, 1985.

Mayhew, M. A., and Galliher, S. C., An equivalent source layer magnetization model for the United States derived from Magsat data, *Geophysical Research Letters*, **9**, 311–313, 1982.

Mayhew, M. A., Thomas, H. H., and Wasilewski, P. J., Satellite and surface geophysical expression of anomalous crustal structure in Kentucky and Tennessee, *Earth and Planetary Science Letters*, **58**, 395–405, 1982.

Mayhew, M. A., Johnson, B. D., and Wasilewski, P. J., A review of problems and progress in studies of satellite magnetic anomalies, *Journal of Geophysical Research*, **90**, B3, 2511–2522, 1985.

Meyer, J., Hufen, J-H., and Siebert, M., On the identification of Magsat anomaly charts as crustal part of the internal field, *Journal of Geophysical Research*, **90**, B3, 2537–2541, 1985.

Morley, L. W., and Larochelle, A., Paleomagnetism as a means of dating geological events, *Royal Society of Canada Special Publication 8*, 39–50, 1964.

Ness, N. F., The interplanetary medium, in *Introduction to Space Science* (Second Edition), W. N. Hess and G. D. Mead, Editors, Gordon and Breach, New York, pp. 345–371, 1968.

Ness, N. F., The magnetic fields of Mercury, Mars, and Moon, *Annual Review of Earth and Planetary Science*, **7**, 249–288, 1979.

Ness, N. F., Behannon, K. W., Lepping, R. P., Whang, Y. C., Schatten, K. H., Magnetic field observations near Mercury: Preliminary results from Mariner 10, *Science*, **185**, 151–160, 1974.

Olsen, N., and 25 others, Oersted initial field model, *Geophysical Research Letters*, **27**, 3607–3610, 2000.

Pierce, J. R., *Electrons, Waves, and Messages*, Hanover House, Garden City, New York, 1956.

Pilkington, M., and Grieve, R. A. F., The geophysical signature of terrestrial impact craters, *Reviews of Geophysics*, **30**, 161–181, 1992.

Pohl, J., On the origin of the magnetization of impact breccias on Earth, *Zeitschrift für Geophysik*, **37**, 549–555, 1971.

Purucker, M. E., and Clark, D., Exploration geophysics on Mars: Lessons from magnetics, *The Leading Edge*, pp. 484–487, March 2000.

Purucker, M. E., and Dyment, J., Satellite magnetic anomalies related to seafloor spreading in the South Atlantic Ocean, *Geophysical Research Letters*, **27**, 2765–2768, 2000.

Purucker, M. E., Langel, R. A., Rajaram, M., and Raymond, C., Global magnetization models with a priori information, *Journal of Geophysical Research*, **103**, B2, 2563–2584, 1998.

Purucker, M. E., Ravat, D., Frey, H. V., Voorhies, C., Sabaka, T., and Acuna, M., An altitude-normalized magnetic map of Mars and its interpretation, *Geophysical Research Letters*, **27**, 2449–2452, 2000.

Raff, A. D., and Mason, R. G., Magnetic survey off the west coast of N. America, 40 deg. N to 52 deg. N latitude, *Geological Society of America Bulletin*, **72**, 1267–1270, 1961.

Ratcliffe, J. A., The development of electricity, in *A Short History of Science*, BBC Radio Symposium, Doubleday Anchor Books, Garden City, New York, pp. 110–117, 1951.

Ravat, D. N., Hinze, W. J., and Taylor, P. T., European tectonic features observed by Magsat, *Tectonophysics*, **220**, 157–173, 1993.

Ravat, D. N., Langel, R. A., Purucker, M. E., Arkani-Hamed, J., and Alsdorf, D. E., Global vector and scalar Magsat magnetic anomaly maps, *Journal of Geophysical Research*, **100**, B10, 20,111–20,136, 1995.

Regan, R. D., and Marsh, B. D., The Bangui magnetic anomaly: Its geological origin, *Journal of Geophysical Research*, **87**, B2, 1107–1120, 1982.

Regan, R. D., Cain, J. C., and Davis, W. M., A global magnetic anomaly map, *Journal of Geophysical Research*, **80**, 794–802, 1975.

Rudnick, R. L., Restites, Eu anomalies, and the lower continental crust, *Geochimica et Cosmochimica Acta*, **56**, 963–970, 1992.

Runcorn, S. K., On the interpretation of lunar magnetism, *Physics of the Earth and Planetary Interiors*, **10**, 327–335, 1975.

Russell, C. T., The magnetosphere, *Annual Review of Earth and Planetary Science*, **19**, 169–182, 1991.

Russell, C. T., and Luhmann, J. G., Venus, magnetic fields, in *The Astronomy and Astrophysics Encyclopedia*, S. P. Maran, Editor, Van Nostrand Reinhold, New York, pp. 948–950, 1992.

Russell, C. T., Elphic, R. C., and Slavin, J. A., Initial Pioneer Venus magnetic field results: Nightside observations, *Science*, **205**, 114–116, 1979.

Schnetzler, C. C., An estimation of continental crust magnetization and susceptibility from Magsat data in the conterminous United States, *Journal of Geophysical Research*, **90**, B3, 2617–2620, 1985.

Schnetzler, C. C., Satellite measurements of the Earth's crustal magnetic field, *Advances in Space Research*, **9**, (1) 5–12, 1989.

Shive, P. N., Can remanent magnetization in the deep crust contribute to long-wavelength magnetic anomalies? *Geophysical Research Letters*, **16**, 89–92, 1989.

Stacey, F. D., Paleomagnetism of meteorites, *Annual Review of Earth and Planetary Science*, **4**, 147–157, 1976.

Stern, D. P., A brief history of magnetospheric physics before the spaceflight era, *Reviews of Geophysics*, **27**, 103–114, 1989.

Sykes, L. R., Intraplate seismicity, reactivation of pre-existing zones of weakness, alkaline magmatism, and other tectonism postdating continental fragmentation, *Reviews of Geophysics and Space Physics*, **16**, 621–688, 1978.

Taylor, P. T., Investigation of plate boundaries in the eastern Indian Ocean using Magsat data, *Tectonophysics*, **192**, 153–158, 1991.

Taylor, P. T., and Ravat, D., An interpretation of the Magsat anomalies of central Europe, *Applied Geophysics*, **34**, 83–91, 1995.

Taylor, P. T., Hinze, W. J., and Ravat, D. N., The search for crustal resources: Magsat and beyond, *Advances in Space Research*, **12**, 7, 5–15, 1992.

Thomas, H. H., Petrologic model of the northern Mississippi Embayment based on satellite magnetic and ground-based geophysical data, *Earth and Planetary Science Letters*, **70**, 115–121, 1984.

Thomas, H. H., A model of ocean basin crustal magnetization

appropriate for satellite elevation anomalies, *Journal of Geophysical Research*, **92**, B11, 11,609–11,613, 1987.

Toft, P. B., Taylor, P. T., Arkani-Hamed, J., and Haggerty, S. E., Interpretation of satellite magnetic anomalies over the West African Craton, *Tectonophysics*, **212**, 21–32, 1992.

Van Allen, J. A., Why radiation belts exist, *EOS*, *Transactions*, *American Geophysical Union*, **72**, 361–363, 1991.

Vestine, E. H., Laporte, L., Lange, I., and Scott, W. E., *The Geomagnetic Field: Its Description and Analysis*, Carnegie Institution of Washington Publication No. 580, Washington, 1947.

Vine, F. J., and Matthews, D. H., Magnetic anomalies over oceanic ridges, *Nature*, **199**, 947–949, 1963.

Voorhies, C. V., and Benton, E. R., Pole-strength of the Earth from Magsat and magnetic determination of the core radius, *Geophysical Research Letters*, **9**, 258–261, 1982.

Warner, R. D., and Wasilewski, P. J., Magnetic petrology of arc xenoliths from Japan and Aleutian Islands, *Journal of Geophysical Research*, **102**, 22,0225–20,243, 1997.

Wasilewski, P. J., and Fountain, D. M., The Ivrea Zone as a model for the distribution of magnetization in the continental crust, *Geophysical Research Letters*, **9**, 333–336, 1982.

Wasilewski, P. J., and Mayhew, M. A., The Moho as a magnetic boundary revisited, *Geophysical Research Letters*, **19**, 2259–2262, 1992.

Wasilewski, P. J., Thomas, H. H., and Mayhew, M. A., The Moho as a magnetic boundary, *Geophysical Research Letters*, **6**, 541–544, 1979.

Wyllie, P. J., *The Way the Earth Works*, John Wiley and Sons, New York, 1976.

Young, R. E., The Galileo probe mission to Jupiter: Science overview, *Journal of Geophysical Research*, **103**, E10, 22,775–22,790, 1998.

Zietz, I., Andreasen, G. E., and Cain, J. C., Magnetic anomalies from satellite magnetometer, *Journal of Geophysical Research*, **75**, 4007–4015, 1970.

Chapter 4

Abrams, M. J., Ashley, R. P., Rowan, L. C., Goetz, A. F. H., and Kahle, A. B., Mapping of hydrothermal alteration in the Cuprite mining district, Nevada, using aircraft scanner images for the spectral region 0.46 to 2.36 micrometers, *Geology*, **5**, 713–718, 1977.

Agar, R. A., and Villaneuva, R., Satellite, airborne, and ground spectral data applied to mineral exploration in Peru, *Proceedings of the Twelfth International Conference on Applied Geologic Remote Sensing*, I-13–20, 1997.

Allen, M. B., Windley, B. F., Chi, Z., and Jinghui, G., Evolution of the Turfan Basin, Chinese central Asia, *Tectonics*, **12**, 889–896, 1993.

Amsbury, D. L., United States manned observations of Earth before the Space Shuttle, *Geocarto International*, **4**, 7–14, 1989.

Apt. J., Helfert, M., and Wilkinson, J., *Orbit*, National Geographic Society, Washington, DC, 1996.

Atwater, T., Implications of plate tectonics for the Cenozoic tectonic evolution of North America, *Geological Society of America Bulletin*, **81**, 3513–3536, 1970.

Bates, R. L., and Jackson, J. A., *Glossary of Geology* (Second Edition), American Geological Institute, Falls Church, 1980.

Best, M. G., *Igneous and Metamorphic Petrology*, W. H. Freeman, San Francisco, CA, 1982.

Bhattacharya, A., Reddy, C. S. S., and Srivastava, S. K., Remote sensing for active volcano monitoring in Barren Island, India, *Photogrammetric Engineering & Remote Sensing*, **59**, 1293–1297, 1993.

Burchfiel, B. C., Plate tectonics and the continents: a review, in *Continental Tectonics, Studies in Geophysics*, National Academy of Sciences, Washington DC, pp. 15–25, 1980.

Cary, T., Visions of the information industry – dreams or nightmares?, *Photogrammetric Engineering & Remote Sensing*, **63**, 767–775, 1997.

Casadevall, T. J., *Volcanic Ash and Aviation Safety*, US Geological Survey Bulletin 2047, 450 pp., 1994.

Choudhury, B. J., Desertification, in *Atlas of Satellite Observations Related to Global Change*, R. J. Gurney, J. L. Foster, and C. L. Parkinson, Editors, Cambridge University Press, Cambridge, UK, pp. 313–326, 1993.

Coleman, J. M., Roberts, H. H., and Huh, O. K., Deltaic landforms, in *Geomorphology from Space*, N. M. Short and R. W. Blair, Editors, Special Publication 486, National Aeronautics and Space Adminstration, Washington, DC, pp. 317–352, 1986.

Davis, G. H., *Structural Geology of Rocks and Regions*, John Wiley, New York, 1984.

De Silva, S. L., and Francis, P. W., *Volcanos of the Central Andes*, Springer-Verlag, Berlin, 1991.

Dean, K., Servilla, M., Roach, A., Foster, B., and Engle, K., Satellite monitoring of remote volcanoes improves study efforts in Alaska, *EOS, Transactions, American Geophysical Union*, **79**, 413–423, 1998.

Dewey, J. F., and Bird, J. M., Mountain belts and the new global tectonics, *Journal of Geophysical Research*, **75**, 2624–2647, 1970.

Donovan, T. J., Friedman, I., and Gleason, J. D., Recognition of petroleum-bearing traps by unusual isotopic compositions of carbonate-cemented surface rocks, *Geology*, **2**, 351–354, 1974.

Dwivedi, R. S., Remote sensing in desertification studies: How to monitor and assess desertification, in *Applications of Remote Sensing in Asia and Oceania*, S. Murai, Editor, Asian Association on Remote Sensing, Tokyo, pp. 180–194, 1991.

Elachi, C., and 15 others, Shuttle imaging radar experiment, *Science*, **218**, 996–1003, 1992.

Engelder, T., Is there a genetic relationship between selected regional joints and contemporary stress within the lithosphere of North America?, *Tectonics*, **1**, 161–177, 1982.

Estes, J. F., and Senger, L. W., *Remote Sensing: Techniques for Environmental Analysis*, Hamilton Publishing Company, Santa Barbara, 1974.

Estes, J. E., Crippen, R. E., and Star, J. L., Natural oil seep detection in the Santa Barbara Channel, California, with Shuttle Imaging Radar, *Geology*, **13**, 282–284, 1985.

Fahrig, W. F., and West, T. D., Diabase dyke swarms of the Canadian Shield, *Geological Survey of Canada Map 1627A*, 1986.

Francis, P. W., Remote sensing of volcanos, *Advances in Space Research*, **1** (1), 89–92, 1989.

Gill, J. R., and Gerathewohl, S. J., The Gemini science program, *Astronautics and Aeronautics*, Nov., 58–65, 1964.

Goetz, A. F. H., Rock, B. N., and Rowan, L. C., Remote sensing for exploration: An overview, *Economic Geology*, **78**, 573–590, 1983.

Goward, S. N., and Williams, D. L., Landsat and earth systems science: Development of terrestrial monitoring, *Photogrammetric Engineering & Remote Sensing*, **63**, 887–900, 1997.

Groth, F. H., and Rivera, H. P., Primary and secondary impacts associated with colonization and oil exploration in the Amazon rain forest, *Proceedings of the Twelfth International Conference on Applied Geologic Remote Sensing*, I-59–66, 1997.

Halbouty, M. T., Application of Landsat imagery to petroleum and mineral exploration, *American Association of Petroleum Geologists Bulletin*, **60**, 745–793, 1976.

Hall, D. K., Williams, R. S., Jr., and Bayr, K. J., Glacier recession in Iceland and Austria, *EOS, Transactions, American Geophysical Union*, **72**, 129–135, 141, 1992.

Hall, D. K., Benson, C. S., and Field, W. O., Changes of glaciers in Glacier Bay, Alaska, using ground and satellite measurements, *Physical Geography*, **16**, 27–41, 1995.

Hancock, P. L., Determining contemporary stress directions from neotectonic joint systems, *Philosophical Transactions, Royal Society of London*, **337**, 29–40, 1991.

Harris, J. L., and 11 others, Real-time satellite monitoring of volcanic hot spots, in *Remote Sensing of Active Volcanism*, P. J. Mouginis-Mark, J. A. Crisp, and J. H. Fink, Editors, *Geophysical Monograph 116*, American Geophysical Union, pp. 139–159, 2000.

Harris, J. R., Murray, R., and Hirose, T., IHS transform for the integration of radar imagery with other remotely sensed data, *Photogrammetric Engineering & Remote Sensing*, **56**, 1631–1641, 1990.

Hill, J., Megier, J., and Mehl, W., Land degradation, soil erosion, and desertification monitoring in Mediterranean ecosystems, *Remote Sensing Reviews*, **12**, 107–133, 1995.

Hobbs, W. H., Repeating patterns in the relief in the structure of the land, *Geological Society of America Bulletin*, **22**, 123–176, 1911.

Hodgson, R. A., Review of significant early studies in lineament tectonics, in *Proceedings of the First International Conference on the New Basement Tectonics, Utah Geological Association Publication No. 5*, Utah Geological Association, Salt Lake City, pp. 1–10, 1976.

Hodgson, R. A., Gay, S. P., Jr., and Benjamins, J. Y. (Editors), *Proceedings of the First International Conference on the New Basement Tectonics, Utah Geological Association Publication No. 5*, Utah Geological Association, Salt Lake City, 1976.

Huang, T. K. (Editor), *Geotectonic Evolution of China*, Springer-Verlag, Berlin, 1987.

Kasturirangan, K., The evolution of satellite-based remote sensing capabilities in India, *International Journal of Remote Sensing*, **6**, 388–400, 1985.

Krueger, A. J., Schaefer, S. J., Krotkov, N., Bluth, G., and Barker, S., Ultraviolet remote sensing of volcanic emissions, in *Remote Sensing of Active Volcanism*, P. J. Mouginis-Mark, J. A. Crisp, and J. H. Fink, Editors, Geophysical Monograph 116, American Geophysical Union, pp. 25–43, 2000.

Kumarapeli, P. D., and Saull, V. A., The St. Lawrence Valley System: A North American equivalent of the East African Rift Valley System, *Canadian Journal of Earth Sciences*, **3**, 639–658, 1966.

Kusaka, T., Ootuka, M., and Shikada, M., Estimation of landslide areas using the multiple regression analysis, *Proceedings of the Twelfth International Conference on Applied Geologic Remote Sensing*, II-4–9, 1997.

Kusky, T. M., Lowman, P. D., Jr., Masuoka, P., and Blodget, H. W., Analysis of Seasat L-band radar imagery of the West Bay–Indian Lake Fault System, Northwest Territories, *Journal of Geology*, **101**, 623–632, 1993.

Kutina, J., and Hildenbrand, T. G., Ore deposits of the western United States in relation to mass distribution in the crust and mantle, *Geological Society of America Bulletin,* **99**, 30–41, 1987.

Lowman, P. D., Jr., Space photography – a review, *Photogrammetric Engineering*, January, 76–86, 1965.

Lowman, P. D., Jr., Geologic orbital photography: Experience from the Gemini Program, *Photogrametria*, **24**, 77–106, 1969.

Lowman, P. D., Jr., Vertical displacement of the Elsinore Fault of southern California, *Journal of Geology*, **88**, 414–432, 1980.

Lowman, P. D., Jr., A global tectonic activity map, *Bulletin of the International Association of Engineering Geology*, **23**, 257–272, 1981.

Lowman, P. D., Jr., A more realistic view of global tectonism, *Journal of Geological Education*, **30**, 97–107, 1982.

Lowman, P. D., Jr., Radar geology of the Canadian Shield: A 10-year review, *Canadian Journal of Remote Sensing*, **20**, 198–209, 1994.

Lowman, P. D., Jr., T plus twenty-five years: A defense of the Apollo Program, *Journal of the British Interplanetary Society*, **49**, 71–79, 1996.

Lowman, P. D., Jr., Apollo and Landsat: The forgotten legacy, *Photogrammetric Engineering & Remote Sensing*, **65**, 1143–1147, 1999.

Lowman, P. D., Jr., and Tiedemann, H. A., *Terrain Photography from Gemini Spacecraft: Final Geologic Report*, Goddard Space Flight Center, X-644–71–15, 75 pp., 1971.

Lowman, P. D., Jr., McDivitt, J. A., and White, E. H., II, *Terrain Photography on the Gemini IV Mission: Preliminary Report*, NASA TN D-3982, 15 pp., 1966.

Lowman, P. D., Jr., Whiting, P. J., Short, N. M., Lohmann, A. M., and Lee, G., Fracture patterns on the Canadian Shield: A lineament study with Landsat and orbital radar imagery, *Basement Tectonics*, **7**, 139–159, 1992.

Lu, Z., Mann, D., and Freymueller, J., Satellite radar interferometry measures deformation at Okmok volcano, *EOS, Transactions, American Geophysical Union*, **79**, 461–468, 1998.

Helfert, M. R., and Lulla, K. P., Analysis of seasonal characteristics of Sambhar Salt Lake, India, from digitized Space Shuttle photography, *Geocarto International*, **4**, 69–74, 1989.

Lulla, K., Helfert, M., Evans, C., Wilkinson, M. J., Pitts, D., and Amsbury, D., Global geologic applications of the Space Shuttle Earth Observations Photography database, *Photogrammetric Engineering & Remote Sensing*, **LIX**, 1225–1231, 1993.

McKee, E. D. (Editor), *A Study of Global Sand Seas*, Professional Paper 1052, US Geological Survey, Washington, DC, 429 pp., 1979.

McKelvey, V. E., *Subsea Mineral Resources*, US Geological Survey Bulletin 1689-A, 106 pp., 1986.

McKenzie, D. P., Speculations on the consequences and causes of plate motions, *Geophysical Journal of the Royal Astronomical Society*, **18**, 1–32, 1969.

Mahmood, A., Crawford, J. P., Michaud, R., and Jezek, K. C., Mapping the world with remote sensing, *EOS, Transactions, American Geophysical Union*, **79**, 17–23, 1998.

Massonet, D., and 6 others, The displacement field of the Landers earthquake mapped by radar interferometry, *Nature*, **364**, 138–142, 1993a.

Massonet, D., Briole, P., and Arnaud, A., Deflation of Mount Etna monitored by spaceborne radar interferometry, *Nature*, **375**, 567–570, 1993b.

Massonet, D., and Sigmundsson, F., Remote sensing of volcano deformation by radar interferometry from various satellites, in *Remote Sensing of Active Volcanism*, P. J. Mouginis-Mark,

J. A. Crisp, and J. H. Fink, Editors, *Geophysical Monograph 116*, American Geophysical Union, pp. 207–221, 2000.

Merifield, P. M., and 7 others, Satellite imagery of the Earth, *Photogrammetric Engineering*, **XXXV**, 654–668, 1969.

Mollard, J. D., Fracture lineament research and application on the western Canadian plains, *Canadian Geotechnical Journal*, **25**, 749–767, 1988.

Molnar, P., and Tapponier, P., Cenozoic tectonics of Asia: Effects of a continental collision, *Science*, **189**, 419–426, 1975.

Mouginis-Mark, P. J., Pieri, D. C., and Francis, P. W., Volcanos, in *Atlas of Satellite Observations Related to Global Change*, R. J. Gurney, J. L. Foster, and C. L. Parkinson, Editors, Cambridge University Press, Cambridge, UK, pp. 341–357, 1993.

Mouginis-Mark, P. J., Crisp, J. A., and Fink, J. H., *Remote Sensing of Active Volcanism, Geophysical Monograph 116*, American Geophysical Union, Washington, DC, 272 pp., 2000.

Murai, S. (Editor), *Applications of Remote Sensing in Asia and Oceania*, Asian Association on Remote Sensing, Tokyo, 372 pp., 1991.

Nagaraja, R., Gautam, N. C., and Rao, D. P., Role of remote sensing for inventory of wastelands in India, in *Applications of Remote Sensing in Asia and Oceania*, S. Murai, Editor, Asian Association on Remote Sensing, Tokyo, pp. 86–99, 1991.

Naisbitt, J., *Megatrends: Ten New Directions Transforming Our Lives*, Warner Books, New York, 1984.

Ni, J., and York, J. E., Late Cenozoic tectonics of the Tibetan Plateau, *Journal of Geophysical Research*, **83**, 5377–5384, 1978.

Nickelsen, R. P., "New basement tectonics" evaluated in Salt Lake City, *Geotimes*, October, pp. 16–17, 1975.

Nur, A., The origin of tensile fracture lineaments, *Journal of Structural Geology*, **4**, 31–40, 1982.

O'Keefe, J. A., Dunkelman, L., Soules, S. D., Huch, W. F., and Lowman, P. D., Jr., Observations of space phenomena, in *Mercury Project Summary, Including Results of the Fourth Manned Orbital Flight*, Special Report 45, National Aeronautics and Space Administration, Washington, DC, 445 pp., 1963.

Onasch, C. M., and Kahle, C. F., Recurrent tectonics in a cratonic setting: An example from northwestern Ohio, *Geological Society of America Bulletin*, **103**, 1259–1269, 1991.

Otterman, J., Lowman, P. D., Jr., and Salomonson, V. V., Surveying earth resources by remote sensing from satellites, *Geophysical Surveys*, **2**, 431–467, 1976.

Pecora, W. T., Earth resource observations from an orbiting spacecraft, in *Manned Laboratories in Space*, S. F. Singer, Editor, Springer-Verlag, New York, pp. 75–87, 1969.

Peltier, W. R. (Editor), *Mantle Convection, Plate Tectonics, and Earth Dynamics*, Gordon and Breach, New York, 1989.

Penner, L. A., and Mollard, J. D., Correlated photolineament and geoscience data on eight petroleum and potash study projects in southern Saskatchewan, *Canadian Journal of Remote Sensing*, **17**, 174–184, 1991.

Quershy, M. N., and Hinze, W. J., *Regional Geophysical Lineaments, Memoir 12*, Geological Society of India, Bangalore, 305 pp., 1989.

Rose, W. I., and Schneider, D. J., Satellite images offer aircraft protection from volcanic ash clouds, *EOS, Transactions, American Geophysical Union*, **77**, 529–532, 1996.

Rothery, D. A., Francis, P. W., and Wood, C. A., Volcano monitoring using short wavelength infrared data from satellites, *Journal of Geophysical Research*, **93**, 7993–8008, 1988.

Rowan, L. C., Application of satellites to geologic exploration, *American Scientist*, **63**, 393–403, 1975.

Rowan, L. C., and Wetlaufer, P. H., *Geologic Evaluation of Major Landsat Lineaments in Nevada and Their Relationship to Ore Districts*, US Geological Survey Open File Report 79-544, 64 pp., 1979.

Rubincam, D. P., Information theory lateral density distribution for Earth inferred from global gravity field, *Journal of Geophysical Research*, **87**, B7, 5541–5552, 1982.

Sabins, F. F., *Remote Sensing: Principles and Interpretation* (Third Edition), W. H. Freeman, New York, 1997.

Sabins, F. F., Remote sensing for petroleum exploration, Part 1, *The Leading Edge*, 467–470, April, 1998, and Part 2, 623–626, May, 1998.

Salomonson, V. V., and Stuart, L., Thematic Mapper research in the earth sciences, *Remote Sensing of the Environment*, **28**, 5–7, 1989.

Schick, R., and Van Haaften, J., *The View from Space: American Astronaut Photography, 1962–1972*, Clarkson N. Potter, New York, 1988.

Schneider, D. J., Dean, K. G., Dehn, J., Miller, T. P., and Kirianov, V. Yu., Monitoring and analyses of volcanic activity using remote sensing data at the Alaska Volcano Observatory: Case study for Kamchatka, Russia, December 1997, in *Remote Sensing of Active Volcanism, Geophysical Monograph 116*, P. J. Mouginis-Mark, J. A. Crisp, and J. H. Fink, Editors, American Geophysical Union, pp. 65–85. 2000.

Schultejann, P. A., Structural trends in Borrego Valley, California: Interpretations from SIR-A and Seasat SAR, *Photogrammetric Engineering & Remote Sensing*, **51**, 1615–1624, 1985.

Searle, M. P., Geological evidence against large-scale pre-Holocene offsets along the Karakorum Fault: Implications for the limited extrusion of the Tibetan Plateau, *Tectonics*, **15**, 171–186, 1996.

Settle, M., and Taranik, J. V., Use of the Space Shuttle for remote sensing research: Recent results and future prospects, *Science*, **218**, 993–995, 1982.

Short, N. M., and Blair, R. W., Jr. (Editors), *Geomorphology from Space, Special Publication 486*, National Aeronautics and Space Administration, Washington, DC, 717 pp., 1986.

Shurr, G. W., Nelson, C. L., and Watkins, I. W., Recent tectonics in the northern Great Plains on Landsat and fractal dimensions of drainage networks, *Proceedings*, *Pecora 12 Symposium*, American Society for Photogrammetry and Remote Sensing, Bethesda, MD, pp. 89–97, 1994.

Singhroy, V. H., Remote sensing in global geoscience processes: Introductory remarks, *Episodes*, **15**, 3–6, 1992a.

Singhroy, V. H., Radar geology: Techniques and results, *Episodes*, **15**, 15–20, 1992b.

Singhroy, V. H., and Lowman, P. D., Jr., *Geology* (Chap. 8) in *Manual of Photographic Interpretation*, (Second Edition), W. R. Philipson, Editor, American Society for Photogrammetry and Remote Sensing, pp. 311–333, 1997.

Singhroy, V. H., Kenny, F. M., and Barnett, P. J., Radar imagery for Quaternary geological mapping in glaciated terrains, *Canadian Journal of Remote Sensing*, **18**, 112–117, 1992.

Singhroy, V. H., Slaney, R., Lowman, P., Harris, J., and Moon, W., Radarsat and radar geology in Canada, *Canadian Journal of Remote Sensing*, **19**, 338–351, 1993.

Sonder, R. A., Discussion of shear patterns of the Earth's crust, *Transactions, American Geophysical Union*, **28**, 939–945, 1947.

Strain, P., and Engle, F., *Looking at Earth*, Turner Publishing, Atlanta, 1993.

Supplee, C., Earth's high-tech checkup, *The Washington Post*, Aug. 20, 2000.

Townshend, J. R. G., Tucker, C. J., and Goward, S. N., Global vegetation mapping, in *Atlas of Satellite Observations Related to Global Change*, R. J. Gurney, J. L. Foster, and C. L. Parkinson, Editors, Cambridge University Press, Cambridge, UK, pp. 301–312, 1993.

Todd, V. R., and Hoggatt, W. C., Vertical tectonics in the Elsinore Fault Zone, *Geological Society of America, Annual Meeting, Abstracts with Program*, **11**, 528, 1989.

Vadon, H., and Sigmundsson, F., Crustal deformation from 1992 to 1995 at the Mid-Atlantic Ridge, Southwest Iceland, mapped by satellite radar interferometry, *Science*, **275**, 193–197, 1997.

Vincent, R. K., *Fundamentals of Geological and Environmental Remote Sensing*, Prentice Hall, Upper Saddle River, NJ, 1997.

Walker A., Aeolian landforms, in *Geomorphology from Space*, N. M. Short and R. W. Blair, Jr., Editors, *Special Publication 486*, National Aeronautics and Space Adminstration, Washington, DC, pp. 447–520, 1986.

White, K., Walden, J., Drake, N., Eckhardt, F., and Settle, J., Mapping the iron oxide contents of dune sands, Namib Sand Sea, Namibia, using Landsat Thematic Mapper data, *Remote Sensing of the Environment*, **62**, 30–39, 1997.

Williams, R. S., Jr., Glaciers and glacial landforms, in *Geomorphology from Space*, N. M. Short and R. W. Blair, Jr., Editors, *Special Publication 486*, National Aeronautics and Space Administration, Washington, DC, pp. 521–596, 1986.

Williams, R. S., Jr., Hall, D. K., Sigurdsson, O., and Chien, J. Y. L., Comparison of satellite-derived with ground-based measurements of the fluctuations of the margins of Vatnajökull, Iceland, 1973–1992, *Annals of Glaciology*, **24**, 72–80, 1997.

Wood, C., Geologic applications of Space Shuttle photography, *Geocarto International*, **4**, 49–54,1989.

Zebker, H. A., and Goldstein, R., Topographic mapping from interferometric SAR observations, *Journal of Geophysical Research*, **91**, 4993–4999, 1986.

Zebker, H. A., Amelung, F., and Jonsson, S., Remote sensing of volcano surface and internal processes using radar interferometry, in *Remote Sensing of Active Volcanism*, P. J. Mouginis-Mark, J. A. Crisp, and J. H. Fink, Editors, *Geophysical Monograph 116*, American Geophysical Union, pp. 179–205, 2000.

Zhang, Y. Q., Vergely, P., and Mercier, J., Active faulting in and along the Qinling Range (China) inferred from SPOT imagery analysis and extrusion tectonics of south China, *Tectonophysics*, **243**, 69–95, 1995.

Zhenda, Z., and Yimou, W., Desertification environment prediction through remote sensing technique, in *Applications of Remote Sensing in Asia and Oceania*, S. Murai, Editor, Asian Association on Remote Sensing, Tokyo, pp. 170–174, 1991.

Chapter 5

Alt, D., Sears, J. M., and Hyndman, D. W., Terrestrial maria: The origins of large basalt plateaus, hotspot tracks, and spreading ridges, *Journal of Geology*, **96**, 647–662, 1988.

Alvarez, L. W., Alvarez, W., Asaro, F., and Michel, H. V., Extraterrestrial cause for the Cretaceous–Tertiary extinction, *Science*, **208**, 1095–1108, 1980.

Alvarez, W., The gentle art of scientific trespassing, *GSA Today*, **1**, 29–34, 1991.

Baldwin, R. B., *The Face of the Moon*, University of Chicago Press, Chicago, 1949.

Baldwin, R. B., *The Measure of the Moon*, University of Chicago Press, Chicago, 1963.

Barnes, V. E., North American Tektites, in *University of Texas Publication 3945*, pp. 477–656, 1940.

Barnes, V. E., and Barnes, M. A., *Tektites, Benchmark Papers in Geology*, Dowden, Hutchinson, and Ross, Stroudsburg, PA, 1973.

Beals, C. S., Innes, M. J. S., and Rottenberg, J. A., Fossil meteorite craters, in *The Moon, Meteorites, and Comets*, B. M. Middlehurst and G. P. Kuiper, Editors, University of Chicago Press, Chicago, pp. 235–284, 1963.

Becker, L., Poreda, R. J., Hunt, A. G., Bunch, T. E., and Rampino, M., Impact event at the Permian–Triassic boundary: Evidence from extraterrestrial noble gases in fullerenes, *Science*, **291**, 1530–1533, 2001.

Bernal, J. D., *The Social Function of Science*, M.I.T. Press, Cambridge, MA, 1967 (original publication 1939).

Chao, E. C. T., Shock effects in certain rock-forming minerals, *Science*, **156**, 192–202, 1967.

Chao, E. C. T., Shoemaker, E. M., and Madsen, B. M., First natural occurrence of coesite, *Science*, **132**, 220–222, 1960.

Chapman, D. R., On the unity and origin of Australasian tektites, *Geochimica et Cosmochimica Acta*, **28**, 841–880, 1964.

Chyba, C., and Sagan, C., Endogenous production, exogenous delivery and impact-shock synthesis of organic molecules: An inventory for the origins of life, *Nature*, **355**, 125–132, 1992.

Cooper, H. F., Jr., A summary of explosion cratering phenomena relevant to meteor impact events, in *Impact and Explosion Cratering*, D. J. Roddy, R. O. Pepin, and R. B. Merrill, Editors, Pergamon Press, New York, pp. 11–44, 1977.

Coughlon, J. P., and Denney, P. P., The Ames Structure and other North American cryptoexplosion structures: Evidence for endogenic emplacement, in *Ames Structure in Northwest Oklahoma and Similar Features: Origin and Petroleum Production*, K. S. Johnson and J. A. Campbell, Editors, *Circular 100*, Oklahoma Geological Survey, pp. 133–152, 1997.

Dence, M. R., The extraterrestrial origin of Canadian craters, *Annals of the New York Academy of Sciences*, **123**, 941–969, 1965.

Dence, M. R., Impact melts, *Journal of Geophysical Research*, **76**, 5521–5565, 1971.

Dietz, R. S., The meteoritic impact origin of the Moon's surface features, *Journal of Geology*, **LIV**, 359–375, 1946.

Dietz, R. S., Meteorite impact suggested by shatter cones in rock, *Science*, **131**, 1781–1784, 1960.

Dietz, R. S., Sudbury structure as an astrobleme, *Journal of Geology*, **72**, 412–434, 1964.

Donofrio, R. R., Survey of hydrocarbon-producing impact structures in North America: Exploration results to date and potential for discovery in Precambrian basement rock, in *Ames Structure in Northwest Oklahoma and Similar Features: Origin and Petroleum Production*, K. S. Johnson and J. A. Campbell, Editors, *Circular 100*, Oklahoma Geological Survey, pp. 17–29, 1997.

Dressler, B. O., The effects of the Sudbury event and the intrusion of the Sudbury Igneous Complex on the footwall rocks of the Sudbury Structure, in *The Geology and Ore Deposits of the Sudbury Structure*, E. G. Pye, A. J. Naldrett, and P. E. Giblin, Editors, Ministry of Natural Resources, Toronto, pp. 97–136, 1984.

Dressler, B. O., Grieve, R. A. F., and Sharpton, V. L. (Editors), *Large Meteorite Impacts and Planetary Evolution, Special Paper 293*, Geological Society of America, 348 pp., 1994.

Flessa, K. W., The "facts" of mass extinctions, in *Global Catastrophes in Earth History*, V. L. Sharpton and P. D. Ward, Editors, *Special Paper 247, Geological Society of America*, 1–7, 1990.

Florentin, J-M., Maurrasse, R., and Sen, G., Impacts, tsunamis, and the Haitian Cretaceous–Tertiary boundary layer, *Science*, **252**, 1690–1693, 1991.

French, B. M., Sudbury Structure, Ontario: Some petrographic evidence for origin by meteoritic impact, *Science*, **156**, 1094–1098, 1967.

French, B. M., 25 years of the impact-volcanic controversy, *EOS, Transactions, American Geophysical Union*, **71**, 411–414, 1990.

French, B. M., *Traces of Catastrophe, Contribution No. 954*, Lunar and Planetary Institute, Houston, 120 pp., 1998.

French, B. M., and Short, N. M. (Editors), *Shock Metamorphism of Natural Materials*, Mono Press, Baltimore, 644 pp., 1967.

Frey, H. V., Origin of the Earth's ocean basins, *Icarus*, **32**, 235–250, 1977.

Frey, H.V., Crustal evolution of the early Earth: The role of major impacts, *Precambrian Research*, **10**, 195–216, 1980.

Galvin, C., The Great Dinosaur Extinction Controversy and the K–T research program in the late 20th century, Essay Review, *Earth Sciences History*, **17**, 41–55, 1998.

Garvin, J. B., Schntezler, C. C., and Grieve, R. A. F., Characteristics of large terrestrial impact structures as revealed by remote sensing studies, *Tectonophysics*, **216**, 45–62, 1992.

Gault, D. E., Quaide, W. L., and Oberbeck, V. R., Impact cratering mechanics and structures, in *Shock Metamorphism of Natural Materials*, B. M. French and N. M. Short, Editors, Mono Book Corp., Baltimore, pp. 87–99, 1966.

Gehrels, T., A proposal to the United Nations regarding the international discovery programs of near-earth asteroids, in *Near-Earth Objects: The United Nations International Conference*, J. L. Remo, Editor, *Annals of the New York Academy of Sciences*, **822**, 603–605, 1997.

Gilbert, G. K., The Moon's face, *Philosophical Society of Washington Proceedings*, **12**, 241–202, 1893.

Girdler, R. W., Taylor, P. T., and Frawley, J. J., A possible impact origin for the Bangui magnetic anomaly (Central Africa), *Tectonophysics*, **212**, 45–58, 1992.

Glass, B. P., *Introduction to Planetary Geology*, Cambridge University Press, Cambridge, UK, 1982.

Glass, B. P., Tektites and microtektites: Key facts and inferences, *Tectonophysics*, **171**, 393–404, 1990.

Glikson, A. Y., Asteroid/comet mega-impacts may have triggered major episodes of crustal evolution, *EOS, Transactions, American Geophysical Union*, **76**, 49–55, 1995.

Goodwin, A. M., Giant impacting and the development of continental crust, in *The Early History of the Earth*, B. F. Windley, Editor, John Wiley, New York, pp. 77–95, 1976.

Green, D. H., Archaean greenstone belts may include terrestrial equivalents of lunar maria?, *Earth and Planetary Science Letters*, **15**, 263–270, 1972.

Green, J., Copernicus as a lunar caldera, *Journal of Geophysical Research*, **76**, 5179–5732, 1971.

Grieve, R. A. F., Impact bombardment and its role in proto-continental growth on the early Earth, *Precambrian Research*, **10**, 217–247, 1980.

Grieve, R. A. F., Terrestrial impact: The record in the rocks, *Meteoritics*, **26**, 175–194, 1991.

Grieve, R. A. F., Terrestrial impact structures: Basic characteristics and economic significance, with emphasis on hydrocarbon production, in *Ames Structure in Northwest Oklahoma and Similar Feature: Origin and Petroleum Production*, K. S. Johnson and J. A. Campbell, Editors, *Circular 100*, Oklahoma Geological Survey, pp. 3–16, 1997.

Grieve, R. A. F., Stoffler, D., and Deutsch, A., The Sudbury Structure: Controversial or misunderstood, *Journal of Geophysical Research*, **96**, E5, 22,753–22,764, 1991.

Hamilton, W., Silicic differentiates of lopoliths, *International Geologic Congress, Report of 21st Session*, Norden, part 8, 59–67, 1960.

Hammel, H. B., and 14 others, HST imaging of atmospheric phenomena created by the impact of comet Shoemaker–Levy-9, *Science*, **267**, 1288–1296, 1995.

Hildenbrand, A. R., and 6 others, Chicxulub crater: A possible Cretaceous–Tertiary boundary impact crater on the Yucatan Peninsula, Mexico, *Geology*, **19**, 867–871, 1991.

Horz, F., Grieve, R. A. F., Heiken, G., Spudis, P., and Binder, A., Lunar surface processes, in *Lunar Sourcebook: A User's Guide to the Moon*, G. Heiken, D. Vaniman, and B. M. French, Editors, Cambridge University Press, Cambridge, UK, pp. 61–120, 1991.

Izett, G. A., Tektites in Cretaceous–Tertiary boundary rocks on Haiti and their bearing on the Alvarez impact extinction hypothesis, *Journal of Geophysical Research*, **96**, E4, 20,879–20,905, 1991.

Johnson, K. S., and Campbell, J. A. (Editors), *Ames Structure in Northwest Oklahoma and Similar Features: Origin and Petroleum Production, Circular 100*, Oklahoma Geological Survey, 396 pp., 1997.

King, E. A., The origin of tektites: A brief review, *American Scientist*, **65**, 212–218, 1977.

Koeberl, C., Tektite origin by hypervelocity asteroidal or cometary impact: Target rocks, source craters, and mechanisms, in *Large Meteorite Impacts and Planetary Evolution, Special Paper 293*, Geological Society of America, B. O. Dressler, R. A. F. Grieve, and V. L. Sharpton, Editors, pp. 133–151, 1994.

Koeberl, C., Impact cratering: The mineralogical and geochemical evidence, in *Ames Structure in Northwest Oklahoma and Similar Features: Origin and Petroleum Production*, K. S. Johnson and J. A. Campbell, Editors, Oklahoma Geological Survey, *Circular 100*, pp. 30–54, 1997.

Lewis, J. S., *Rain of Iron and Ice*, Addison-Wesley, Reading, MA, 1996.

Lowman, P. D., Jr.,The relation of tektites to lunar igneous activity (Abstract), *Journal of Geophysical Research*, **67**, 1646, 1962.

Lowman, P. D., Jr., The relation of tektites to lunar igneous activity, *Icarus*, **2**, 35–48, 1963.

Lowman, P. D., Jr., Magnetic reconnaissance of Sierra Madera, Texas, and nearby igneous intrusions, *Annals of the New York Academy of Sciences*, **123**, Article 2, 1182–1197, 1965.

Lowman, P. D., Jr., *Lunar Panorama*, Weltflugbild, Zürich, 1969.

Lowman, P. D., Jr., Crustal evolution in silicate planets: Implications for the origin of continents, *Journal of Geology*, **84**, 1–26, 1976.

Lowman, P. D., Jr., Original shape of the Sudbury Structure, Canada: A study with airborne imaging radar, *Canadian Journal of Remote Sensing*, **17**, 152–161, 1991.

Lowman, P. D., Jr., The Sudbury Structure as a terrestrial mare basin, *Reviews of Geophysics*, **30**, 227–243, 1992.

Lowman, P. D., Jr., *Lunar Limb Observatory: An Incremental Plan for the Utilization, Exploration, and Settlement of the Moon*, NASA Technical Memorandum 4757, 92 pp., 1996.

Lowman, P. D., Jr., Extraterrestrial impact craters, in *Ames Structure in Northwest Oklahoma and Similar Features: Origin and Petroleum Production*, K. S. Johnson and J. A. Campbell, Editors, *Circular 100*, Oklahoma Geological Survey, pp. 55–81, 1997.

McCall, G. J. H. (Editor), *Astroblemes – Cryptoexplosion Structures*, Dowden, Hutchinson, and Ross, Stroudsburg, PA, 1979.

McLaren, D. J., and Goodfellow, W. D., Geological and biological consequences of giant impacts, *Annual Reviews of Earth and Planetary Sciences*, **18**, 123–171, 1990.

Margulis, L., and West, O., Gaia and the colonization of Mars, *GSA Today*, **3**, 277–291, 1993.

Marsh, B. D., and Zieg, M. J., Melt sheet madness: Superheated emulsion differentiation (Abstract), *Geological Association of Canada, Sudbury 1999, Abstract Volume 24*, pp. 78–79, 1999.

Marvin, U. B., Impact and its revolutionary implications for geology, in *Global Catastrophes in Earth History*, V. L. Sharpton and P. D. Ward, Editors, *Special Paper 247*, Geological Society of America, pp. 147–154, 1990.

Melosh, H. J., *Impact Cratering: A Geologic Process*, Oxford University Press, New York, 1989.

Melosh, H. J., Moon, Origin and Evolution, in *The Astronomy and Astrophysics Encyclopedia*, S. P. Maran, Editor, Van Nostrand Reinhold, New York, pp. 456–459, 1992.

Morrison, D. *The Spaceguard Survey: Report of the NASA International Near-Earth-Object Detection Workshop*, National Aeronautics and Space Adminstration, 52 pp., 3 appendices, 1992.

Oberbeck, V. R., Marshall, J. R., and Aggarwal, H., Impacts, tillites, and the breakup of Gondwanaland, *Journal of Geology*, **101**, 1–19, 1993.

Officer, C. B., and Page, J., *The Great Dinosaur Extinction Controversy*, Addison-Wesley Longman, Inc., Reading, MA, 1996.

O'Keefe, J. A., *Tektites*, University of Chicago Press, Chicago, 1963.

O'Keefe, J. A., *Tektites and Their Origin*, Elsevier, New York, 1976.

O'Keefe, J. A., The coming revolution in planetology, *EOS, Transactions, American Geophysical Union*, **66**, 89–90, 1985.

O'Keefe, J. A., and Cameron, W. S., Evidence from the Moon's surface for the production of lunar granites, *Icarus*, **1**, 271–285, 1962.

Oliver, B. M., The rationale for a preferred frequency band: the water hole, Chapter 4 in *The Search for Extraterrestrial Intelligence*, P. Morrison, J. Billingham, and J. Wolfe, Editors, Dover Publications, New York, pp. 38–47, 1979.

Poag, C. W., The Chesapeake Bay bolide impact: A convulsive event in Atlantic Coastal Plain evolution, *Sedimentary Geology*, **108**, 45–90, 1997.

Pye, E. G., Naldrett, A. J., and Giblin, P. E. (Editors), *The Geology and Ore Deposits of the Sudbury Structure*, Ministry of Natural Resources, Toronto, 1984.

Rampino, M. R., Volcanic hazards, in *Understanding the Earth*, G. C. Brown, C. J. Hawkesworth, and R. C. L. Wilson, Editors, Cambridge University Press, pp. 506–522, 1992.

Rondot, J., Recognition of eroded astroblemes, *Earth-Science Reviews*, **35**, 331–365, 1994.

Rondot, J., *Les Impacts Météoritiques*, Publications MNH, Beauport, Quebec, 1995.

Russell, D. A., The enigma of the extinction of the dinosaurs, *Annual Review of Earth and Planetary Sciences*, **7**, 163–182, 1979.

Salisbury, J. W., and Ronca, L. B., The origin of continents, *Nature*, **210**, 669–670, 1966.

Sharpton, V. L., and Ward, P. D. (Editors), *Global Catastrophes in Earth History, Special Paper 247*, Geological Society of America, 631, pp., 1990.

Sharpton, V. I., Dalrymple, G. B., Marin, I. E., Ryder, G., Schuraytz, B. D., and Urrutia-Fucugauchi, J., New links between the Chicxulub impact structure and the Cretaceous–Tertiary boundary, *Nature*, **359**, 819–821, 1992.

Shaw, H. R., *Craters, Cosmos, and Chronicles*, Stanford University Press, Stanford, 1994.

Shoemaker, E. M., Interpretation of lunar craters, in *Physics and Astronomy of the Moon*, Z. Kopal, Editor, Academic Press, New York, pp. 283–359, 1962.

Shoemaker, E. M., Why study impact craters? in *Impact and Explosion Cratering*, D. J. Roddy, R. O. Pepin, and R. B. Merrill, Editors, Pergamon Press, New York, pp. 1–10, 1977.

Shoemaker, E. M., and Chao, E. C. T., New evidence for the impact origin of the Ries Basin, Bavaria, Germany, *Journal of Geophysical Research*, **66**, 3371–3378, 1961.

Shoemaker, E. M., and Uhlherr, H. R., Stratigraphic relations of australites in the Port Campbell Embayment, Victoria, *Meteoritics and Planetary Science*, **34**, 369–384, 1999.

Shoemaker, E. M., Wolfe, R. F., and Shoemaker, C. S., Asteroid and comet flux in the neighborhood of Earth, in *Global Catastrophes in Earth History*, Sharpton, V. L., and Ward, P. D., Editors, *Special Paper 247*, Geological Society of America, pp. 155–170, 1990.

Short, N. M., Impact and nuclear explosion craters, in *Geological Problems in Lunar Research*, J. Green, Editor, *Annals of the New York Academy of Science*, **CXXIII**, pp. 573–616, 1965.

Short, N. M., *Planetary Geology*, Prentice-Hall, Englewood Cliffs, NJ, 1975.

Silver, L. T., and Schultz, P. H. (Editors), *Geological Implications of Impacts of Large Asteroids and Comets on the Earth, Special Paper 190*, Geological Society of America, 528 pp., 1982.

Smit, J., and 8 others, Tektite-bearing, deep-water clastic unit at the Cretaceous–Tertiary boundary in northeastern Mexico, *Geology*, **20**, 99–103, 1991.

Spray, J. G., Localized shock- and friction-induced melting in response to hypervelocity impact, in *Meteorites: Flux with Time and Impact Effects*, M. M. Grady, R. Hutchinson, G. J. H. McCall, and D. A. Rothery, Editors, *Special Publication 140*, Geological Society of London, London, pp. 195–204, 1998.

Spudis, P. D., *The Geology of Multi-ring Basins*, Cambridge University Press, New York, 1993.

Stevenson, D. J., Origin of the Moon – the collision hypothesis, *Annual Reviews of Earth and Planetary Sciences*, **15**, 271–315, 1987.

Stoffler, D., Coesite and stishovite in shocked crystalline rocks, *Journal of Geophysical Research*, **76**, 5474–5488, 1971.

Tarter, J., *SETI-List of Searches*, www manuscript, SETI Institute, Mountain View, California, 1996.

Taylor, S. R., *Solar System Evolution: A New Perspective*, Cambridge University Press, Cambridge, UK, 1992.

Tonks, W. B., and Melosh, J., Magma ocean formation due to giant impacts, *Journal of Geophysical Research*, **98**, E3, 5319–5333, 1993.

Ward, P. D., and Brownlee, D., *Rare Earth: Why Complex Life is Uncommon in the Universe*, Springer-Verlag, New York, 2000.

Warren, P. H., The magma ocean as an impediment to lunar plate tectonics, *Journal of Geophysical Research*, **98**, E3, 5335–5345, 1993.

Watkis, N. C., *The Western Front From the Air*, Sutton Publishing, Ltd, Phoenix Mill, England, 1999.

Wilhelms, D. D., *The Geologic History of the Moon, US Geological Survey Professional Paper 1348*, 302 pp., 1987.

Wilshire, H. G., Offield, T. W., Howard, K. A., and Cummings, D., *Geology of the Sierra Madera Cryptoexplosion Structure, Pecos County, Texas*, US Geological Survey, Professional Paper 599-H, 42pp., 1972.

Zubrin, R., *The Case for Mars*, Simon and Schuster, New York, 1996.

Chapter 6

Adler, I., and 13 others, Apollo 16 geochemical X-ray fluorescence experiment: Preliminary report, *Science*, **177**, 256–259, 1972.

Anderson, D. L., Hotspots, basalts, and the evolution of the mantle, *Science*, **213**, 83–89, 1981.

Anderson, D. L., The Earth as a planet: Paradigms and paradoxes, *Science*, **223**, 347–355, 1984.

Armstrong, R. L., A model for the evolution of strontium and lead isotopes in a dynamic earth, *Reviews of Geophysics*, **6**, 175–199, 1968.

Armstrong, R. L., Radiogenic isotopes: The case for crustal recycling on a near-steady-state no-continental-growth Earth, *Philosophical Transactions of the Royal Society of London, Series A*, **301**, 443–457, 1981.

Bandfield, J. L., Hamilton, V. E., and Christensen, P. R., A global view of martian surface compositions from MGS-TES, *Science*, **287**, 1626–1630, 2000.

Barrell, J., Geologic bearing of the moon's surficial features, *American Journal of Science, 5th Series*, **13**, 306–314, 1927.

Bartholomew, M. J., and Lewis, S. E., Evolution of Grenville massifs in the Blue Ridge geologic province, southern and central Appalachians, in *The Grenville Event in the Appalchians and Related Topics, Special Paper 194*, M. J. Bartholomew, E. R. Force, A. K. Sinha, N. Herz, Editors, Geological Society of America, pp. 229–254, 1984.

Bartholomew, M. J., Force, E. R., Sinha, A. K., Krishna, A. K., and Herz, N. (Editors), *The Grenville Event in the Appalachians and Related Topics, Special Paper 194*, Geological Society of America, 287 pp., 1984.

Basilevsky, A. T., and Head, J. W., III, The geologic history of Venus: A stratigraphic view, *Journal of Geophysical Research*, **103**, E11, 8531–8544, 1998.

Basu, A. R., Ray, S. I., Saha, A. K., and Sarkar, S. N., Eastern Indian 3800- million-year-old crust and early mantle differentiation, *Science*, **212**, 1502–1506, 1981.

Beatty, J. K., O'Leary, B., and Chaikin, A., *The New Solar System* (Second Edition), Cambridge University Press, Cambridge (UK), and Sky Publishing Corporation, Cambridge (MA), 1982.

Best, M. G., *Igneous and Metamorphic Petrology*, W. H. Freeman, San Francisco, 1982.

Bickford, M. E., The formation of continental crust, *Geological Society of America Bulletin*, **100**, 1375–1391, 1988.

Binder, A. B., Lunar Prospector: Overview, *Science*, **281**, 1475–1476, 1998.

Bird, J. M., and Dewey, J. F., Lithosphere plate continental margin tectonics and the evolution of the Appalachian orogen, *Geological Society of America Bulletin*, **81**, 1031–1060, 1970.

Bowring, S. A., and Housh, T., The Earth's early evolution, *Science*, **269**, 1535–1540, 1995.

Bowring, S. A., Housh, T. B., and Isachsen, C. E., The Acasta gneisses: Remnant of Earth's early crust, in *Origin of the Earth*, H. E. Newsom and J. H. Jones, Editors, Oxford University Press, New York, pp. 319–339, 1990.

Brown, M., Rushmer, T., and Sawyer, E. W., Introduction to special section: Mechanisms and consequences of melt segregation from crustal protoliths, *Journal of Geophysical Research*, **100**, B8, 15,551–15,563, 1995.

Burke, K. C., and Dewey, J. F., Plume-generated triple junctions: Key indicators in applying plate tectonics to old rocks, *Journal of Geology*, **81**, 406–433, 1973.

Burke, K. C., Dewey, J. F., and Kidd, W. S. F., World distribution of sutures – the sites of former oceans, *Tectonophysics*, **40**, 69–99, 1977.

Bursnall, J. T., Leclair, A. D., Moser, D. E., and Percival, J. A., Structural correlation within the Kapuskasing uplift, *Canadian Journal of Earth Sciences*, **31**, 1081–1095, 1994.

Cameron, A. G. W., Higher-resolution simulations of the giant impact, in *Origin of the Earth and Moon*, R. M. Canup and K. Righter, Editors, University of Arizona Press, Tucson, pp. 133–144, 2000.

Cameron, A. G. W., and Truran, J. W., The supernova trigger for the formation of the solar system, *Icarus*, **30**, 447–461, 1977.

Canup, R. M., and Righter, K. (Editors), *Origin of the Earth and Moon*, University of Arizona Press, Tucson, 2000.

Card, K. D., *Geology of the Sudbury–Manitoulin Area, Report 166*, Ontario Geological Survey, 238 pp., 1978.

Card, K. D., A review of the Superior Province of the Canadian Shield, a product of Archean accretion, *Precambrian Research*, **48**, 99–156, 1990.

Card, K. D., and Innes, D. G., *Geology of the Benny Area, District of Sudbury, Report 206*, Ontario Geological Survey, 117 pp., 1981.

Carlson, R. W., and Lugmair, G. W., Timescales of planetesimal formation and differentiation based on extinct and extant radioisotopes, in *Origin of the Earth and Moon*, R. M. Canup, and K. Righter, Editors, University of Arizona Press, Tucson, pp. 25–44, 2000.

Cattermole, P., *Planetary Volcanism: A Study of Volcanic Activity in the Solar System* (Second Edition), Praxis Publishing, Chichester, UK, 1996.

Choi, D. R., Late Permian–Early Triassic paleogeography of northern Japan: Did Pacific microplates accrete to Japan?, *Geology*, **12**, 728–831, 1984.

Christensen, P. R., Bandfield, J. L., Smith, M. D., Hamilton, V. R., and Clark, R. N., Identification of a basaltic component on the martian surface from Thermal Emission Spectrometer data, *Journal of Geophysical Research*, **105**, E3, 9609–9621, 2000.

Clark, B. C., Baird, A. K., Weldon, R. J., Tsusaki, D. M., Schnabel, L., and Candelaria, M. P., Chemical composition of Martian fines, *Journal of Geophysical Research*, **87**, B12, 10,059–10,067, 1982.

Clayton, D. D., *Principles of Stellar Evolution and Nucleosynthesis*, New York, McGraw-Hill, 1968.

Clayton, D. D., Origin of Xe-HL and Supernova 1987A, *Lunar and Planetary Science XX*, Lunar and Planetary Institute, Houston, 165–166, 1989.

Clowes, R. M., Variations in continental crustal structure in Canada from LITHOPROBE seismic reflection and other data, *Tectonophysics*, **219**, 1–27, 1993.

Cobbing, E. J., The Coastal Batholith and other aspects of Andean magmatism in Peru, in *Understanding Granites: Integrating New and Classical Techniques*, A. Castro, C. Fernandez, and J. L. Vigneresse, Editors, Geological Society of London, *Special Publication No. 168*, pp. 111–122, 1999.

Compston, W., and Pidgeon, R. T., Jack Hills, evidence of more very old detrital zircons, *Nature*, **321**, 766–769, 1986.

Condie, K. C., *Archean Greenstone Belts*, Elsevier Scientific Publishing Company, Amsterdam, 1981.

Condie, K. C., *Plate Tectonics and Crustal Evolution* (Third Edition), Pergamon Press, New York, 1989a.

Condie, K. C., Origin of the Earth's crust, *Global and Planetary Change*, **75**, 57–81, 1989b.

Consolmagno, G. J., and Schaefer, M. W., *Worlds Apart: A Textbook in Planetary Science*, Prentice Hall, Englewood Cliffs, NJ, 1994.

Cook, F. A., Brown, L. D., and Oliver, J. E., The southern Appalachians and the growth of continents, *Scientific American*, **243**, 156–168, 1980.

Corriveau, L., and van Breeman, O., Docking of the Central Metasedimentary Belt to Laurentia in geon 12: evidence from the 1.17–1.16 Ga Chevreuil intrusive suite and host gneisses, Quebec, *Canadian Journal of the Earth Sciences*, **37**, 253–269, 2000.

Dana, J. D., On the plan of development in the geological history of North America, *American Journal of Science, 2nd series*, **22**, 335–344, 1856.

Davidson, A., An overview of Grenville Province geology, Canadian Shield, in *Geology of the Precambrian Superior and Grenville Provinces and Precambrian Fossils in North America*, S. B. Lucas and M. R. St-Onge, Editors, Geological Survey of Canada, *Geology of Canada, No. 7*, pp. 205–270, 1998.

Dence, M. R., Hartung, J. B., and Sutter, J. F., Old K–Ar mineral ages from the Grenville Province, Ontario, *Canadian Journal of Earth Sciences*, **8**, 1495–1498, 1971.

DePaolo, D. K., Sources of continental crust: neodymium isotope evidence from the Sierra Nevada and Peninsular Ranges, *Science*, **209**, 684–687, 1980.

Dewey, J. F., Evolution of the Appalachian/Caledonian orogen, *Nature*, **222**, 124–129, 1969.

Dewey, J. F., Suture zone complexities: A review, *Tectonophysics*, **40**, 53–67, 1977.

Dewey, J. F., Extensional collapse of orogens, *Tectonics*, **7**, 1123–1139, 1988.

Dewey, J. F., and Burke, K. C. A., Tibetan, Variscan, and Precambrian basement reactivation: products of continental collision, *Journal of Geology*, **81**, 683–692, 1973.

Dewey, J. F., and Windley, B. F., Growth and differentiation of the continental crust, *Philosophical Transactions of the Royal Society of London, Series A*, **301**, 189–206, 1981.

De Wit, M. J., and Ashwal, L. D. (Editors), *Greenstone Belts*, Clarendon Press, Oxford, 1997.

Dickin, A. P., Crustal formation in the Grenville Province: Nd-isotope evidence, *Canadian Journal of the Earth Sciences*, **37**, 165–181, 2000.

Dickin, A. P., and McNutt, R. H., Nd model-age mapping of Grenville lithotectonic domains: Mid-Proterozoic crustal evolution in Ontario, in *Mid-Proterozoic Laurentia-Baltica*, C. F. Gower, T. Rivers, and B. Ryan, Editors, *Special Paper 38*, Geological Association of Canada, pp. 79–94, 1990.

Donn, W. L., Donn, B. D., and Valentine, W. G., On the early history of the Earth, *Geological Society of America Bulletin*, **76**, 287–306, 1965.

Drury, S. A., Structures induced by granite diapirs in the Archaean greenstone belt at Yellowknife, Canada: Implications for Archaean geotectonics, *Journal of Geology*, **85**, 345–358, 1977.

Dudas, F. O., Davidson, A., and Bethune, K. M., Age of the Sudbury diabase dikes and their metamorphism in the Grenville Province, Ontario, in *Radiogenic Age and Isotopic Studies: Report 8*, Geological Survey of Canada, pp. 97–106, 1994.

Dymek, R. F., Supracrustal rocks, polymetamorphism, and evolution of the SW Greenland Archean gneiss complex, in *Patterns of Change in Earth Evolution*, H. D. Holland and A. F. Trendall, Editors, Springer-Verlag, Berlin, pp. 313–343, 1984.

Engel, A. E. J., Geological evolution of North America, *Science*, **140**, 143–152, 1963.

Ernst, W. G., Metamorphic terranes, isotopic provinces, and implications for crustal growth of the western United States, *Journal of Geophysical Research*, **93**, B7, 7634–7642, 1988.

Ernst, W. G., *The Dynamic Planet*, Columbia University Press, New York, 1990.

Fahrig, W. R., and Wanless, R. K., Age and significance of diabase dike swarms of the Canadian Shield, *Nature*, **200**, 934–937, 1963.

Fahrig, W. F., and West, T. D., Diabase dike swarms of the Canadian Shield, *Map 1627A*, Geological Survey of Canada, 1986.

Faure, G., *Principles of Isotope Geology*, John Wiley, New York, 1986.

Feldman, W. C., and 6 others, Major compositional units of the Moon: Lunar Prospector thermal and fast neutrons, *Science*, **281**, 1489–1493, 1998.

Frey, H. V., Crustal evolution of the early Earth: the role of major impacts, *Precambrian Research*, **10**, 195–216, 1980.

Froude, D. O., and 6 others, Ion microprobe identification of 4,100–4,200 Myr old terrestrial zircons, *Nature*, **304**, 616–618,1983.

Fyfe, W. S., The evolution of the Earth's crust: Modern plate tectonics to ancient hot spot tectonics, *Chemical Geology*, **23**, 89–114, 1978.

Fyfe, W. S., Hot spots, magma underplating, and modification of continental crust, *Canadian Journal of the Earth Sciences*, **30**, 908–912, 1993.

Gamow, G., *Biography of the Earth*, Mentor Books, New York, 1948.

Gehrels, G. E., and Stewart, J. H., Detrital zircon U–Pb geochronology of Cambrian to Triassic miogeoclinal and eugeoclinal strata of Sonora, Mexico, *Journal of Geophysical Research*, **103**, 2471–2492, 1998.

Glikson, A. Y., Geochemical, isotopic, and palaeomagnetic tests of early sial-sima patterns: the Precambrian crustal enigma revisited, *Geological Society of America Memoir 161*, pp. 95–118, 1981.

Goodwin, A. M., *Precambrian Geology*, Academic Press, London, 1990.

Green, M. G., Sylvester, P. J., and Buick, R., Growth and recycling of early Archean continental crust: Geochemical evidence from the Coonterunah and Warrawoona Groups, Pilbara Craton, Australia, *Tectonophysics*, **322**, 69–88, 2000.

Halls, H. C., The importance and potential of mafic dike swarms in studies of geodynamic processes, *Geoscience Canada*, **9**, 145–154, 1982.

Halls, H. C., and Bates, M. P., The evolution of the 2.45 Ga Matachewan dike swarm, Canada, in *Mafic Dykes and Emplacement Mechanisms, Proceedings of the Second International Dyke Conference*, A. J. Parker, P. C. Rickwood, and D. H. Tucker, Editors, A. A. Balkema, Rotterdam, pp. 237–248, 1990.

Hamilton, W. B., *Tectonics of the Indonesian Region*, *Professional Paper 1078*, US Geological Survey, 345 pp., 1979.

Hamilton, W. B., Evolution of Archean mantle and crust, in *The Geology of North America*, C-2, *Precambrian: Conterminous U.S.*, J. C. Reed, Jr., T. T. Ball, G. L. Farmer, and W. B. Hamilton, Editors, Geological Society of America, pp. 597–614, 1993.

Hamilton, W. B., Archean tectonics and magmatism, *International Geology Review*, **40**, 1–39, 1998.

Hanel, R., and 12 others, Investigation of the martian environment by infrared spectroscopy on Mariner 9, *Icarus*, **17**, 423–442, 1972.

Hanks, T. C., and Anderson, D. L., The early thermal history of the earth, *Physics of the Earth and Planetary Interiors*, **2**, 19–29, 1979.

Hanmer, S., Corrigan, D., Pehrsson, S., and Nadeau, L., SW Grenville Province, Canada: the case against post-1.4 Ga accretionary tectonics, *Tectonophysics*, **319**, 33–51, 2000.

Hargraves, R. B., Precambrian geologic history, *Science*, **193**, 363–370, 1976.

Hartmann, W. K., Phillips, R. J., and Taylor, G. J. (Editors), *Origin of the Moon*, Lunar and Planetary Institute, Houston, 1986.

Head, J. W., III, Lunar volcanism in time and space, *Reviews of Geophysics and Space Physics*, **14**, 265–300, 1976.

Head, J.W., III, and Crumpler, L. S., Evidence for divergent plate-boundary characteristics and crustal spreading on Venus, *Science*, **238**, 1380–1385, 1987.

Head, J. W., III, Crumpler, L. S., Aubele, J. C., Guest, J. E., and Saunders, R. S., Venus volcanism: Classification of volcanic features and structures, associations, and global distribution from Magellan data, *Journal of Geophysical Research*, **97**, E8, 13,153–13,197, 1992.

Hoffman, P., United plates of America, *Annual Reviews of Earth and Planetary Sciences*, **16**, 543–603, 1988.

Hofmann, A. W., Early evolution of continents, *Science*, **275**, 498–499, 1997.

Howell, D. G., *Principles of Terrane Analysis: New Applications for Global Tectonics* (Second Edition), Chapman and Hall, London, 1995.

Hunt, C. W., Collins, L. G., and Skobelin, E. A., *Expanding Geospheres*, Polar Publishing, Calgary, 1992.

Hurley, P. M., and Rand, J. R., Pre-drift continental nuclei, *Science*, **164**, 1229–1242, 1969.

Ireland, T. R., Crustal evolution of New Zealand: Evidence from age distributions of detrital zircons, *Geochimica et Cosmochimica Acta*, **56**, 911–920, 1992.

Jacobs, J. A., When did the Earth's core form?, in *The Continental Crust and Its Mineral Deposits*, D. W. Strangway, Editor, Geological Association of Canada, *Special Paper 20*, pp. 35–48, 1980.

Jacobsen, S. B., and Dymek, R. F., Nd and Sr isotope systematics of clastic metasediments from Isua, West Greenland: Identification of pre-3.8 Ga differentiated crustal components, *Journal of Geophysical Research*, **93**, 338–354, 1988.

Jones, D. L., Silberling, N. J., and Hillhouse, J., Wrangellia – A displaced terrane in northwestern North America, *Canadian Journal of the Earth Sciences*, **14**, 2565–2577, 1977.

Kaula, W. M., The beginning of the Earth's thermal evolution, in *The Continental Crust and Its Mineral Deposits*, D. W. Strangway, Editor, Geological Association of Canada, *Special Paper 20*, pp. 25–34, 1980.

Kaula, W. M., Venus: A contrast in evolution to Earth, *Science*, **247**, 1191–1196, 1990.

Kerr, A. C., Tarney, J., Marriner, G. F., Nivia, A., and Saunders, A. D., The Caribbean–Colombian Igneous Province: The internal anatomy of an oceanic plateau, in *Large Igneous Provinces: Continental, Oceanic, and Planetary Flood Volcanism*, J. J. Mahoney and M. F. Coffin, Editors, *Geophysical Monograph 100*, American Geophysical Union, pp. 123–144, 1997.

Kerr, R. A., Suspect terranes and continental growth, *Science*, **222**, 36–38, 1983.

Krogh, T. E., Davis, D. W., and Corfu, F., Precise U–Pb zircon and baddeleyite ages for the Sudbury area, in *The Geology and Ore Deposits of the Sudbury Structure, Special Volume 1*, E. G. Pye, A. J. Naldrett, and P. E. Giblin, Editors, Ontario Geological Survey, pp. 431–446, 1984.

Kroner, A., Evolution of the Archean continental crust, *Annual Reviews of Earth and Planetary Sciences*, **13**, 49–74, 1985.

Kushiro, I., Effect of water on the composition of magmas formed at high pressures, *Journal of Petrology*, **13**, 311–334, 1972.

Kusky, T. M., Accretion of the Archean Slave Province, *Geology*, **17**, 63–57, 1989.

Lee, D., Halliday, A. N., Snyder, G. A., and Taylor, L. A., Age and origin of the Moon, *Science*, **278**, 1089–1103, 1997.

Lowman, P. D., Jr., Composition of the lunar highlands: Possible implications for evolution of the Earth's Crust, *Journal of Geophysical Research*, **74**, 495–504, 1969.

Lowman, P. D., Jr., Crustal evolution in silicate planets: Implications for the origin of continents, *Journal of Geology*, **84**, 1–26, 1976.

Lowman, P. D., Jr., Formation of the earliest continental crust: Inferences from the Scourian Complex of northwest Scotland and geophysical models of the lower continental crust, *Precambrian Research*, **24**, 199–215, 1984.

Lowman, P. D., Jr., Plate tectonics with fixed continents: A testable hypothesis, *Journal of Petroleum Geology*, **8**, 373–388, 1985, and **9**, 71–88, 1986.

Lowman, P. D., Jr., A comparison of three fronts: Grenville, Nelson, and Allegheny (Abstract), *Saskatoon 87*, *Annual Meeting*, Geological Association of Canada, 1987.

Lowman, P. D., Jr., Comparative planetology and the origin of continental crust, *Precambrian Research*, **44**, 171–195, 1989.

Lowman, P. D., Jr., The Sudbury Structure as a terrestrial mare basin, *Reviews of Geophysics*, **30**, 227–243, 1992.

Lowman, P. D., Jr., Andesites on Mars: Implications for the origin of terrestrial continental crust, *Lunar and Planetary Science Conference*, **XXXIX**, Abstract #1227, Lunar and Planetary Institute, Houston, 1998.

Lowman, P. D., Jr., Whiting, P. J., Short, N. M., Lohman, A. M., and Lee, G., Fracture patterns on the Canadian Shield: A lineament study with Landsat and orbital radar imagery, *Basement Tectonics*, **7**, 139–159, 1992.

Luais, B., and Hawkesworth, C. J., The generation of continental crust: An integrated study of crust-forming processes in the Archaean of Zimbabwe, *Journal of Petrology*, **35**, 43–94, 1994.

Ludden, J., and Hynes, A., The Abitibi–Grenville LITHOPROBE transect part III: Introduction, *Canadian Journal of the Earth Sciences*, **37**, 115–116, 2000a.

Ludden, J., and Hynes, A., The LITHOPROBE Abitibi–Grenville transect: two billion years of crust formation and recycling in the Precambrian Shield of Canada, *Canadian Journal of the Earth Sciences*, **37**, 459–476, 2000b.

MacDonald, G. J. F., The internal constitutions of the inner planets and the Moon, *Space Science Reviews*, **2**, 473–557, 1963.

McGregor, V. R., Archean grey gneisses and the origin of the continental crust: Evidence from the Godthab region, West Greenland, in *Trondhjemites, Dacites, and Related Rocks*, F. Barker, Editor, Elsevier, New York, pp. 169–204, 1979.

McKee, B., *Cascadia*, McGraw-Hill Book Company, New York, 1972.

McKenzie, D., McKenzie, J. M., and Saunders, R.S., Dike emplacement on Venus and Earth, *Journal of Geophysical Research*, **97**, 15,977–154,990, 1992.

McPhee, J., *In Suspect Terrain*, Farrar, Strauss, and Giroux, New York, 1982.

McSween, H. Y., Jr., SNC meteorites: clues to Martian petrologic evolution?, *Reviews of Geophysics*, **23**, 391–415, 1985.

McSween, H. Y., Jr., Rocks at the Mars Pathfinder landing site, *American Scientist*, **87**, 36–45, 1998.

McSween, H. Y., *Meteorites and their Parent Planets* (Second Edition), Cambridge University Press, Cambridge, UK, 1999.

Marsh, B. D., Some Aleutian andesites: Their nature and source, *Journal of Geology*, **84**, 27–46, 1976.

Meissner, R., *The Continental Crust, A Geophysical Approach*, Academic Press, New York, 1986.

Melosh, H. J., Moon, Origin and Evolution, in *The Astronomy and Astrophysics Encyclopedia*, S. P. Maran, Editor, Van Nostrand Reinhold, New York, pp. 456–459, 1992.

Meszaros, S. P., *Planetary Size Comparisons: A Photographic Study*, NASA Technical Memorandum 85017, 40 pp., 1983.

Meyerhoff, A. A., Taner, I., Morris, A. E. L., Martin, B. D., Agocs, W. B., and Meyerhoff, H. A., Surge tectonics: A new hypothesis of Earth dynamics, in *New Concepts in Global Tectonics*, Texas Tech University Press, Lubbock, TX, pp. 309–409, 1992.

Minster, J. B., and Jordan, T., Present-day plate motions, *Journal of Geophysical Research*, **83**, 5331–5354, 1978.

Moorbath, S., Evolution of Precambrian crust from strontium isotopic evidence, *Nature*, **254**, 395–398, 1975.

Moore, J. M., Davidson, A., and Baer, A. J. (Editors), *The Grenville Province, Special Paper 31*, Geological Association of Canada, 358 pp., 1986.

Muehlberger, W. E., Denison, R. E., and Lidiak, E. G., Basement rocks in continental interior of United States, *American Association of Petroleum Geologists Bulletin,* **51**, 2351–2380, 1967.

Nisbet, E. G., *The Young Earth*, Allen and Unwin, Boston, 1987.

Nur, A., and Ben-Avraham, Z., Oceanic plateaus, the fragmentation of continents, and mountain building, *Journal of Geophysical Research*, **87**, B5, 3644–3661, 1982.

Phillips, R. J., and Malin, M. C., The interior of Venus and tectonic implications, in *Venus*, D. M. Hunten, L. Colin, T. M. Donahue, and M. I. Moroz, Editors, University of Arizona Press, Tucson, pp. 159–214, 1983.

Pieters, C. M., Staid, M. I., Fischer, E. M., Tompkins, S., and He, G., A sharper view of impact craters from Clementine data, *Science*, **266**, 1844–1848, 1994.

Poldervaart, A., Chemistry of the earth's crust, in *Crust of the Earth*, A. Poldervaart, Editor, *Special Paper 62*, Geological Society of America, pp. 119–144, 1955.

Raia, F., and Spera, F. J., Simulations of crustal anatexis: Implications for the growth and differentiation of continental crust, *Journal of Geophysical Research*, **102**, B10, 22,629–22,648, 1997.

Redfern, M., *Journey to the Centre of the Earth*, Broadside Books, London, 1991.

Richardson, R. M., Ridge forces, absolute plate motions, and the intraplate stress field, *Journal of Geophysical Research*, **97**, B8, 11,739–11,748, 1992.

Rieder, R., and 7 others, The chemical composition of martian soil and rocks returned by the mobile Alpha Proton X-ray Spectrometer: Preliminary results from the X-ray mode, *Science*, **278**, 1771–1774, 1997.

Rousell, D. H., Gibson, H. L., and Jonasson, I. R., The tectonic, magmatic, and mineralization history of the Sudbury Structure, *Exploration and Mining Geology*, **6**, 1–22, 1997.

Rudnick, R. L., Restites, Eu anomalies, and the lower continental crust, *Geochimica et Cosmochimica Acta*, **56**, 963–970, 1992.

Rudnick, R. L., Making continental crust, *Nature*, **378**, 571–578, 1995.

Saunders, R. S., and Carr, M. H., Venus, in *The Geology of the Terrestrial Planets*, M. H. Carr, R. S. Saunders, R. G. Strom, D. E. Wilhelms, Editors, *NASA Special Publication 469*, National Aeronautics and Space Adminstration, Washington, DC, pp. 57–77, 1984.

Shaw, D. M., Development of the early continental crust, in *The Early History of the Earth*, B. F. Windley, Editor, John Wiley, New York, pp. 33–53, 1976.

Shaw, D. M., Evolutionary tectonics of the Earth in the light of early crustal structure, in *The Continental Crust and Its Mineral Deposits*, D. W. Strangway, Editor, Geological Association of Canada *Special Paper 20*, pp. 65–73, 1980.

Shirey, S. B., and Hanson, G. N., Mantle-derived Archaean monzodiorites and trachyandesites, *Nature*, **310**, 222–224, 1984.

Siever, R., Comparison of Earth and Mars as differentiated planets, *Icarus*, **22**, 312–324, 1974.

Sleep, N. H., Martian plate tectonics, *Journal of Geophysical Research*, **99**, E4, 5639–5655, 1994.

Smith, J. V., Development of the Earth–Moon system: implications for the geology of the early Earth, in *The Early History of the Earth*, B. F. Windley, Editor, John Wiley, New York, pp. 3–19, 1976.

Solomon, S. C., Differentiation of crusts and cores of the terrestrial planets: Lessons for the early Earth?, *Precambrian Research*, **10**, 177–194, 1980.

Solomon, S. C., and Head, J. W., Evolution of the Tharsis Province of Mars: The importance of heterogeneous lithospheric thickness and volcanic construction, *Journal of Geophysical Research*, **87**, B12, 9755–9774, 1982.

Solomon, S. C., and Head, J. W., Fundamental issues in the geology and geophysics of Venus, *Science*, **252**, 252–260, 1991.

Solomon, S. C., and 10 others, Venus tectonics: An overview of Magellan observations, *Journal of Geophysical Research*, **97**, E8, 13,199–13,255, 1992.

Spudis, P. D., and Davis, P. A., A chemical and petrological model of the lunar crust and implications for lunar crustal origin, *Proceedings of the Lunar and Planetary Science Conference 17*, *Journal of Geophysical Research*, **91**, 84–90, 1986.

Suppe, J., and Connors, C., Critical taper wedge mechanics of fold-and-thrust belts on Venus: Initial results from Magellan, *Journal of Geophysical Research*, **97**, E8, 13,545–13,561, 1992.

Surkov, Yu. A., Studies of Venus rocks by Veneras 8, 9, and 10, in *Venus*, D. M. Hunten, L. Colin, T. M. Donahue, and M. I. Moroz, Editors, University of Arizona Press, Tucson, pp. 154–158, 1983.

Sylvester, P. J., Continent formation, growth and reycling, *Tectonophysics*, **322**, vii–viii, 2000.

Taylor, S. R., *Planetary Science: A Lunar Perspective*, Lunar and Planetary Institute, Houston, 1982.

Taylor, S. R., *Solar System Evolution: A New Perspective*, Cambridge University Press, Cambridge, UK, 1992; and (Second Edition) 2001.

Taylor, S. R., and McLennan, S. M., *The Continental Crust: Its Composition and Evolution*, Blackwell Scientific Publications, London, 1985.

Thomas, M. D., Gravity studies of the Grenville province: Significance for Precambrian plate collision and the origin of anorthosite, in *The Utility of Regional Gravity and Magnetic Anomaly Maps*, W. J. Hinze, Editor, Society of Exploration Geophysicists, Tulsa, OK, pp. 109–123, 1985.

Trombka, J. I., and 23 others, The elemental composition of asteroid 433 Eros: Results of the NEAR–Shoemaker X-ray spectrometer, *Science*, **289**, 2101–2105, 2000.

Turner, F. J., and Verhoogen, J., *Igneous and Metamorphic Petrology* (Second Edition), McGraw-Hill Book Company, 1960.

Urey, H. C., *The Planets*, Yale University Press, New Haven, 1952.

Walker, D., Lunar and terrestrial crust formation, *Proceedings of the Fourteenth Lunar and Planetary Science Conference, Journal of Geophysical Research*, **88**, 17–25, 1983.

Walker, D., Stolper, E. M., and Hays, J. F., *Proceedings of the 10th Lunar Science Conference*, Pergamon Press, New York, pp. 1995–2015, 1979.

Ward, P. D., and Brownlee, D., *Rare Earth*, Copernicus Springer-Verlag, New York, 2000.

Wasserburg, G. J., Crustal history and the Pre-Cambrian time scale, *Annals of the New York Academy of Sciences*, **91**, (2), 583–594, 1961.

Wasserburg, G. J., and Papanastassiou, D. A., Some short-lived nuclides in the early solar system – a connection with the placental ISM (interstellar medium), in *Essays in Nuclear Astrophysics*, C. A. Barnes, D. D. Clayton, and D. N. Schramm, Editors, Cambridge University Press, London, UK, pp. 77–135, 1982.

Weaver, B. L., and Tarney, J., Major and trace element composition of the continental lithosphere, in *Structure and Evolution of the Continental Lithosphere*, H. N. Pollack and V. R. Murthy, Editors, Pergamon Press, New York, pp. 39–68, 1984.

Wetherill, G. W., The beginning of continental evolution, *Tectonophysics*, **13**, 31–45, 1972.

White, D. J., and 7 others, Seismic images of the Grenville Orogen in Ontario, *Canadian Journal of the Earth Sciences*, **31**, 203–307, 1994.

Williams, H., and Hatcher, R. D., Suspect terranes and accretionary history of the Appalachian Orogen, *Geology*, **10**, 530–536, 1982.

Williams, H., Hoffman, P. F., Lewry, J. F., Monger, J. W. H., and Rivers, R., Anatomy of North America: thematic geologic portrayals of the continent, *Tectonophysics*, **187**, 117–134, 1991.

Wilson, J. T., The development and structure of the crust, in *The Earth as a Planet*, G. P. Kuiper, Editor, University of Chicago Press, Chicago, pp. 138–214, 1954.

Windley, B. F., *The Evolving Continents*, John Wiley, New York, 1984.

Windley, B. F., Bishop, F. C., and Smith, J. V., Metamorphosed layered igneous complexes in Archean granulite–gneiss belts, *Annual Reviews of Earth and Planetary Sciences*, **9**, 175–198, 1981.

Wise, D. U., Freeboard of continents through time, *Geological Society of America Memoir 132*, pp. 87–100, 1972.

Wood, J. A., Fragments of terra rock in the Apollo 12 soil samples and a structural model of the moon, *Icarus*, **16**, 462–501, 1972.

Wood, J., and Wallace, H., *Volcanology and Mineral Deposits, Miscellaneous Paper 129*, Ontario Geological Survey, 183 pp., 1986.

Wright, F. E., Polarization of light reflected from rough surfaces with special reference to light reflected by the moon, *National Academy of Sciences Proceedings*, **13**, 535–540, 1927.

Wyllie, P. J., *The Way the Earth Works*, John Wiley, New York, 1976.

Wyllie, P. J., Magma genesis, plate tectonics, and chemical differentiation of the Earth, *Reviews of Geophysics*, **26**, 370–415, 1988.

Wyllie, P. J., Huang, W., Stern, C. R., and Maaloe, S., Granitic magmas: possible and impossible sources, water contents, and crystallization sequences, *Canadian Journal of the Earth Sciences*, **13**, 1007–1019, 1976.

Wynne-Edwards, H. R., Tectonic overprinting in the Grenville Province, southwestern Quebec, in *Age Relations in High-grade Metamorphic Terrains, Special Paper 5*, Geological Association of Canada, H. R. Wynne-Edwards, Editor, pp. 163–182, 1969.

Yoder, H. S., Jr., Calcalkalic andesites: experimental data bearing on the origin of their assumed characteristics, in *Proceedings of the Andesite Conference*, A. R. McBirney, Editor, Oregon Department of Geology and Mineral Industries, *Bulletin 65*, 77–89, 1969.

Yoder, H. S., Jr., *Generation of Basaltic Magma*, National Academy of Sciences, Washington, DC, 1976.

Yoder, H. S., Jr., The great basaltic "floods," *South African Journal of Geology*, **91**, 139–156, 1988.

Zoback, M. L., First- and second-order patterns of stress in the lithosphere: The World Stress Map Project, *Journal of Geophysical Research*, **97** B8, 11,703–11,728, 1992.

Chapter 7

Allegre, C. J., and Schneider, S. H., The evolution of the Earth, in *Life in the Universe, Scientific American Special Issue*, W. H. Freeman, New York, pp. 29–40, 1995.

Anderson, D. L., The Earth as a planet : Paradigms and paradoxes, *Science*, **223**, 347–355, 1984.

Campbell, I. H., and Taylor, S. R., No water, no granites – No oceans, no continents, *Geophysical Research Letters*, **10**, 1061–1064, 1983.

DiGregorio, B. E., *Mars: The Living Planet*, North Atlantic Books, Berkeley, CA, 1997.

Falkowski, P. G., Barber, R. T., and Smetaček, V., Biogeochemical controls and feedbacks on ocean primary production, *Science*, **281**, 200–206, 1998.

Frederickson, J. K., and Onstott, T. C., Microbes deep within the Earth, *Scientific American*, October, 68–79, 1996.

Gilliland, R. L., Solar evolution, *Global and Planetary Change*, **75**, 35–55, 1989.

Jacobson, M. C., Charlson, R. J., Rodhe, H., and Orians, G. H., *Earth System Science: From Biogeochemical Cycles to Global Change*, Academic Press, San Diego, 2000.

Kasting, J. F., Long-term stability of the Earth's climate, *Global and Planetary Change*, **75**, 83–95. 1989.

Kellogg, L. H., Hager, B. H., and van der Hilst, R. D., Compositional stratification in the deep mantle, *Science*, **283**, 1881–1884, 1999.

Lenton, T. M., Gaia and natural selection, *Nature*, **394**, 439–447, 1998.

Lovelock, J. E., *Gaia: A New Look at Life on Earth*, Oxford University Press, Oxford, UK, 1979.

Lovelock, J. E., *The Ages of Gaia: A Biography of Our Living Earth*, Bantam Books, New York, 1988.

Lowman, P. D., Jr., Twelve key 20th-century discoveries in the geosciences, *Journal of Geoscience Education*, **44**, 485–502, 1996.

McKay, C. P., Terraforming: Making an Earth of Mars, *Planetary Report*, **7**, 26–27, 1987.

McKay, D. S., and 8 others, Search for past life on Mars: Possible relic biogenic activity in martian meteorite ALH 84001, *Science*, **273**, 924–930, 1996.

Margulis, L., *Symbiotic Planet*, Basic Books, New York, 1998.

Margulis, L., and Lovelock, J. E., Biological modulation of the Earth's atmosphere, *Icarus*, **21**, 471–489, 1974.

Margulis, L., and West, O., Gaia and the colonization of Mars, *GSA Today*, **3**, 277–291, 1993.

Nisbet, E. G., *The Young Earth*, Allen and Unwin, Boston, 1987.

Seyfert, C. K., Plastic deformation of rocks and rock-forming minerals, in *The Encyclopedia of Structural Geology and Plate Tectonics*, C. K. Seyfert, Editor, Van Nostrand Reinhold Company, New York, pp. 495–502, 1987a.

Seyfert, C. K., Fluid pressure and the formation of overthrusts and gravity slides, in *The Encyclopedia of Structural Geology and Plate Tectonics*, C. K. Seyfert, Editor, Van Nostrand Reinhold Company, New York, pp. 239–249, 1987b.

Stein, C. A., and Stein, S., A model for the global variation in oceanic depth and heat flow with lithospheric age, *Nature*, **359**, 123–129, 1992.

van der Hilst, R. B., and Karason, H., Compositional heterogeneity in the bottom 1000 kilometers of Earth's mantle: Toward a hybrid convection model, *Science*, **283**, 1885–1888, 1999.

Walker, J. C. G., Hays, P. B., and Kasting, J. F., A negative feedback mechanism for the long-term stabilization of Earth's surface temperature, *Journal of Geophysical Research*, **86**, C10, 9776–9782, 1981.

Watson, A. J., and Lovelock, J. E., Biological homeostasis of the global environment: The parable of the "daisy" world, *Tellus*, **35B**, 284–289, 1983.

Yoder, H. S., Jr., *The Role of Water in Metamorphism, Special Paper 62*, pp. 505–524, Geological Society of America, 1955.

Afterword

Mencken, H. L., The cult of hope, in *Prejudices, A Selection*, J. T. Farrell, Editor, Vintage Books, New York, pp. 84–89, 1958.

Glossary

Dewey, J. F., and Windley, B. F., Growth and differentiation of the continental crust, *Philosophical Transactions of the Royal Society of London, Series A*, **301**, 189–206, 1981.

Lowman, P. D., Jr., Crustal evolution in silicate planets: Evidence from comparative planetology, *Journal of Geology*, **84**, 1–26, 1976.

INDEX

This index includes terms from the body of the text only; captions are not included

Printed in the United States
By Bookmasters